Differential Algebra
and Related Topics

Proceedings of the International Workshop

Differential Algebra and Related Topics

Newark Campus of Rutgers, The State University of New Jersey
2–3 November 2000

Editors

Li Guo
Rutgers University, Newark, NJ, USA

Phyllis J. Cassidy
Smith College, Northampton, MA, USA

William F. Keigher
Rutgers University, Newark, NJ, USA

William Y. Sit
City College of New York, New York, NY, USA

World Scientific
New Jersey • London • Singapore • Hong Kong

Published by

World Scientific Publishing Co. Pte. Ltd.

P O Box 128, Farrer Road, Singapore 912805

USA office: Suite 1B, 1060 Main Street, River Edge, NJ 07661

UK office: 57 Shelton Street, Covent Garden, London WC2H 9HE

British Library Cataloguing-in-Publication Data
A catalogue record for this book is available from the British Library.

ISBN 981-02-4703-6

Printed in Singapore by Uto-Print

Foreword

Differential algebra was founded by Joseph Fels Ritt, who, at the end of the first third of the twentieth century, succeeded in making both rigorous and algorithmic the classical theory of algebraic differential equations. Although in 1957, differential algebra was described by Irving Kaplansky as being 99% the work of Ritt and his student, Ellis R. Kolchin, it has in recent decades become a lively research area. This new subject has not only begun to influence related fields such as the theory of Hamiltonian systems and the analysis of systems of differential equations with symmetries, but has also had surprising interconnections with diophantine geometry over function fields, computer algebra, and model theory. The organizers of the Workshop on Differential Algebra and Related Topics aimed to bring together mathematicians in branches of mathematics that both influence and have been influenced by differential algebra in order to stimulate further discourse and research. We thank the forty-eight participants and twelve invited speakers, all of whom made the conference a success. In the spirit of the meeting, nine of the speakers agreed to write introductory tutorial or survey articles for these proceedings. We thank the authors and the referees for making this volume possible.

The speakers have extended their talks into didactic expositions that contain numerous examples, and, in some cases, exercises for the reader. Sit's paper is an introduction to the development of the commutative algebra that underlies the new geometry known as differential algebraic geometry. His article describes recent work in computer algebra on elimination theory that demonstrates the power and effectiveness of the Ritt-Kolchin approach. The objects of affine differential algebraic geometry are solution sets of systems of differential polynomial equations. Until recently, the language of schemes has not been widely accepted in differential algebra, and there has been no universally accepted definition of abstract differential algebraic variety. One of the reasons for this is, perhaps, the challenge presented by the fact that the coordinate ring of an affine differential algebraic variety may not be isomorphic to the ring of global differential rational functions, and, indeed, may be much smaller. Kovacic's paper addresses this problem by developing a theory of differential schemes, restricting his attention to an appropriate class of differential rings, which includes the coordinate rings of affine differential algebraic varieties. An explicit link of differential algebraic geometry with algebraic geometry is provided by Gillet in his paper on the Kolchin irreducibility theorem, which states that the differential algebraic variety associated with an integral domain of finite type over a differential field is irreducible. Gillet's article marks the first time that "modern" or "Grothendieck style" algebraic geometry is applied to understand conceptually and to give a new proof of this important theorem.

The intertwining of contemporary symbolic logic and differential algebra is the focus of Scanlon's paper on the model theory of differentially closed fields. After introducing the reader to the language of model theory, Scanlon discusses recent deep work on differential algebraic varieties equipped with an exotic structure first introduced by Hrushovski and Zilber, called a *Zariski geometry*. In differential algebra, *trivial* Zariski geometries first surfaced in work by Shelah, Rosenlicht and Kolchin on the nature of minimal differentially closed fields; *modular* Zariski geometries were exhibited later in Buium's papers on Manin kernels on abelian varieties. By describing the remarkable symbiosis of model theory and differential algebra, Scanlon makes clear the role that differential fields play as "proving grounds for pure model theory."

Recently, there has been a renewal of interest in the Galois theory of linear differential operators, a theory that was initiated at the end of the 19th century by Emile Picard and Ernest Vessiot, and revived in 1948 by Ellis Kolchin. Building on Kovacic's solution, for connected solvable groups, of the inverse problem in differential Galois theory, Singer and Mitschi solved the problem for connected algebraic groups over an algebraically closed constant field. In his article on the inverse problem, Magid gives a brief history of the problem, and describes his own approach, illustrated in the case of the two-dimensional affine group. He also presents a variant of the inverse problem, called the lifting problem. Van der Put recalls briefly the basic ideas of differential Galois theory, and introduces a universal Picard-Vessiot ring for a class of differential equations defined over differential fields of convergent and formal Laurent series. The differential Galois groups of these universal differential rings are certain affine group schemes. Churchill and Kovacic turn to a classical result in the differential Galois theory of Picard-Vessiot extensions. They present two proofs of the existence of cyclic vectors for differential structures on an n-dimensional vector space over an ordinary differential field of any characteristic of degree at least n over its field of constants. The first, a constructive version of a non-constructive proof by Deligne, is supplemented by a MAPLE implementation; the second is a generalized version of Cope's constructive proof, which was specific to the field of rational functions.

A beautiful and surprising application of differential Galois theory is to the problem of determining the non-integrability of Hamiltonian systems. Because he assumes a working knowledge of differential Galois theory, but not of Hamiltonian systems, Churchill's article includes an introduction to the classical study of Hamilton's equations, which is illustrated by carefully chosen examples such as the two-body problem and the problem of spinning tops. The link between differential algebra and the study of differential equations with symmetries is explored by Mansfield in an article that emphasizes the technique of moving frames initiated by Elie Cartan and recently generalized by Olver and Fels. This instructive tutorial points the way to an algorithmic approach to the problem of constructing invariants as well as invariant differential operators.

While differential algebra focuses mainly on differential equations, applied mathematicians are often interested in the more general integral equations, or integral-differential equations. Baxter algebras provide a natural setting for an algebraic theory. The paper by Li Guo not only introduces us to this fascinating topic, but relates it to the umbral calculus of combinatorics.

We, the editors, are responsible for the final camera-ready copy of these proceedings. The cover design originated from a joint effort of the editors and Mr. Chun S. Park, who offered expert technical assistance as well (thanks from all of us, Chun). Our thanks also go to Jerald J. Kovacic who contributed LaTeX macros that helped to give the articles a more uniform look. We thank World Scientific Publishing Company, Singapore and River Edge, N.J., U.S.A. for publishing this volume.

We sincerely hope that the readers find this volume useful and that in the near future, a sequel to DART will be organized to cover the many topics still omitted.

> Li Guo (Chief Editor)
> Phyllis J. Cassidy
> William F. Keigher
> William Y. Sit

Acknowledgements

The organizers of the Workshop gratefully acknowledge funding from the National Security Agency, the Kolchin Seminar in Differential Algebra Fund of The City College of New York, and the University Research Council, the Office of the Provost, the Office of the Dean of the Faculty of Arts and Sciences, and the Department of Mathematics and Computer Science at the Newark campus of Rutgers, The State University of New Jersey.

Workshop Participants

Beth Arnold, Texas A&M University, USA.
Teijo T. Arponen, Helsinki University of Technology, Finland.
Alberto Baider, Hunter College (CUNY), USA.
Peter Berman, North Carolina State University, USA.
Raphaël Bomboy, INRIA, France.
Mireille Boutin, University of Minnesota, USA.
Manuel Bronstein, INRIA, France.
Giuseppa Carrá-Ferro, University of Catania, Italy.
Phyllis J. Cassidy, Smith College, USA.
Richard Churchill, Hunter College (CUNY), USA.
Peter A. Clarkson, University of Kent, UK.
Richard Cohn, Rutgers University, New Brunswick, USA.
Henri Gillet, University of Illinois at Chicago, USA.
Li Guo, Rutgers University at Newark, USA.
Mark van Hoeij, Florida State University, USA.
Evelyne Hubert, INRIA, France.
Joseph L. Johnson, Rutgers University, New Brunswick, USA.
Lourdes Juan, Mathematical Sciences Research Institute, USA.
William Keigher, Rutgers University at Newark, USA.
Irina Kogan, Yale University, USA.
Ilias Kotsireas, University of Western Ontario, Canada.
Jerald Kovacic, The City College of New York (CUNY), USA.
Yueh-er Kuo, University of Tennessee, USA.
Peter Landesman, Graduate Center (CUNY), USA.
Jaewoo Lee, Graduate Center (CUNY), USA.
Alexander B. Levin, The Catholic University of America, USA.
Ian Levitt, Federal Aviation Administration, USA.
Robert H. Lewis, Fordham University, USA.
Joseph Lewittes, Lehman College (CUNY), USA.
Anton Leykin, University of Minnesota, USA.
Gennady Lyubeznik, University of Minnesota, USA.
Andy Magid, University of Oklahoma, USA.
Elizabeth Mansfield, University of Kent, UK.
Dave Marker, University of Illinois at Chicago, USA.
Laura Felicia Matusevich, University of California, Berkeley, USA.
Wai Yan Pong, University of Illinois, USA.
F. Leon Pritchard, York College (CUNY), USA.

Marius van der Put, University of Groningen, the Netherlands.
Greg Reid, University of Western Ontario, Canada.
Thomas Scanlon, University of California, Berkeley, USA.
Michael Singer, North Carolina State University, USA.
William Sit, The City College of New York (CUNY), USA.
Earl Taft, Rutgers University, New Brunswick, USA.
William Traves, U.S. Naval Academy, USA.
Harrison Tsai, Cornell University, USA.
Felix Ulmer, Université de Rennes I, France.
Uli Walther, University of Minnesota, USA.
Yang Zhang, University of Western Ontario, Canada.

Workshop Program

Thursday, November 2, 2000

8:30–9:00		Registration and welcoming remarks
9:00–10:00	Phyllis Cassidy	The evolution of differential algebraic geometry
10:15–11:15	Marius van der Put	Differential Galois theory
11:30–12:30	Andy Magid	Inverse differential Galois theory
2:00–3:00	Michael Singer	Galois theory of difference equations
3:15–4:15	Richard Churchill	Applications of differential Galois theory in Hamiltonian mechanics
4:30–5:30	William Sit	The Ritt-Kolchin theory for differential polynomials
5:30–6:30		Reception

Friday, November 3, 2000

9:00–10:00	Manuel Bronstein	Differential algebra and symbolic integration
10:15–11:15	Elizabeth Mansfield	Applications of differential algebra to symmetries and vice versa
11:30–12:30	Jerald Kovacic	Differential schemes
2:00–3:00	Henri Gillet	Differential algebra and schemes
3:15–4:15	Thomas Scanlon	Model theory and differential algebra
4:30–5:30	Li Guo	Baxter algebras and differential algebra

Contents

Differential Algebra and Related Topics, pp. 1–70
Proceedings of the International Workshop
Eds. L. Guo, P. J. Cassidy, W. F. Keigher & W. Y. Sit
© 2002 World Scientific Publishing Company

THE RITT–KOLCHIN THEORY
FOR DIFFERENTIAL POLYNOMIALS

WILLIAM Y. SIT

Department of Mathematics,
The City College of The City University of New York,
New York, NY 10031, USA
E-mail: wyscc@cunyvm.cuny.edu

This is a tutorial article to introduce the Ritt-Kolchin theory for differential polynomials. This theory has recently been further developed and successfully applied
to many computational problems where the main object of study is described by
a set of algebraic differential equations. A basic knowledge of the Ritt-Kolchin
theory has become a necessary prerequisite for leading-edge research in many pure
as well as applied disciplines. The goal of this paper is to present the fundamental
concepts and results with an emphasis on the relationship between the classical
theory and the modern symbolic computation approach.

Preface

This tutorial will be an exposition of the computational aspects of differential
polynomials as given in Kolchin [23]. It is both a much abridged and a much
expanded version. Abridged, because I have omitted many equally important
topics due to lack of time and space, and expanded, because I have filled in
details and included some (though by no means, all) recent developments.
I shall introduce the main concepts most relevant to symbolic computation.
Kolchin's treatment is very general and he developed the theory for differential polynomial rings and their ideals as much as possible for the partial case
over differential fields of arbitrary characteristics. To make the text more
accessible to the general reader, I have restricted these notes to the characteristic zero case only, and I have rearranged the order of presentation to provide
a tighter grouping of related topics. Compared to an earlier version of this
tutorial, where I skipped many technical lemmas but covered more ground,
this paper is almost self-contained and includes all that is essential to a thorough understanding of the main results. The highlight of the paper will be
the algorithm for decomposing a radical differential ideal into an intersection
of prime differential ideals or radical differential ideals with special properties
(for example, with known characteristic sets).

I have assumed the readers are familiar with ideal theory for the algebraic
case and have a working knowledge of Gröbner basis, and except for these,
I will discuss analogous concepts and constructions in detail. My goal is

to provide an easy but rigorous introduction to this rich theory. Readers whose appetites have been whetted should of course refer to Kolchin's text for generalizations and further results, and to Ritt [32] and original papers for authenticity. For an excellent historical commentary on differential algebra, see Buium and Cassidy [11].

These notes grew out of tutorial lectures I gave at the Mathematical Sciences Research Institute at Berkeley in 1998, the Mathematical Mechanization Research Center in Beijing in 2000, and the Workshop on Differential Algebra and Related Topics at Rutgers University at Newark in 2000. They are now divided into twelve sections and cover the basic notions and reduction methods which form a main staple for modern symbolic algorithms on algebraic differential equations. A large portion, which appeared in the version distributed at the Workshop, has been removed to make the paper more focused. As a consequence, the following topics are omitted: the geometric aspects of the theory, finitely generated differential field extensions, universal differential fields, differential dimension, differential dimension polynomials, systems of bounded orders, differential algebraic groups, and (old) results on the intersection of differential algebraic varieties that have yet to appear in book form. This paper also does not cover differential Galois theory.

A brief synopsis of the sections now follows. Section 1 introduces differential rings, differential ideals, and differential polynomial rings. Section 2 defines triangular sets of polynomials and pseudo-division. This is followed by a section on invertibility of polynomials, in particular, of initials of polynomials. Both Sections 2 and 3 are for a polynomial ring setting. In Section 4, ranking and reduction concepts in the differential polynomial setting are given. Characteristic sets are developed in Section 5, and this is followed by the reduction algorithm of Ritt and Kolchin in Section 6. The Rosenfeld properties, which provide crucial links between certain differential polynomial ideals and polynomial ideals, are studied in detail in Section 7. The algorithmic aspects for testing Rosenfeld properties and their relationship to coherence are developed in Section 8, leading to the famous lemma of Rosenfeld. In Section 9, we study the Ritt-Raudenbush basis theorem for differential polynomial rings (analogue of the Hilbert basis theorem in polynomial rings). Section 10 highlights the main steps and difficulties for the decomposition of a radical differential polynomial ideal into its prime components. The component theorems for a single differential polynomial are the focus in Section 11. The last section covers the preparation equation and the Low Power Theorem, which solves the prime decomposition problem for a single differential polynomial. Examples and exercises are given throughout the paper. Solutions to selected exercises are given in an Appendix.

1 Basic Definitions

In this paper, all non-zero rings are commutative with 1 and contain the ring of integers, $\mathbb{Z}$. All fields are assumed to have characteristic zero. The set of natural numbers, rational numbers, real numbers, and complex numbers are denoted respectively by $\mathbb{N}, \mathbb{Q}, \mathbb{R}$, and $\mathbb{C}$.

Definition 1.1 A *derivation* δ on a ring $\mathcal{R}$ is an additive map $\delta: \mathcal{R} \to \mathcal{R}$, such that for all $a, b \in \mathcal{R}$, $\delta(ab) = a \cdot \delta(b) + b \cdot \delta(a)$. Two derivations δ_1, δ_2 on $\mathcal{R}$ are said to *commute* if for all $a \in \mathcal{R}$, we have $\delta_1(\delta_2(a)) = \delta_2(\delta_1(a))$. We customarily write δa for $\delta(a)$ and inductively, write $\delta^h a$ for $\delta(\delta^{h-1}a)$. A ring $\mathcal{R}$ with a given single derivation δ is called an *ordinary differential ring* or *δ-ring*, and a ring $\mathcal{R}$ with a given family $\Delta = \{\delta_i\}_{1 \leq i \leq m}$ of m ($m \geq 0$) pairwise commuting, derivations $\delta_1, \cdots, \delta_m$ is called a *partial differential ring* or *Δ-ring* (which may be an ordinary differential ring or even just a ring). A Δ-ring $\mathcal{R}$ which is a field is a *differential field* or *Δ-field*. An element $x \in \mathcal{R}$ such that $\delta x = 0$ for all $\delta \in \Delta$ is called a *constant* or *Δ-constant* of $\mathcal{R}$. The set $\mathcal{C}$ of constants of $\mathcal{R}$ is a field if $\mathcal{R}$ is a Δ-field. A subset J of $\mathcal{R}$ is an *ideal* of a Δ-ring $\mathcal{R}$ if for all $a \in \mathcal{R}$ and $x, y \in J$, we have $ax \in J$ and $x + y \in J$. An ideal $\mathfrak{a}$ is a *differential ideal* or *Δ-ideal* of $\mathcal{R}$ if for all $x \in \mathfrak{a}$, $\delta \in \Delta$, we have $\delta x \in \mathfrak{a}$. A differential ideal $\mathfrak{a}$ of $\mathcal{R}$ is said to be *proper* if $\mathfrak{a} \neq \mathcal{R}$ (thus, the zero differential ideal is considered proper if $\mathcal{R}$ is not the zero ring). A differential ideal $\mathfrak{a}$ of $\mathcal{R}$ is *radical* (or *perfect*) if $\mathfrak{a}$ contains an element $F \in \mathcal{R}$ whenever it contains some positive integral power of F.

The intersection of an arbitrary family of (differential) ideals, the union of an ascending chain of (differential) ideals, and the sum of a finite number of (differential) ideals, are all (differential) ideals. In particular, any set $S \subset \mathcal{R}$ can generate a differential ideal, denoted by $[S]$, which is the smallest differential ideal containing S. The differential ideal $[S]$ is simply the set of all elements in $\mathcal{R}$ that can be obtained through a finite number of operations on the set S using addition, multiplication by elements of $\mathcal{R}$ and differentiation.

Example 1.2 Any ring $\mathcal{R}$ can become a Δ-ring with $\delta a = 0$ for all $a \in \mathcal{R}$ and all $\delta \in \Delta$. If $\mathbb{C}$ denotes the (differential) field of complex numbers, then $\mathcal{F} = \mathbb{C}(\sin^2 x, \sin x \cos x)$ is an ordinary differential field with respect to the derivation $\delta = d/dx$.

Example 1.3 If $\mathcal{F}$ is a field (or even just a ring), the polynomial ring $\mathcal{R} = \mathcal{F}[x_1, \cdots, x_m]$ is a partial differential ring with $\Delta = \{\partial/\partial x_1, \cdots, \partial/\partial x_m\}$. Here m, and hence Δ, may even be infinite. If $\mathbb{Z}$ is the ring of integers, then $\mathcal{R} = \mathbb{Z}[x]$ is an ordinary differential ring with respect to $\delta = d/dx$, the ideal

$\mathfrak{a} = (2, 2x)$ is a differential ideal, and the differential ideal $[x]$ generated by the single element x is the entire ring $\mathcal{R}$.

Example 1.4 The ring $\mathbb{Z}[x, e^x]$ of polynomials in x and e^x with coefficients over the integers $\mathbb{Z}$ is an ordinary differential ring with $\delta = \partial/\partial x$. The ideal (e^x) generated by e^x is the set of all elements in $\mathbb{Z}[x, e^x]$ that have e^x as a factor, and is a differential ideal.

Proposition 1.5 *Let $\mathcal{M}$ be the set of radical differential ideals in $\mathcal{R}$. Under set inclusion, $\mathcal{M}$ is an ordered set with the following properties:*

(a) *The intersection of any set of ideals in $\mathcal{M}$ is in $\mathcal{M}$.*

(b) *The union of any non-empty totally ordered set of ideals in $\mathcal{M}$ is in $\mathcal{M}$.*

(c) *If $\mathfrak{a} \in \mathcal{M}$ and $H \in \mathcal{R}$, then $\mathfrak{a} \colon H \in \mathcal{M}$.*

Definition 1.6 Given a Δ-ring $\mathcal{R}$, the free multiplicative monoid Θ generated by the set Δ is the set of *derivative operators* $\theta = \delta_1^{e_1} \cdots \delta_m^{e_m}$, where $e_1, \cdots, e_m$ are natural numbers. The *order* of θ is the natural number $e_1 + \cdots + e_m$. The set Θ acts on $\mathcal{R}$ by $\theta a = \delta_1^{e_1} \cdots \delta_m^{e_m} a$ for $a \in \mathcal{R}$. For any natural number s, we denote by $\Theta(s)$ the set of derivative operators of order $\leq s$.

Definition 1.7 Let $\mathcal{F}$ be a Δ-ring. Let $n > 0$ be given and let $\Theta Y = \{ y_{\theta,j} \}_{1 \leq j \leq n, \theta \in \Theta}$ be a family of indeterminates over $\mathcal{F}$. The polynomial ring $\mathcal{F}[\Theta Y]$ has a unique structure of a Δ-ring extending the given Δ-structure on $\mathcal{F}$ such that $\delta(y_{\theta,j}) = y_{\delta\theta,j}$ (where $\delta\theta$ is the product of δ and θ in Θ) for all $\delta \in \Delta$, $\theta \in \Theta$, and $1 \leq j \leq n$. The ring $\mathcal{F}[\Theta Y]$ equipped with this structure is called the *differential polynomial ring in n differential indeterminates* $y_1, \cdots, y_n$ (where $y_j = y_{1,j}$) and is denoted by $\mathcal{R} = \mathcal{F}\{ y_1, \cdots, y_n \}$ or $\mathcal{F}\{ y_1, \cdots, y_n \}_\Delta$. From now on, we shall replace $y_{\theta,j}$ by θy_j, which is a partial derivative of y_j. The order of a derivative $u = \theta y_j$ is denoted by $\mathrm{ord}\, u$ and is defined to be the order of θ. A *differential monomial* M is a finite power product of derivatives of the form θy_j and may be written as $\prod_k (\theta_k y_{j_k})^{e_k}$. In this product, the derivatives $\theta_k y_{j_k}$ are usually distinct, but neither the derivative operators θ_k nor the indices j_k need be distinct. A *differential polynomial* or Δ-*polynomial* is an element of $\mathcal{R}$ and is a finite linear combination $\sum_M a_M M$ of differential monomials M with coefficients a_M in $\mathcal{F}$. For any natural number s, the set of differential polynomials in $\mathcal{R}$ of order $\leq s$ is denoted by $\mathcal{R}_s$, which is simply the polynomial subring $\mathcal{F}[\{ \theta y_j \}_{\theta \in \Theta(s), 1 \leq j \leq n}]$ of $\mathcal{R}$.

Remark 1.8 The differential polynomial ring $\mathcal{R}$ is also a differential ring with respect to the infinite family of derivations $\{ \frac{\partial}{\partial(\theta y_j)} \}_{\theta \in \Theta, 1 \leq j \leq n}$.

Example 1.9 An example of a differential monomial of order 7 and degree 5 in 2 variables is $(\delta_1^4 \delta_2^3 y_1)(\delta_2^4 y_1)(\delta_1^3 \delta_2 y_2)^2 y_2$. An example of a δ-polynomial in n differential indeterminates is the Wronskian determinant

$$W(y_1, \ldots, y_n) = \begin{vmatrix} y_1 & y_2 & \cdots & y_n \\ \delta y_1 & \delta y_2 & \cdots & \delta y_n \\ \vdots & \vdots & \vdots & \vdots \\ \delta^{n-1} y_1 & \delta^{n-1} y_2 & \cdots & \delta^{n-1} y_n \end{vmatrix}.$$

It is well-known that n elements $\eta_1, \ldots, \eta_n$ of a δ-field $\mathcal{F}$ are linearly dependent over its field $\mathcal{C}$ of constants if and only if $W(\eta_1, \ldots, \eta_n)$ is zero (see Theorem 1, p. 86 of Kolchin [23] for a generalization to the partial case). For $n = 2$, the Wronskian determinant is $W = y_1 \delta y_2 - y_2 \delta y_1$. The derivative of W is $\delta W = y_1 \delta^2 y_2 - y_2 \delta^2 y_2$, the partial derivative of W with respect to δy_2 is y_1, and the partial derivative of δW with respect to δy_2 is zero.

Exercise 1.10 How many derivatives θy_j (with $\theta \in \Theta$, $1 \leq j \leq n$) are there in $\mathcal{R} = \mathcal{F}\{y_1, \cdots, y_n\}$ that are of order less than or equal to s? How many differential monomials are there in $\mathcal{R}$ of degree equal to 2?

Exercise 1.11 In the differential polynomial ring $\mathcal{R} = \mathcal{F}\{y_1, y_2\}$ with $\Delta = \{\delta_1, \delta_2\}$, show that for any natural number $h > 0$, the differential ideal generated in $\mathcal{R}$ by the four *linear* differential polynomials

$$P_{1,1} = \delta_1^h y_1, \; P_{1,2} = \delta_1 \delta_2^{h-1} y_2 + y_1, \; P_{2,1} = \delta_2 \delta_1^{h-1} y_1 + y_2, \; P_{2,2} = \delta_2^h y_2$$

is $[y_1, y_2]$.

From now on, unless explicitly stated otherwise, $\mathcal{F}$ will be a differential field of characteristic zero with a fixed set Δ of m commuting derivation operators. This tutorial will be mainly about differential polynomials and differential ideals in a differential polynomial ring with n differential indeterminates, which we henceforth denote by $\mathcal{R} = \mathcal{F}\{y_1, \cdots, y_n\}$. A differential polynomial $F \in \mathcal{R}$ is associated with a differential equation $F = 0$, where the y_j's are viewed as "infinitely differentiable unknown functions". A finite subset $\Phi \subset \mathcal{R}$ is associated with a system E_Φ of partial differential equations $F = 0, F \in \Phi$. The differential ideal $[\Phi]$ is associated with the system of all partial differential equations obtainable from E_Φ through a finite number of operations using addition, multiplication by differential polynomials (including by elements from $\mathcal{F}$), and differentiation.

2 Triangular Sets and Pseudo-Division

The notion of triangularity seems not to be explicitly used by Ritt or Kolchin, even though the property is inherent in an autoreduced set (to be defined later). One of the goals in "solving" a system of algebraic equations is to put the system into triangular form, or to replace it by systems that are triangular (for example, the resulting forms after applying the Gaussian elimination method to linear algebraic systems; or see Lazard [24] for zero-dimensional non-linear systems). This weaker notion is often sufficient for many applications in solving algebraic systems, in geometric theorem proving using the Ritt-Wu [39] method, in the Kalkbrener [19] algorithm for the unmixed-dimensional decomposition of an algebraic variety, and in the Rosenfeld-Gröbner algorithm of Boulier, *et al.* [5]. Since these will involve reduction procedures which are based on pseudo-division[1], we give a brief review here for the purpose of setting up the necessary notations. Details on pseudo-division may be found in Knuth [21], Mishra [28], or Chou [14], for example.

Definition 2.1 A subset **A** of differential polynomials is *triangular* if its elements can be rearranged as $A_1, A_2, \ldots, A_k, \ldots$ such that each A_k involves at least one derivative $\theta_k y_{j_k}$ which does not appear in $A_1, \ldots, A_{k-1}$ (in particular, $A_1 \notin \mathcal{F}$). When such a rearrangement is chosen, we write $\mathbf{A}: A_1, \ldots, A_k, \ldots$ and say **A** is in *triangular form with respect to* $\theta_1 y_{j_1}, \ldots, \theta_k y_{j_k}, \ldots$.

Example 2.2 The set $\mathbf{A}: A_1, A_2, A_3$ where $A_1 = \delta^2 y_2 + y_2$, $A_2 = \delta^2 y_2 + y_2^2 + y_3$, $A_3 = \delta^2 y_2 + y_1$ is in triangular form with respect to y_2, y_3, y_1, and *also* in triangular form with respect to $\delta^2 y_2, y_3, y_1$. The set **A** is triangular, but $\mathbf{A}: A_2, A_1, A_3$ is *not* in triangular form.

Remark 2.3 For a recent tutorial and comparative study in the theories and implementations of triangular sets, see the articles by Aubry *et al.* [1] and by Aubry and Maza [2]. However, Definition 2.1, while given in a differential algebra setting, differs from the one given by Aubry *et al.* (as well as by others, for example Lazard [24], Morrison [29]), even in the cases when Δ is specialized to the empty set, or when the derivatives are viewed as algebraic indeterminates, in that *no ordering of the variables* $y_1, \ldots, y_n$ *or of their derivatives are required*, but some ordering of the *members* of the set **A** *and a corresponding distinguished set of variables* must be given. When an ordering

---

[1]Kolchin described the process as a generalized Euclidean algorithm, even though the term pseudo-division already appeared in Knuth [21] in 1969. The modern reader is likely to be more familiar with the latter.

is given, elements of a triangular set in the sense of Definition 2.1 need not satisfy the defining property of triangularity according to Aubry *et al.*, who require that the highest variable appearing in A_k (called its main variable) be distinct for each k. In Example 2.2, when each A_k is viewed as a *polynomial*, and $\delta^2 y_2$ is viewed as y_4, with an ordering of $y_1 < y_2 < y_3 < y_4$, the main variable of A_k is equal to y_4 for every k and hence $\mathbf{A}$ is not triangular in the sense of Aubry *et al.* for *this* ordering. In the algebraic case, if a triangular set is finite, it is always possible to find an ordering (not necessarily uniquely) of the variables $y_1, \ldots, y_n$ so that the main variables are distinct. If we use an ordering $y_4 < y_1 < y_2 < y_3$, the main variables of A_k would all be distinct[2]. Unfortunately, this is generally not the case for differential polynomials because, as we shall see, the ordering (called a ranking) must be compatible with differentiations. In any ranking, we must have $y_2 < \delta^2 y_2$. A careful study of what results in computational differential algebra will remain valid for triangular sets (instead of autoreduced sets) would be of significant interest. Such studies are already evident in recent literature, for example, Morrison [29] and Sadik [36].

Proposition 2.4 *Let $\mathbf{S}$ be an integral domain, and v an indeterminate over $\mathbf{S}$. Let F, A be two polynomials in $\mathbf{S}[v]$ of respective degrees d_F and d_A, and suppose $A = I_{d_A} v^{d_A} + \cdots + I_1 v + I_0 \neq 0$, where $I_k \in \mathbf{S}$ for $0 \leq k \leq d_A$. Let $e = \max(d_F - d_A + 1, 0)$. Then we can compute unique polynomials $Q, R \in \mathbf{S}[v]$ such that*

$$I_{d_A}^e F = QA + R, \quad \text{and } \deg(R) < \deg(A). \quad (2.5)$$

Definition 2.6 The leading coefficient $I_{d_A} \in \mathbf{S}$ of A is called the *initial of A with respect to the variable v*. The unique polynomial Q (resp. R) is called[3] the *pseudo-quotient* (resp. *pseudo-remainder*) of F with respect to A and will be denoted by $Q(F, A, \mathbf{S}, v)$ (resp. $R(F, A, \mathbf{S}, v)$). We will call e in Proposition 2.4 the *pseudo-exponent* and denote this by $E(F, A, \mathbf{S}, v)$. *Any* triple (e, Q, R) which satisfies Equation (2.5) will be called a *pseudo-division triple* of F by A over $\mathbf{S}$ with respect to v.

Remark 2.7 In modern literature on symbolic computation, $R(F, A, \mathbf{S}, v)$ is often simply written as $\text{prem}(F, A)$. For reasons that will be clear later, we

---

[2]In this sense, Definition 2.1 generalizes that of Aubry *et al.*

[3]The prefix "pseudo" used in these notes are different from that used in Section 11 of Chapter I, p. 83, Kolchin [23], where its use is to generalize concepts of leader, partial reduction and reduction to the non-zero characteristic case.

emphasize the coefficient ring $\mathbf{S}$ and the univariate variable v in our notation. Observe that when A is linear in v, $E(F, A, \mathbf{S}, v) = \deg F$. In general, $E(F, A, \mathbf{S}, v)$ need not be the least exponent such that an equation of the form (2.5) holds. For example, $E(A, A, \mathbf{S}, v) = 1$, not 0. The uniqueness of Q, R are thus subject to the predefined e in Proposition 2.4. If (e, Q, R) and (e', Q', R') are two pseudo-division triples and $e' > e$, then $Q' = I_A^{e'-e} Q$ and $R' = I_A^{e'-e} R$. The pseudo-quotient, pseudo-remainder, and pseudo-exponent are easily computable (see Algorithm R, p. 369 of Knuth [21]). The reason for not using the least possible exponent is because the pseudo-division algorithm requires no division or greatest common divisor calculation in $\mathbf{S}$ and thus presumably runs faster, but at the expense of space. Hubert [17], for example, uses gcd computations to control coefficient and expression growth. In some parts of the theory that follows, only the existence of a pseudo-division triple (e, Q, R) plays a role, but in other parts, the specific one from Proposition 2.4 will be needed.

Example 2.8 In $\mathbb{Z}[v]$, let $F = 5v^2 + 3v$ and $A = 2v$. Performing pseudo-division of F by A, we have first $2F = 10v^2 + 6v = 5v(2v) + 6v = 5vA + 6v$ and then pseudo-dividing $6v$ by A, we have $4F = 10vA + 6(2v) + 0 = (10v + 6)A + 0$. So $Q(F, A, \mathbb{Z}, v) = 10v + 6$, $R(F, A, \mathbb{Z}, v) = 0$ and $E(F, A, \mathbb{Z}, v) = 2 = \deg F$. Note that we could also have $2F = (5v + 3)A + 0$. Thus $(2, 10v + 6, 0)$ and $(1, 5v + 3, 0)$ are both pseudo-division triples of F by A over $\mathbb{Z}$ with respect to v. If A were replaced by $2v + 1$, we would have $(2, 10v + 1, -1)$ as a pseudo-division triple instead.

For simultaneous pseudo-divisions of several polynomials $F_1, \cdots, F_q$ by A, we can obtain a pseudo-division triple (e, Q_i, R_i) of F_i for each i with a common exponent e since the initial of A does not involve v. We now generalize Proposition 2.6 to successive pseudo-divisions using a sequence of divisors.

Proposition 2.9 Let $\mathbf{S}_0$ be an integral domain, $V = \{v_1, \cdots, v_p\}$ be a set of indeterminates over $\mathbf{S}_0$. For $1 \leq k \leq p$, let $\mathbf{S}_k = \mathbf{S}_0[v_1, \ldots, v_k]$. Let $\mathbf{A}: A_1, \ldots, A_p$ be a subset of $\mathbf{S}_p$ in triangular form with respect to $v_1, \ldots, v_p$. For $1 \leq k \leq p$, let d_k be the degree of A_k in v_k and let $I_k \in \mathbf{S}_{k-1}$ be the initial of A_k with respect to v_k. Then we can compute natural numbers $e_1, \cdots, e_p$ and polynomials $Q_1, \cdots, Q_p$ and R_1 in $\mathbf{S}_0[V]$ such that

$$I_1^{e_1} \cdots I_p^{e_p} F = Q_1 \cdot A_1 + \cdots + Q_p \cdot A_p + R_1, \tag{2.10}$$

where $\deg_{v_k} R_1 < d_k$ for $1 \leq k \leq p$.

Proof. With $R_{p+1} = F$, we can compute inductively for $k = p, \ldots, 1$ pseudo-division triples (e_k, Q'_k, R_k) of R_{k+1} by A_k over $\mathcal{S}_{k-1}$ with respect to v_k so that $I_k^{e_k} R_{k+1} = Q'_k A_k + R_k$, with $\deg_{v_i} R_k < d_i$ for $k \leq i \leq p$. Then let $Q_k = I_1^{e_1} \cdots I_{k-1}^{e_{k-1}} Q'_k$. $\qquad\square$

Definition 2.11 If the pseudo-division triples (e_k, Q'_k, R_k) in the proof of Proposition 2.9 are taken to be the corresponding pseudo-exponents, quotients, and remainders, the sequence $R_p, \ldots, R_1$ is known as the *pseudo-remainder sequence* of F with respect to the set $\mathbf{A}$. These particular choices of pseudo-exponents e_k (resp. quotients Q_k, resp. remainders R_k) are denoted by $E_k(F, \mathbf{A}, \mathcal{S}_0, V)$ (resp. $Q_k(F, \mathbf{A}, \mathcal{S}_0, V)$, resp. $R_k(F, \mathbf{A}, \mathcal{S}_0, V)$). We call *any* R_1 satisfying the property in Proposition 2.9 a *pseudo-remainder* of F with respect to $\mathbf{A}$.

Remark 2.12 The reader should be aware that there is nothing unique in Equation (2.10). Even for the pseudo-remainder sequence, the exponents $e_1, \ldots, e_p$ do not just depend on F and the degrees d_k, but also on R_k and hence on the degrees in the initials I_k. Many authors used $\mathrm{prem}(F, \mathbf{A})$ to mean some kind of pseudo-remainder R_1 without giving a specific sequence of pseudo-divisions or the pseudo-division algorithm. If that is the case, it is the responsibility of the authors to prove that their results are independent of such choices. While for a single pseudo-division operation, one can easily minimize the exponent used through a gcd computation, it is far from clear how to choose a sequence of pseudo-division operations to compute some pseudo-remainder R_1 satisfying Equation (2.10) with overall efficiency in time and space. This becomes even more complex when the set $\mathbf{A}$ is not triangular (repeated pseudo-division is still possible). Morrison [29] studied the theoretical aspect for such cases. A comparative study of efficiency with various sequencing strategies would be of great interest.

3 Invertibility of Initials

In this section, we study a useful notion and an important lemma (Lemma 3.6) which appeared as part of Lemma 13(a), page 36 of Kolchin [23]. Our version is for a triangular set rather than an autoreduced set (see later for definition), and this generalization was observed by Sadik [36]. This notion has been studied afresh recently under the term *invertibility* (and a closely related notion: *regularity*). The results of this section, while presented differently and with new examples, have all appeared in one or more of the papers by Audry *et al.* [1], Kandri Rody *et al.* [18], Hubert [17],

Boulier *et al.* [5,6,7,8], Bouziane *et al.* [10], and Sadik [36]. By replacing irreducibility with invertibility or regularity, the constructive treatments by these authors allow a significant improvement in the Ritt-Kolchin algorithm for expressing a radical differential ideal as an intersection of prime (or radical) differential ideals with special properties that we shall explore later. Many of these recent modifications depend on the special properties of Gröbner bases of zero-dimensional polynomial ideals. In this tutorial, we present a simpler theory without going into implementation, and will be satisfied merely with demonstrating the effectiveness.

Throughout this section, S_0 is an integral domain containing $\mathbb{Q}$ with quotient field K_0, and $v_1, \ldots, v_p$ are indeterminates over K_0. Let k be an integer with $1 \leq k \leq p$. Let S_k be the polynomial ring $S_0[v_1, \ldots, v_k]$ and let K_k be the polynomial ring $K_0[v_1, \ldots, v_k]$. Let $A: A_1, \ldots, A_p$ be a subset in S_p that is triangular with respect to $v_1, \ldots, v_p$ over S_0. Let I_k be the initial of A_k, d_k be the degree in v_k of A_k, and $I = I_A = I_1 \cdots I_p$ be the product of the initials. Let (A_k) denote the ideal generated by $A_k: A_1, \ldots, A_k$ in either S_k or K_k depending on context. If need be, for any subset or ideal Σ of S_k, we shall use $S_k \cdot \Sigma$ or Σ_S for the ideal (Σ) in S_k, and Σ_K for the one in K_k. We adopt the convention that $(\Sigma) = (0)$ if $\Sigma = \emptyset$. We denote $K_p/(A)$ by V_A, which is both a ring (perhaps the zero ring, which happens if and only if $(A)_S \cap S_0 \neq (0)$) and a vector space over K_0. Let $\eta : K_p \longrightarrow V_A$ be the canonical quotient homomorphism. For any $F \in S_p$, let $\eta_F : V_A \longrightarrow V_A$ denote the multiplication map by F, that is, $\eta_F(\eta(G)) = \eta(FG)$ for any $G \in K_p$. Clearly η_F is a vector space endomorphism of V_A. In any commutative ring (including the zero ring), an element x is *multiplicatively invertible* if there exists an element y such that $xy = 1$; a *zero divisor* is a non-zero element x such that there is a non-zero y with $xy = 0$.

Proposition 3.1 *Let $F \in S_p$. Then the following conditions are equivalent:*

(a) *η_F is surjective.*

(b) *$\eta(F)$ is multiplicatively invertible in the ring V_A.*

(c) *There exist $L \in S_0$, $L \neq 0$ and $M \in S_p$ such that $L \equiv MF \bmod (A)_S$.*

(d) *$(A, F)_S \cap S_0 \neq (0)$.*

Proof. When V_A is the zero ring (equivalently, when $(A) \cap S_0 \neq (0)$ or when $(A) = S_p$), conditions (a) through (d) are trivially satisfied by any F. Thus we need only prove the case when V_A is not trivial.

(a) $\Rightarrow$ (b): When η_F is surjective, there exists $G \in K_p$ such that $\eta_F(\eta(G)) = \eta(1)$, that is, $\eta(F)\eta(G) = \eta(1)$.

(b) $\Rightarrow$ (c): If $\eta(F)$ is multiplicatively invertible, let its inverse be $\eta(G)$, where $G = N/D$, $N \in \mathbf{S}_p$, $D \in \mathbf{S}_0$, $D \neq 0$. We have $FG \equiv 1 \mod (\mathbf{A})_K$ and hence $FG - 1 = \sum_{i=1}^p C_i A_i$ where $C_i = N_i/D_i$, $N_i \in \mathbf{S}_p$, $D_i \in \mathbf{S}_0$, $D_i \neq 0$. Then $L = D \prod_{i=1}^p D_i$ and $M = N \prod_{i=1}^p D_i$ satisfy (b).

(c) $\Rightarrow$ (d) is clear.

(d) $\Rightarrow$ (a): Let $L \in \mathbf{S}_0$, $L \neq 0$, and $L = MF + \sum_{k=1}^p C_k A_k$, where $M, C_k \in \mathbf{S}_p$. Since $V_\mathbf{A}$ is not trivial, $\eta(L) \neq 0$ and $\eta(1) = \eta(MF/L)$. Let $G \in K_p$. We have $\eta(G) = \eta(MFG/L) = \eta_F(\eta(MG/L))$ and hence η_F is surjective. $\qquad\square$

Definition 3.2 We say $F \in \mathbf{S}_p$ is *invertible with respect to the triangular set* $\mathbf{A}: A_1, \ldots, A_p$ *and corresponding variables* $v_1, \ldots, v_p$ *over* $\mathbf{S}_0$ (or simply that F is *invertible* when the context is clear) if any one of the conditions (a) through (d) in Proposition 3.1 holds.

Corollary 3.3 *Let $F = F_1 F_2 \in \mathbf{S}_p$. Then F is invertible if and only if F_1, F_2 are invertible.*

Corollary 3.4 *Let $\mathbf{S}_0'$ be an integral domain with quotient field K_0', $\mathbf{S}_0$ be a subdomain of $\mathbf{S}_0'$, and $v_1, \ldots, v_p$ be indeterminates over K_0'. Let $\mathbf{A}: A_1, \ldots, A_p$ be a subset of $\mathbf{S}_p$ triangular with respect to $v_1, \ldots, v_p$ over $\mathbf{S}_0$. If $F \in \mathbf{S}_p$ is invertible with respect to $\mathbf{A}$ over $\mathbf{S}_0$, then $F \in \mathbf{S}_0'[v_1, \ldots, v_p]$ is invertible with respect to $\mathbf{A}$ over $\mathbf{S}_0'$.*

Definition 3.5 If $\mathbf{A}: A_1, \ldots, A_p$ is triangular with respect to $v_1, \ldots, v_p$, and if for $1 \leq k \leq p$, the initial I_k of A_k is invertible with respect to $A_1, \ldots, A_{k-1}$ and corresponding variables $v_1, \ldots, v_{k-1}$ over $\mathbf{S}_0$, we say the triangular set $\mathbf{A}$ *has invertible initials*, or more precisely, *has invertible initials with respect to* $v_1, \ldots, v_p$ *over* $\mathbf{S}_0$. If $p = 1$, then I_1 is always invertible (with respect to the empty set) over $\mathbf{S}_0$.

Lemma 3.6 *Suppose $\mathbf{A}: A_1, \ldots, A_p$ has invertible initials with respect to $v_1, \ldots, v_p$ over an integral domain $\mathbf{S}_0$. Let I_k be the initial and d_k be the degree in v_k of A_k. Let $I = I_1 \cdots I_p$. Let $F \in (A_1, \ldots, A_p): I^\infty$ be such that for $1 \leq k \leq p$, the degree of F in v_k is $< d_k$. Then $F = 0$.*

Proof. We prove the lemma by induction on p. For $p = 1$, if $F \in (A_1): I_1^\infty \subset \mathbf{S}_1$ were non-zero with degree in v_1 less than d_1, then for some $e_1 \in \mathbb{N}$, $I_1^{e_1} F$ would be divisible by A_1, which would be a contradiction since $I_1 \in \mathbf{S}_0$ has degree 0 in v_1. Suppose the lemma has been proved for the case $p \geq 1$ and suppose the hypothesis of the lemma holds when p is replaced by $p + 1$.

Let $F \in (A_1, \ldots, A_{p+1}) \colon (I_1 \cdots I_{p+1})^\infty \subseteq \mathbf{S}_{p+1}$ have degree $< d_k$ in v_k for $1 \leq k \leq p+1$. Let $I_{p+1}^h I^h F = \sum_{i=1}^{p+1} C_i A_i$, where $h \in \mathbb{N}$, $C_i \in \mathbf{S}_{p+1}$ and $I = I_1 \cdots I_p$. Let (e, Q_i, R_i), where $Q_i, R_i \in \mathbf{S}_{p+1}$, be a pseudo-division triple of C_i by A_{p+1} for $1 \leq i \leq p$. Then

$$I_{p+1}^{e+h} I^h F - \sum_{i=1}^{p} R_i A_i = \left(I_{p+1}^e C_{p+1} + \sum_{i=1}^{p} Q_i A_i \right) A_{p+1}.$$

Each side of the above equation must be identically zero, since otherwise, the left hand side would have a lower degree in v_{p+1} than the right hand side. Let $L = MI_{p+1} + N \in (A_1, \ldots, A_p, I_{p+1}) \cap \mathbf{S}_0$, where $L \neq 0$, $L \in \mathbf{S}_0$, $M \in \mathbf{S}_p$, and $N \in (A_1, \ldots, A_p)$. Letting $R_i' = M^{e+h} R_i$, we obtain,

$$(L - N)^{e+h} I^h F = \sum_{i=1}^{p} R_i' A_i,$$

showing that $L^{e+h} F \in \mathbf{S}_{p+1} \cdot (A_1, \ldots, A_p) \colon I^\infty$. Letting $\mathbf{S}_0' = \mathbf{S}_0[v_{p+1}]$ and $\mathbf{S}_p' = \mathbf{S}_{p+1} = \mathbf{S}_0'[v_1, \ldots, v_p]$, the induction hypothesis shows that $L^{e+h} F = 0$ and hence $F = 0$. $\qquad\square$

Definition 3.7 The ideal $(\mathbf{A}) \colon I^\infty = (A_1, \ldots, A_p) \colon I^\infty$ (in $\mathbf{S}_p$) in Lemma 3.6 is called a *saturation ideal of* $\mathbf{A}$ *with respect to its initials* and henceforth will be denoted by J_I whenever $\mathbf{A}$ is clear from the context.

Remark 3.8 If $\mathbf{A}$ has invertible initials, then by Lemma 3.6, $J_I \cap \mathbf{S}_0 = (0)$ and also $\mathbf{S}_k \cdot (A_1, \ldots, A_k) \cap \mathbf{S}_0 = (0)$ for $1 \leq k \leq p$. It follows that $V_{\mathbf{A}}$ is a non-trivial vector space. For any non-zero $L \in \mathbf{S}_0$, we have $\eta(L) = L\eta(1) \neq 0$.

Corollary 3.9 *Assume the hypotheses are as in Lemma 3.6. Let K_0 be the quotient field of $\mathbf{S}_0$. Then $V_{\mathbf{A}} = K_0[v_1, \ldots, v_p]/(\mathbf{A})$ is a non-trivial finite dimensional vector space over K_0 with dimension $\prod_{k=1}^{p} d_k$.*

Proof. By Remark 3.8, $V_{\mathbf{A}}$ is non-trivial. Let $\mathbf{A}_k \colon A_1, \ldots, A_k$, and let $\eta_k \colon K_k \longrightarrow K_k/(\mathbf{A}_k)$ be the canonical homomorphism. Let

$$B_k = \{\, \eta_k(v_1^{e_1} \cdots v_k^{e_k}) \mid 0 \leq e_i < d_i \text{ for all } 1 \leq i \leq k \,\}. \qquad (3.10)$$

Any linear dependence relation $\sum C_j \eta_k(D_j) = 0$, where $C_j \in K_0$, D_j is a monomial in $v_1, \ldots, v_k$ with degree in v_i less than d_i, yields, after clearing denominators, an $F = L \sum C_j D_j \in \mathbf{S}_k \cdot (\mathbf{A}_k)$ for some non-zero $L \in \mathbf{S}_0$ such that $LC_j \in \mathbf{S}_0$. By Lemma 3.6, $LC_j = 0$ and hence $C_j = 0$. It thus suffices to prove by induction on k that B_k also generates $K_k/(\mathbf{A}_k)$ as a vector space over K_0. For $k = 0$, this is trivial. For $k \geq 1$, let $G \in K_k = K_{k-1}[v_k]$ be written as

$\sum_{i=1} G_i v_k^i$ where $G_i \in K_{k-1}$. Clearly, we have a natural embedding ι_k of vector spaces from $K_{k-1}/(\mathbf{A}_{k-1})$ into $K_k/(\mathbf{A}_k)$. Since $\eta_k(G_i) = \iota_k(\eta_{k-1}(G_i))$, by the induction hypothesis, it suffices to show that for any $i \in \mathbb{N}$, $\eta_k(v_k^i)$ belongs to the subspace generated by B_k. Let (e, Q_k, R_k) be a pseudo-remainder triple of v_k^i with respect to A_k over $\mathbf{S}_{k-1}$, so that $I_k^e v_k^i = Q_k A_k + R_k$. Since I_k is invertible respect to $\mathbf{A}_{k-1}$, let $L_k = M_k I_k + N_k \in \mathbf{S}_{k-1} \cdot (\mathbf{A}_{k-1}, I_k) \cap \mathbf{S}_0$, where $L_k \neq 0$, $N_k \in \mathbf{S}_{k-1} \cdot (\mathbf{A}_{k-1})$, and $M_k \in \mathbf{S}_{k-1}$. By Remark 3.8, $\eta_k(L_k) \neq 0$. It follows that $\eta_k(v_k^i) = \eta_k((M_k I_k/L_k)^e v_k^i) = \eta_k((M_k/L_k)^e R_k)$. Since the degree in v_k of $(M_k/L_k)^e R_k$ is $< d_k$, the proof is complete by induction. $\square$

Remark 3.11 The proof of Corollary 3.9 is constructive, that is, given any $F \in K_p$, we can compute a unique representation of $\eta(F)$ as a linear combination of the basis B_p with coefficients in K_0. Notice that when it is known that $\mathbf{A}$ has invertible initials, we can compute, using Gröbner basis methods, an expression representing 1 as a linear combination of $A_1, \ldots, A_{k-1}, I_k$ over K_0. By clearing denominators, we can compute L_k, M_k, N_k with the properties as stated in the proof. Since pseudo-division is also effective, this allows us to replace any v_k^i occuring in F where $k \geq d_k$ by $(M_k/L_k)^e R_k$ (here e and R_k both depend also on i as well as k), that is, by a polynomial in K_k where the degree in v_k is $< d_k$. Proceeding with this replacement inductively for $k = p, \ldots, 1$ eventually yields the representation for $\eta(F)$.

Corollary 3.12 *Suppose $\mathbf{A}$ has invertible initials and $F \in \mathbf{S}_p$. Then we can compute a non-zero polynomial $P(X) \in \mathbf{S}_0[X]$ of minimal degree such that $\eta(P(F)) = 0$.*

Proof. If d is the dimension of the K_0 vector space $V_{\mathbf{A}}$, then the elements $\eta(1), \eta(F), \ldots, \eta(F^d)$ must be linearly dependent over K_0. By Corollary 3.9 and Remark 3.11, such a dependence relation involving a minimal power $F^{d'}$ of F can be obtained by linear algebra. If $\sum_{i=0}^{d'} N_i \eta(F^i)/D_i = 0$, where $N_i, D_i \in \mathbf{S}_0, D_i \neq 0$, then clearing the denominators yields the polynomial $P(X)$ of degree d' satisfying the requirements. $\square$

Definition 3.13 A non-zero polynomial $P \in \mathbf{S}_0[X]$ of minimal degree in X such that $\eta(P(F)) = 0$ will be called a *minimal polynomial of F modulo $\mathbf{A}$ over $\mathbf{S}_0$* and denoted by $P_F(X)$. Note that for some $L \neq 0$ in $\mathbf{S}_0$, $LP(F) \in (\mathbf{A})_\mathbf{S}$ and $P_F(X)$, if it exists, is unique up to a non-zero factor in $\mathbf{S}_0$.

Example 3.14 Corollaries 3.9 and 3.12 are false if $\mathbf{A}$ is triangular but does not have invertible initials. For example, if $\mathbf{A}: v_1, v_1 v_2$, then $V_{\mathbf{A}} \cong K_0[v_2]$ is non-trivial and has infinite dimension over K_0. The polynomial $F = v_2$ has no minimal polynomial and is not invertible. The converse of Corollary 3.9 is also false: if $\mathbf{A}: v_1, v_1 v_2^2 + v_2$, then $V_{\mathbf{A}} \cong K_0$ has dimension 1 but $\mathbf{A}$ does not have invertible initials.

We now define another notion, closely related to invertibility. This concept of regularity is used by many authors for computing the decomposition of a radical polynomial ideal. We show below that this seemingly weaker notion is, for all practical purposes, equivalent to invertibility.

Definition 3.15 We say $F \in \mathcal{S}_p$ is *regular with respect to the triangular set* $\mathbf{A} : A_1, \ldots, A_p$ *and corresponding variables* $v_1, \ldots, v_p$ *over* $\mathcal{S}_0$ (or simply that F is *regular* when the context is clear) if $((\mathbf{A}): I^\infty, F) \cap \mathcal{S}_0 \neq (0)$, where $I = I_1 \cdots I_p$ is the product of initials. We say $\mathbf{A}$ *has regular initials*[4] with respect to $v_1, \ldots, v_p$ over $\mathcal{S}_0$ if for each k, $1 \leq k \leq n$, the initial I_k of A_k is regular with respect to $\mathbf{A}_{k-1}: A_1, \ldots, A_{k-1}$.

Theorem 3.16 *Suppose* $\mathbf{A}$ *has invertible initials and* $F \in \mathcal{S}_p$. *Then the following are equivalent:*

(a) *F is regular with respect to* $\mathbf{A}$ *over* $\mathcal{S}_0$.
(b) *F is invertible with respect to* $\mathbf{A}$ *over* $\mathcal{S}_0$.
(c) *$\eta(F) \neq 0$ and $\eta(F)$ is not a zero divisor.*
(d) *η_F is injective.*
(e) *Every pseudo-remainder of* $G \in (\mathbf{A}): F^\infty$ *with respect to* $\mathbf{A}$ *is zero.*
(f) *For every minimal polynomial $P_F(X)$ of F, we have $P_F(0) \neq 0$.*

Proof. (a) $\Rightarrow$ (b): Suppose F is regular, and let $L' \in \mathcal{S}_0$, $L' \neq 0$, $M' \in \mathcal{S}_p$ and $N' \in (\mathbf{A}): I^\infty$ be such that $L' = M'F + N'$. Then $N' = L' - M'F$ and there exists some $e \in \mathbb{N}$ such that $I^e N' = I^e(L' - M'F) \in (\mathbf{A})$. Since $\mathbf{A}$ has invertible initials, let $L_k \in \mathcal{S}_0, M_k \in \mathcal{S}_{k-1}, N_k \in (\mathbf{A}_{k-1})$ be such that $L_k \neq 0$ and $L_k = M_k I_k + N_k$. Thus $\prod_{k=1}^p (L_k - N_k)^e N' = \prod_{k=1}^p (L_k - N_k)^e (L' - M'F) \in (\mathbf{A})$. Hence there exist $L \in \mathcal{S}_0$, $L \neq 0$, and $M \in \mathcal{S}_p$, $N \in (\mathbf{A})$ such that $L = MF + N$.

(b) $\Rightarrow$ (c): Let F be invertible and let L, M be as in (c) of Proposition 3.1. Then $\eta(L) = \eta(MF)$. Since $V_{\mathbf{A}}$ is not trivial, $\eta(L) \neq 0$, and hence $\eta(F) \neq 0$ and $\eta(MF/L) = \eta(1)$. Let $G \in K_0[v_1, \ldots, v_p]$ be such that $\eta(FG) = 0$. Then $\eta(G) = \eta(MFG/L) = 0$, showing that $\eta(F)$ cannot be a zero-divisor in $K_0[v_1, \ldots, v_p]/(\mathbf{A})$.

(c) clearly implies (d).

---

[4]Some authors say $\mathbf{A}$ is a *regular chain*. Corollary 3.17 is given as Theorem 1.2.1 in Bouziane *et al.* [10] and as part of Theorem 6.1 of Aubry *et al.* [1].

(d) $\Rightarrow$ (e): Let $G \in (\mathbf{A}): F^\infty$. For some $e \in \mathbb{N}$, $F^e G \in (\mathbf{A})$. Let G_0 be a pseudo-remainder of G with respect to $\mathbf{A}$ and let $I_1^{e_1} \cdots I_p^{e_p} G = Q_1 A_1 + \cdots + Q_p A_p + G_0$. Then $F^e G_0 \in (\mathbf{A})$. Hence $\eta(F)^e \eta(G_0) = 0$. Since η_F is injective, we must have $\eta(G_0) = 0$. This means $L G_0 \in (\mathbf{A})$ for some non-zero $L \in \mathbf{S}_0$, and hence by Lemma 3.6, $G_0 = 0$.

(e) $\Rightarrow$ (f): Suppose (f) is false and suppose there is a minimal polynomial $P(X) \in \mathbf{S}_0[X]$ of F such that $P(0) = 0$. Write $P(X) = X Q(X)$. Then there exists a non-zero $L \in \mathbf{S}_0$ such that $L Q(F) \in (\mathbf{A}): F^\infty$. By hypothesis (e), every pseudo-remainder of $L Q(F)$ with respect to $\mathbf{A}$ is zero. In particular, by the invertibility of initials, there exists a non-zero $L' \in \mathbf{S}_0$ such that $L' L Q(F) \in (\mathbf{A})$. Thus $P'(X) = L' L Q(X)$, which has lower degree than the minimal polynomial $P(X)$, satisfies $\eta(P(F)) = 0$, leading to a contradiction.

(f) $\Rightarrow$ (a): By Corollary 3.12, there exists a minimal polynomial $P_F(X)$ for F. Let $P_F(X) = P_F(0) + X Q(X)$. Since $P_F(0) \in \mathbf{S}_0$, and $\eta(P(F)) = 0$, there exists a non-zero $L \in \mathbf{S}_0$ such that $L P_F(F) \in (\mathbf{A})_{\mathbf{S}}$ and hence $L P_F(0) = L P_F(F) - F Q(F)$ is a non-zero element of $(\mathbf{A}, F) \cap \mathbf{S}_0$ and therefore of $((\mathbf{A}): I^\infty, F) \cap \mathbf{S}_0$. Thus F is regular. $\qquad\square$

Corollary 3.17 *Let $\mathbf{A}$ be a triangular subset of $\mathbf{S}_p$. Then $\mathbf{A}$ has invertible initials if and only if $\mathbf{A}$ has regular initials.*

Proof. Clearly, if $\mathbf{A}$ has invertible initials, it has regular initials. Conversely, suppose $\mathbf{A}$ has regular initials. Since $\mathbf{A}_1 : A_1$ has invertible initial, it follows from Theorem 3.16 that I_2 is invertible with respect to $\mathbf{A}_1$. By induction, $\mathbf{A}$ has invertible initials. $\qquad\square$

Example 3.18 The equivalences in Theorem 3.16 do not always hold without the hypothesis that $\mathbf{A}$ has invertible initials. Let $p = 2$ and let $\mathbf{S}_0$ be the polynomial ring $\mathbb{Z}[u]$ in one variable u. Let $A_1 = u v_1^2$, $A_2 = v_1 v_2 - 1$. Then $\mathbf{A}: A_1, A_2$ is triangular with respect to v_1, v_2 over $\mathbf{S}_0$, and since $1 = v_2 v_1 - A_2$, $F = v_1$ is invertible with respect to $\mathbf{A}$ by (c) of Proposition 3.1. But $u = v_2^2 A_1 - u(v_1 v_2 + 1) A_2 \in (\mathbf{A})$, and hence $V_{\mathbf{A}}$ is trivial. So η_F is injective, and $\eta(F) = 0$ even though $F \notin (\mathbf{A})_{\mathbf{S}} = (u, A_2)$. The minimal polynomial $P_F(X)$ (with respect to $\mathbf{A}$) is 1 since $\eta(P_F(F)) = \eta(1) = 0$. The polynomial $G = u v_1 \in (\mathbf{A}): F^\infty$ has pseudo-remainder G. Thus (b), (d), and (f) do not imply (c) or (e). Note that the initial $I_2 = v_1$ is *not* invertible *with respect to* $\mathbf{A}_1$, its minimal polynomial (with respect to $\mathbf{A}_1$) is $u X^2$, and therefore $\mathbf{A}$ does not have invertible initials.

Example 3.19 Let u, v_1, v_2, v_3 be indeterminates over $\mathbb{Q}$. Let $\mathbf{S}_0 = \mathbb{Q}[u]$ and let $A_1 = u v_1$, $A_2 = u(v_1 + 1) v_2$ and $A_3 = (v_1 + 1) v_3 + u$. Then $\mathbf{A}: A_1, A_2, A_3$

is triangular with respect to v_1, v_2, v_3 over $\mathbf{S}_0$. The initials are $I_1 = u$, $I_2 = u(v_1+1)$, and $I_3 = v_1+1$. Since $u = I_2 - A_1$ and $u = uI_3 - A_1$, $\mathbf{A}$ has invertible initials. Note that the ideal $(\mathbf{A})$ in $K_0[v_1, v_2, v_3]$ is generated by $v_1, v_2, v_3 + u$. Thus $\eta(I_1), \eta(I_2), \eta(I_3)$ are non-zero, and they are not zero-divisors. However, the images of I_1, I_2, I_3 in $\mathbf{S}_0[v_1, v_2, v_3]/(\mathbf{A})$ are all zero-divisors.

Corollary 3.20 Let $\mathbf{A}: A_1, \ldots, A_p$ be a triangular set with respect to $v_1, \ldots, v_p$ in $\mathbf{S}_p$. Let J_I be the saturation ideal of $\mathbf{A}$ with respect to initials. Then $\mathbf{A}$ has invertible initials if and only if every pseudo-remainder of $G \in J_I$ with respect to $\mathbf{A}$ is zero.

Proof. If $\mathbf{A}$ has invertible initials, then I is invertible with respect to $\mathbf{A}$ by Corollary 3.3 and hence by Theorem 3.16, every pseudo-remainder of $G \in J_I$ with respect to $\mathbf{A}$ is zero. To prove the converse, note that I_1 is by definition invertible with respect to the empty set. Suppose by induction, we have proved that for some $2 \leq k < p$, I_j is invertible with respect to $\mathbf{A}_{j-1}$ for $j < k$. Let $G \in \mathbf{S}_{k-1} \cdot ((\mathbf{A}_{k-1}): I_k^\infty)$. Then $G \in J_I$ and let $G_p, \ldots, G_1$ be the pseudo-remainder sequence of G with respect to $\mathbf{A}$ (see Definition 2.11). Then $G_p = \cdots = G_k = G$ and the pseudo-remainder sequence of G with respect to $\mathbf{A}_{k-1}$ is $G_{k-1}, \ldots, G_1$. By hypothesis, $G_1 = 0$. By Theorem 3.16 applied to $\mathbf{A}_{k-1}$, I_k is invertible with respect to $\mathbf{A}_{k-1}$. $\square$

Remark 3.21 Corollary 3.20 shows that the converse of Lemma 3.6 also holds. Note that $\mathbf{A}$ having invertible initials is not equivalent to I invertible with respect to $\mathbf{A}$ (see Example 3.18).

Theorem 3.22 *Suppose $\mathbf{A}$ is triangular as before.*
(a) *The property that $\mathbf{A}$ has invertible initials is decidable.*
(b) *If $\mathbf{A}$ has invertible initials, and $F \in \mathbf{S}_p$, then the invertibility of F with respect to $\mathbf{A}$ is decidable. If F is invertible, we can compute a non-zero $L \in \mathbf{S}_0$, $M \in \mathbf{S}_p$, and $N \in (\mathbf{A})$ such that $L = MF + N$. If F is not invertible, we can compute a $G \in \mathbf{S}_p$, such that the degree of G in v_k is $< d_k$ for all $1 \leq k \leq p$, $GF \in (\mathbf{A})$ and $G \notin (\mathbf{A})$.*

Proof. We prove (a) and simultaneously (b) in the special case when $\mathbf{A}$ is $\mathbf{A}_k$ and $F = I_k$, by induction on k, $k = 1, \ldots, p$. For $k = 1$, $\mathbf{A}_1$ always has invertible initial and for $F = I_1$, we may take $L = I_1, M = 1, N = 0$. For $2 \leq k < p$, assume that the above holds for $j \leq k-1$. We are done if $\mathbf{A}_{k-1}$ does not have invertible initials. So suppose $\mathbf{A}_{k-1}$ has invertible initials. By Corollary 3.12, we can compute a minimal polynomial $P_k(X) = P_k(0) + XQ_k(X)$ of $F = I_k$ with respect to $\mathbf{A}_{k-1}$ and $L_k \neq 0$, $L_k \in \mathbf{S}_0$ such that $L_k P_k(I_k) \in (\mathbf{A}_{k-1})$. If

$P_k(0) \neq 0$, then I_k is invertible with respect to $\mathbf{A}_{k-1}$ by Theorem 3.16, and we may take $L = L_k P_k(0), M = -Q_k(I_k)$ and $N = L_k P_k(I_k)$. If $P_k(0) = 0$, let $V_{k-1} = \{v_1, \ldots, v_{k-1}\}$ and let $I_1^{e_1} \cdots I_{k-1}^{e_{k-1}} Q_k(I_k) = \sum_{j=1}^{k-1} Q'_j A_j + G_k$ be the result of the pseudo-division of $Q_k(I_k)$ by $\mathbf{A}_{k-1}$. Now $G_k \notin (\mathbf{A}_{k-1})$, for otherwise, we would have $I_1^{e_1} \cdots I_{k-1}^{e_{k-1}} Q_k(I_k) \in (\mathbf{A}_{k-1})$; the invertibility of initials of $\mathbf{A}_{k-1}$ would imply that there be a non-zero $L' \in \mathbf{S}_0$ (explicitly, $L' = \prod_{j=1}^{k-1} (L_j P_j(0))^{e_j}$) such that $L' Q_k(I_k) \in (\mathbf{A}_{k-1})$; and this would contradict the minimality of $P_k(X)$. Now $G = L_k G_k$ has the property we need for I_k, that is, $G \notin (\mathbf{A}_{k-1})$ and

$$GI_k = L_k I_k G_k \equiv L_k I_k I_1^{e_1} \cdots I_{k-1}^{e_{k-1}} Q_k(I_k) \equiv 0 \bmod (\mathbf{A}_{k-1}).$$

This completes the induction and proves (a). The proof for (b) for a general $F \in \mathbf{S}_p$ is similar (append $A_{p+1} = F v_{p+1}$, for example). □

The above theorem, due to Kandri Rody *et al.* [18,10], with its constructive proof, not only provides an algorithm for invertibility and for a triangular set to have invertible initials, but more importantly, in the case of a negative determination, the proof shows how to find the polynomial(s) G, which are essential to the decomposition algorithm later. Another possible algorithm, based on linear algebra, decides the invertibility of F by computing a matrix representation of the map η_F and testing its injectivity or surjectivity (see Corollary 3.9, Proposition 3.1 and Theorem 3.16, or Sadik [36]). Hubert [17] (in Proposition 3.4) gave a different version based on the Chinese Remainder Theorem. When $\mathbf{S}_0$ is a polynomial ring in finitely many indeterminates over a field (or more generally, a strongly computable domain, see Mishra [28]), the invertibility of a polynomial F may be easily decided, without the assumption that $\mathbf{A}$ has invertible initials, by the elimination theorem in Gröbner basis since we can test whether the ideal $(\mathbf{A}, F) \cap \mathbf{S}_0$ is trivial, and find the polynomials L, M, N in the theorem in case the ideal is not trivial (F is invertible). However, we would need to find G in the case the ideal is trivial (F is not invertible). A similar result for triangular sets with regular initials was one of the basic tools in the Rosenfeld-Gröbner algorithm of Boulier *et al.* [5,6,7,8].

We end this section with a notion of invertibility for separants.

Definition 3.23 Let $\mathbf{A}: A_1, \ldots, A_p$ be triangular with respect to $v_1, \ldots, v_p$ over $\mathbf{S}_0$. The polynomial $S_k = \partial A_k / \partial v_k$ is called the *separant* of A_k with respect to v_k (or simply the *separant* of A_k). We denote the product of separants of $\mathbf{A}$ by S, and let $H = IS$. We say $\mathbf{A}$ *has invertible separants with respect to* $v_1, \ldots, v_p$ over $\mathbf{S}_0$ (or simply *has invertible separants*) if for $1 \leq k \leq p$, S_k is invertible with respect to the set $\mathbf{A}_k$, when $\mathbf{A}_k$ is considered as a triangular set with respect to $v_1, \ldots, v_k$ over $\mathbf{S}_0$.

Example 3.24 Note that in Example 3.18, the separant $S_1 = 2uv_1$ is not invertible with respect to $\mathbf{A}_1$, but the separant $S_2 = v_1$ is invertible with respect to $\mathbf{A}$. The same polynomial v_1, as the initial I_2, on the other hand, is not invertible with respect to $\mathbf{A}_1$. Thus $\mathbf{A}$ does not have invertible initials or invertible separants, even though *all* its initials and separants *are* invertible as polynomials with respect to $\mathbf{A}$ over $\mathbf{S}_0$.

Corollary 3.25 *Let $\mathbf{A}$ be triangular as before and suppose $\mathbf{A}$ has invertible initials. Then the property that $\mathbf{A}$ has invertible separants is decidable.*

When $\mathbf{A}$ has both invertible initials and invertible separants, it is called *separable* by Sadik [36], and $\mathbf{A}$ is said to be a *squarefree regular chain* by Boulier [8]. When $\mathbf{A}$ is also autoreduced (see next section), $\mathbf{A}$ is said to be *satured* by Bouziane *et al.* [10]. This latter condition, which first appeared in Lemma 13 on page 36 of Kolchin [23], implies that the ideal $J_I = (\mathbf{A}):I^\infty$ is *separable*[5] over $\mathbf{S}_0$. The proof of Lemma 13(b), which works equally well for triangular sets (as observed in Sadik [36]), provides the theoretical basis of the Ritt-Kolchin decomposition algorithm using factorization over algebraic extensions. More importantly, it shows that, under these hypothesis, J_I is radical and $J_I = (\mathbf{A}):H^\infty$ (see Hubert [17], Proposition 3.3, or Sadik [36], Corollary 2.1.1). Bouziane *et al.* and Sadik used the properties to compute the dimension of $J_I = (\mathbf{A}):I^\infty$ and the differential dimension of a related differential ideal $\mathfrak{a}_I = [\mathbf{A}]:I^\infty$. Hubert developed similar algorithms using properties of Gröbner basis of zero-dimensional polynomial ideals. Due to lack of time and space, we shall not investigate these important results.

4 Ranking and Reduction Concepts

In Exercise 1.11, we saw how a system of arbitrarily high order linear partial differential equations can be simplified, in this case, to a trivial system. This is the essense of elimination theory. We are interested in automating this process of simplifying a given system through symbolic computations. The first computational requirement towards this goal is data representation and an important related concept is that of ranking, or a total ordering of derivatives, which affects how the differential monomials are stored, and the order in which algebraic operations and differentiations are performed on them.

Definition 4.1 A *ranking* of $y_1, \cdots, y_n$ is a total ordering $\leq$ on the set ΘY of derivatives such that for all $u, v \in \Theta Y$ and $\delta \in \Delta$, we have $u \leq \delta u$ and

---

[5]See page 8 of Kolchin [23] for a precise definition.

$u \leq v \Rightarrow \delta u \leq \delta v$. A ranking is said to be *orderly* if $\operatorname{ord} u < \operatorname{ord} v$ implies $u < v$. A ranking is said to be *unmixed* if for every $(i,j), 1 \leq i,j \leq n$, $y_i < y_j$ implies $\theta y_i < y_j$ for every $\theta \in \Theta$.

A ranking on ΘY is, if $n = 1$, similar to a term ordering on monomials in a polynomial ring in m variables, and if $n \geq 1$, to a term ordering on a free module of rank n over a polynomial ring in m variables. Rankings are usually given via an embedding $\varphi \colon \Theta Y \longrightarrow \mathbb{N}^s$ (or $\mathbb{R}^s$) for some s. The set $\mathbb{R}^s$ is ordered lexicographically: for two s-tuples $(a_1, \cdots, a_s)$ and $(b_1, \cdots, b_s)$ in $\mathbb{R}^s$, we say $(a_1, \cdots, a_s) <_{lex} (b_1, \cdots, b_s)$ if there exists some $r < s$ such that $a_i = b_i$ for $1 \leq i \leq r$ and $a_{r+1} < b_{r+1}$. This lexicographic order on $\mathbb{R}^s$ induces a ranking on ΘY via φ: that is, $u < v$ if $\varphi(u) <_{lex} \varphi(v)$. In such cases, we say loosely that the ranking is lexicographic with respect to the s-tuple $\varphi(u)$.

Example 4.2 Let $u = \delta_1^{e_1} \cdots \delta_m^{e_m} y_j$ be a typical derivative. Then the lexicographical order $<_{lex}$ with respect to the $(m+1)$-tuple $(\operatorname{ord} u, j, e_1, \cdots, e_{m-1})$ induces a ranking on ΘY. This ranking is clearly orderly. The lexicographical order with respect to $(j, \operatorname{ord} u, e_1, \cdots, e_{m-1})$ is unmixed.

Example 4.3 Let $m = 2, n = 1$ and order the set ΘY lexicographically with respect to the tuple $(2e_1 + e_2, e_1)$. Then $\delta_2^2 y < \delta_1 y$ since $(2,0) <_{lex} (2,1)$. This ranking is not orderly.

Exercise 4.4 Show that every ranking is a well-ordering.

Exercise 4.5 Show that there is a natural bijection between the set of all term orderings on a polynomial ring $\mathcal{F}[x_1, \cdots, x_m]$ and the set of all rankings on a differential polynomial ring $\mathcal{F}\{y\}$ with m commuting derivation operators $\delta_1, \cdots, \delta_m$. Furthermore, show that a term ordering is degree-compatible if and only if the corresponding ranking is orderly.

Exercise 4.6 Characterize the set of all rankings for the case $n = 1$ (see Exercise 4.5). That is: (1) find an algorithm to construct all rankings, (2) find an algorithm to decide if two constructed rankings are equal (and if possible, find a canonical representation of a ranking).

Exercise 4.7 (Difficult) Characterize the set of all rankings for the general case, that is, for m derivation operators and n differential indeterminates.

We now assume that a ranking is fixed and given on ΘY. To avoid verbal description of certain polynomial subrings of $\mathcal{R} = \mathcal{F}\{y_1, \cdots, y_n\}$, it is convenient to adapt the suggestive notation of Morrison [29]. For any $v \in \Theta Y$, let $\mathcal{R}_{[v]} = \mathcal{F}[\{\theta y_j\}_{\theta \in \Theta, 1 \leq j \leq n, \theta y_j \leq v}]$ and let $\mathcal{R}_{(v)} = \mathcal{F}[\{\theta y_j\}_{\theta \in \Theta, 1 \leq j \leq n, \theta y_j < v}]$.

Definition 4.8 The *leader* of a differential polynomial $A \in \mathfrak{R}$, $A \notin \mathfrak{F}$, is the highest ranked derivative in ΘY that appears in A and is denoted by u_A. We often write A as a univariate polynomial in u_A, thus:

$$A = I_d u_A^d + I_{d-1} u_A^{d-1} + + \cdots + I_0, \tag{4.9}$$

where $I_d, \ldots, I_0 \in \mathfrak{R}_{(u_A)}$ and $I_d \neq 0$. The leading coefficient I_d in this unique representation is called the *initial*[6] of A and is denoted by I_A. The partial derivative $\frac{\partial A}{\partial u_A}$ is called the *separant* of A and is denoted by S_A. The *rank*[7] of A is the pair $(u_A, \deg_{u_A} A)$. *From now on, whenever we speak of the leader, initial, separant, or rank of any differential polynomial, it will be implicit that the polynomial is not in $\mathfrak{F}$.* Note that these notions are *not* defined for an element in the ground field $\mathfrak{F}$ since such an element does not have a leader.

Definition 4.10 We say a differential polynomial A *has lower rank than* a differential polynomial B if the rank of A is lexicographically lower than that of B. The relation *has lower rank than* is a pre-order, that is, it is reflexive and transitive, on $\mathfrak{R} \backslash \mathfrak{F}$. It extends to a pre-ordering on $\mathfrak{R}$, denoted by $<$, when we specify that every element of $\mathfrak{F}$ has lower rank than every element of $\mathfrak{R} \backslash \mathfrak{F}$. For any A, the initial and separant of A, which may lie in $\mathfrak{F}$, have lower rank than A.

Example 4.11 Let $m = 1$ and $n = 2$. Denoting the differential indeterminates by w and z, there is a unique orderly ranking on $\mathfrak{F}\{w, z\}$ such that $w < z$. Using the prime notation for differentiation, let

$$A = (w' + w^3)(z''')^2 + w^2 z''' - 3w^2 (z')^3 z''.$$

We have $u_A = z'''$, $I_A = w' + w^3$ and $S_A = 2(w' + w^3)z''' + w^2$. The rank of A is $(z''', 2)$. Both I_A and S_A are of lower rank than A, and this important property holds in general for any differential polynomial $A \in \mathfrak{R} = \mathfrak{F}\{y_1, \cdots, y_n\}$ using any ranking on $\mathfrak{R}$.

Remark 4.12 When $\Delta = \emptyset$, a ranking is an ordering of the differential indeterminates $y_1, \cdots, y_n$ only and not a term-ordering. Using the ranking in Example 4.11, $w^3 z^2$ and $w^2 z^2$ both have the same rank $(z, 2)$.

---

[6]Thus, for a differential polynomial A, the initial of A is the initial of A with respect to its leader (in the sense of Definition 2.6).

[7]In Ritt [32] or Kolchin [23], the "rank" of a differential polynomial is not explicitly defined and is only used in the comparative sense. If y_j is the highest ranked variable in an algebraic polynomial F and if d is the degree of F in y_j, the index j is sometimes called the *class* of F and the degree d is called the *class degree* of f. These terms are commonly used in the literature on geometric theorem proving based on the Ritt-Wu method.

Exercise 4.13 Let $m = 1$ and let the ranking be induced by the lexicographical order with respect to the tuple (j, h), where $u = \delta^h y_j \in \Theta Y$. Show that in every subset of differential polynomials, there is one (which need not be unique) with the lowest rank.

Exercise 4.14 Given any ranking, show that for any $\theta \in \Theta$ and any $A \in \mathcal{R} \backslash \mathcal{F}$ and $\theta \neq 1$, we have $\theta A - S_A \theta u_A < \theta u_A$. In particular, θA has rank $(\theta u_A, 1)$ with initial and separant $I_{\theta A} = S_{\theta A} = S_A$ and we can write $\theta A = S_A \theta u_A - T$ with $T < \theta u_A$.

Definition 4.15 Let $\mathbf{A} \subset \mathcal{R} \backslash \mathcal{F}$ and F in $\mathcal{R}$. We say F is *partially reduced with respect to* $\mathbf{A}$ if for every $A \in \mathbf{A}$, no proper derivative of u_A appears in F. We say F is *reduced with respect to* $\mathbf{A}$ if either $F = 0$ or F is partially reduced and $\deg_{u_A} F < \deg_{u_A} A$ for all $A \in \mathbf{A}$. We say $\mathbf{A}$ of $\mathcal{R}$ is *autoreduced* if every $A \in \mathbf{A}$ is reduced with respect to every other one. In particular, the empty set and the set consisting of a single differential polynomial A, $A \notin \mathcal{F}$, are autoreduced.

Remark 4.16 Ritt and some authors in the symbolic computation community define an autoreduced set differently to allow it to contain non-zero elements from $\mathcal{F}$. This may seem convenient, since during the construction of an autoreduced set, the reduction process (see later) may produce elements of the ground field. Unfortunately, such a liberal extension of the concept creates many theoretical inconveniences. Elements of the ground field do not have leaders and it is meaningless to speak of reduction with respect to such elements (Definition 4.15). For example, Ritt was careful to assume that the autoreduced set contains no elements from the ground field before describing his reduction process (in fact, that assumption was mentioned 5 times on page 164 of Ritt [32], with 3 within 8 lines of each other). In any software designed for computational differential algebra, if such elements are obtained through some elimination process from a given set, the differential ideal generated by the given set may not be proper, and should be handled separately in any case. If desired, we may specifically allow the set $\mathbf{E}$ consisting solely of the unit $1 \in \mathcal{F}$ to be an autoreduced set which has lower rank than every other autoreduced set (including the empty set). In such case, one has to be careful always to qualify that $\mathbf{A} \neq \mathbf{E}$ when a statement involves the rank of elements of $\mathbf{A}$. In his entire book [23], Kolchin used this only once, on page 168, and even there, it is not really necessary.

Example 4.17 Using the same differential polynomial ring and ranking as in Example 4.11, let $A = (z')^3 (z''')^2 - w''$, $B = (z')^3 (z'')^2 - w''$ and $F =$

22

$w^{(4)} - (w')^2 z'''$. Then F is partially reduced and reduced with respect to A, but is not partially reduced with respect to B. Observe also that F is of higher rank than A because the ranking is orderly. Since A is reduced with respect to F, $\{A, F\}$ is autoreduced, but $\{B, F\}$ is not.

Example 4.18 When $\Delta = \emptyset$, every F is partially reduced with respect to every $A \notin \mathcal{F}$, and is reduced with respect to A if either u_A does not appear in F or appears to a lower degree.

Exercise 4.19 When $\Delta = \emptyset$, show that an autoreduced set $\mathbf{A} \subset \mathcal{F}[y_1, \ldots, y_n]$ is in triangular form with respect to its set of leaders when the members are listed in order of increasing rank. In particular, $\mathbf{A}$ can have no more than n elements.

Exercise 4.20 Relative to any ranking of $\mathcal{R} = \mathcal{F}\{y_1, \cdots, y_n\}$, prove that in every non-empty autoreduced set $\mathbf{A}$ of $\mathcal{R}$, distinct members have distinct leaders (and hence also distinct ranks). Show that $\mathbf{A}$ therefore must be finite and triangular with respect to the leaders of its elements when arranged in order of increasing rank.

Remark 4.21 Let $\mathbf{A} \subset \mathcal{R}$. We shall often consider $\mathbf{A}$ as a subset of a polynomial subring $\mathcal{F}[V]$, where $V \subset \Theta Y$ is any set containing all the derivatives that appear in one or more $A \in \mathbf{A}$. We can consider $\mathbf{A}$ either as an autoreduced subset of $\mathcal{R}$ in the sense of Definition 4.15, or as an autoreduced subset of $\mathcal{F}[V]$ in the algebraic sense, that is, treating V as a set of algebraic indeterminates over $\mathcal{F}$ (the "$\Delta = \emptyset$" case). These are not equivalent. An autoreduced subset of $\mathcal{R}$ is an autoreduced subset of $\mathcal{F}[V]$, but the converse is false in general. More generally, if $F \in \mathcal{R}$ is such that $F \in \mathcal{F}[V]$, the notion of F being reduced as a differential polynomial in $\mathcal{R}$ and that as a polynomial in $\mathcal{F}[V]$ are different. In Example 4.17, F is "reduced" with respect to B in the polynomial ring $\mathcal{F}[V] = \mathcal{F}[z', z'', z''', w'', w^{(4)}]$ since the degree in z'' (the leader of B) in F is zero and hence $\{F, B\}$ is "autoreduced" in $\mathcal{F}[V]$. One should be alert of such differences when applying algebraic results from $\mathcal{F}[V]$ to $\mathcal{R}$. To avoid confusion, in this paper, we adopt the convention that **every polynomial in $\mathcal{F}[V]$ is considered as a differential polynomial of $\mathcal{R}$ in matters regarding reduction.**

Exercise 4.22 When $\Delta = \emptyset$, give an example of a Gröbner basis which is not triangular (and hence not autoreduced). Give an example of a triangular set which is not autoreduced.

By Exercise 4.20, we may list the elements of an autoreduced set in order of increasing ranks and for the rest of the paper, we shall assume this is the case for every autoreduced set *if a ranking has been fixed beforehand*. This assumption is implicit in the notation $\mathbf{A}: A_1, A_2, \ldots, A_r$, or emphasized with the notation $\mathbf{A}: A_1 < A_2 < \cdots < A_r$. Ritt [32] (and later, Wu [39]) called such a sequence a *chain* or an *ascending set*. Let $\mathbf{A}: A_1 < \cdots < A_r$ and $\mathbf{B}: B_1 < \cdots < B_s$ be two autoreduced sets in $\mathcal{F}\{y_1, \cdots, y_n\}$. We extend the concept of comparative rank to autoreduced sets.

Definition 4.23 We say $\mathbf{A}$ is *of lower rank than* $\mathbf{B}$ (or $\mathbf{A} < \mathbf{B}$) if either

(a) there exists a $t, 1 \leq t < \min(r, s)$ such that $\operatorname{rank}(A_i) = \operatorname{rank}(B_i)$ for $1 \leq i \leq t$ and $\operatorname{rank}(A_{t+1}) < \operatorname{rank}(B_{t+1})$, or

(b) $r > s$ and $\operatorname{rank}(A_i) = \operatorname{rank}(B_i)$ for $1 \leq i \leq s$.

The relation $\leq$ is a pre-order, being reflexive and transitive.

Proposition 4.24 *Let $\mathcal{A}$ be any non-empty set of autoreduced subsets of $\mathcal{F}\{y_1, \cdots, y_n\}$. Then there exists an autoreduced subset $\mathbf{A} \in \mathcal{A}$ which has the lowest rank.*

Proof. If the empty set is the only set in $\mathcal{A}$, then it is of lowest rank. Assume the opposite. Among all non-empty autoreduced subsets in $\mathcal{A}$, select those whose first element is of lowest rank. Each such first element must have the same leader v_1. If possible, among these which have a second element, select those whose second element is of lowest rank. Let their common leader be $v_2 > v_1$. Suppose we were able to continue this way and obtain an infinite increasing sequence of derivatives. Then there would be an infinite subsequence which were derivatives of the same y_j. There would be a further infinite subsequence in which each element was a derivative of the previous one. This would contradict that the leaders of an autoreduced set cannot be derivatives of another. Thus, at some i-th step, all the selected autoreduced sets have exactly i elements. Clearly, anyone of these is of the lowest rank in $\mathcal{A}$. $\square$

We summarize this section by listing the different orderings and pre-orderings used in differential algebra.

- A ranking is a well-ordering on the set θy_j of derivatives that is compatible with differentiation.

- A term-ordering is a well-ordering on the monomials in a *polynomial* ring such as $\mathcal{R}_s = \mathcal{F}[(\theta y_j)_{\theta \in \Theta(s), 1 \leq j \leq n}]$, or differential monomials in $\mathcal{R} = \mathcal{F}\{y_1, \cdots, y_n\}$, compatible with ranking and multiplication.

- A comparative rank on differential polynomials is a well-pre-ordering, and is induced by a ranking using the lexicographic order on the ranks of differential polynomials. The rank of a differential polynomial F is (u_F, d) where u_F is leader of F and d is the degree of F in u_F.
- A comparative rank on autoreduced sets is a well-pre-ordering, induced by a lexicographic order on the lists of ranks of elements of autoreduced sets when the elements of an autoreduced set are listed in increasing order of rank.

5   Characteristic Sets

We begin with Kolchin's definition (page 82, Kolchin [23]) which, contrary to Ritt's (page 5, Ritt [32]), is given for a differential ideal and not for a set.

Definition 5.1 Let $\mathfrak{a}$ be a differential ideal of $\mathcal{F}\{y_1, \cdots, y_n\}$. An autoreduced subset $\mathbf{A}$ of $\mathfrak{a}$ of lowest rank among all autoreduced subsets $\mathbf{B}$ of $\mathfrak{a}$ with the property that $S_B \notin \mathfrak{a}$ for all $B \in \mathbf{B}$ is called a *characteristic set* of $\mathfrak{a}$.

Since the empty set is an autoreduced set, by Proposition 4.24, every differential ideal has a characteristic set. When $\mathcal{F}$ is a differential field of characteristic zero, Ritt gave two equivalent definitions for a characteristic set of a *set*. Beginning with Boulier [4], Ritt's versions have been followed by researchers (for example Boulier *et al.* [5], Hubert [17], and Bouziane *et al.* [10]) in symbolic computation because they are less cumbersome. To facilitate discussion, let us introduce the notion of a *zero-reduced* set.

Definition 5.2 A subset Σ of $\mathcal{R} = \mathcal{F}\{y_1, \cdots, y_n\}$ is said to be *zero-reduced* with respect to an autoreduced set $\mathbf{A}$ if no non-zero element $F \in \Sigma$ is reduced with respect to $\mathbf{A}$.

Lemma 5.3 *Let $\mathbf{A}$ be an autoreduced subset of a proper differential ideal $\mathfrak{a}$ in $\mathcal{R}$. Then the following are equivalent:*
(a) *$\mathbf{A}$ is of minimal rank among all autoreduced subsets $\mathbf{B}$ of $\mathfrak{a}$.*
(b) *$\mathbf{A}$ is a characteristic set of $\mathfrak{a}$.*
(c) *$\mathfrak{a}$ is zero-reduced with respect to $\mathbf{A}$.*

Proof. We first prove that (a) implies (b). Suppose that $\mathbf{A}$ is of minimal rank. Let $\mathbf{B}$ be a characteristic set of $\mathfrak{a}$. Clearly, $\mathbf{A} \leq \mathbf{B}$. So it needs to be shown that $S_A \notin \mathfrak{a}$ for all $A \in \mathbf{A}$. This is trivially true if $\mathbf{A} = \emptyset$. Otherwise, S_A is non-zero ($\mathcal{F}$ has characteristic zero) and is reduced with respect to $\mathbf{A}$ and $S_A \notin \mathbf{A}$. If for some $C \in \mathbf{A}$, $F = S_C \in \mathfrak{a}$, then F and the elements of $\mathbf{A}$ for which u_A is lower than the leader u_F of F would form an autoreduced set of lower rank than $\mathbf{A}$, which would be a contradiction.

Next, we will prove that (b) implies (c). Suppose there were a non-zero differential polynomial in $\mathfrak{a}$, reduced with respect to $\mathbf{A}$ and let F be one with minimal rank. Then $F \notin \mathcal{F}$ because $\mathfrak{a}$ is proper, and F together with the elements of $\mathbf{A}$ for which u_A is lower than the leader u_F of F would form an autoreduced set of lower rank than $\mathbf{A}$. By the minimality of $\mathbf{A}$, S_F, which is non-zero and reduced with respect to $\mathbf{A}$, would belong to $\mathfrak{a}$, contradicting the choice of F.

To prove (c) implies (a), assume that the $\mathbf{A} \colon A_1 < \cdots < A_r$ given in (c) were not the lowest. Let $\mathbf{B} \colon B_1 < \cdots < B_s$ be an autoreduced subset of $\mathfrak{a}$ of lowest rank. Then $\mathbf{B} < \mathbf{A}$. Either for some $t < \min(r,s)$, rank $(B_i) = $ rank (A_i) for $1 \le i \le t$ and rank $(B_{i+1}) < $ rank (A_{i+1}), or $s > r$, rank $(B_i) = $ rank (A_i) for $1 \le i \le r$. Thus, either B_{t+1} (in the first case) or B_{s+1} (in the second case) would be non-zero and reduced with respect to $\mathbf{A}$. This would be a contradiction. $\qquad\square$

It is clear that the arrangement of elements of an autoreduced set in order of increasing rank is important in computations. Thus we fix, once and for all in this paper, the following notations. For any autoreduced set $\mathbf{A} \colon A_1 < \cdots < A_p$, we shall denote the corresponding leaders by $v_1, \ldots, v_p$, initials by $I_1, \ldots, I_p$, separants by $S_1, \ldots, S_p$. The product of initials (resp. separants, resp. initials and separants) is denoted by $I_\mathbf{A}$ (resp. $S_\mathbf{A}$, resp. $H_\mathbf{A}$) or simply I (resp. S, resp. H) if $\mathbf{A}$ is clear from the context. This entire set of notations will be implicit whenever $\mathbf{A}$ is an autoreduced set. The symbol k, $1 \le k \le p$ will be used as a running index as in $I = \prod_{k=1}^p I_k$. We shall occasionally use the notations I_A, S_A ($A \in \mathbf{A}$) when these are needed in a random order.

Corollary 5.4 Let $\mathbf{A}$ be a characteristic set of a proper differential ideal $\mathfrak{a}$ in $\mathfrak{R}$. Then I_k and S_k are not in $\mathfrak{a}$ for every $A \in \mathbf{A}$. If $\mathfrak{a}$ is prime, then I, S, H are not in $\mathfrak{a}$.

Remark 5.5 If the differential ideal $\mathfrak{a}$ is $\mathfrak{R}$, then the set $\mathbf{A}_0$ consisting of $y_1, \ldots, y_n$ is clearly a lowest autoreduced subset of $\mathfrak{a}$ for any ranking. Any other lowest autoreduced subset $\mathbf{B}$ of $\mathfrak{a}$ must have exactly n elements of the form $b_1 y_1 + c_1, \ldots, b_n y_n + c_n$ where $b_i, c_i \in \mathcal{F}$ and $b_i \ne 0$ for all i. Since any element of $\mathfrak{a}$ which has a separant will have the separant in $\mathfrak{a}$, by Definition 5.1, the only characteristic set of $\mathfrak{a}$ is the empty set. Thus (a) and (b) of Lemma 5.3 are not equivalent when $\mathfrak{a} = \mathfrak{R}$. In this case, (c) is false (whether one uses the autoreduced set $\mathbf{A}_0$ or the empty autoreduced set). On the other hand, the lemma is trivially true for $\mathfrak{a} = (0)$ since the empty set is the only autoreduced subset.

In Ritt [32] and recent works by others on computational differential algebra, a characteristic set is defined for an arbitrary subset $\Sigma \subseteq \mathcal{F}\{\, y_1, \cdots, y_n \,\}$ as an autoreduced subset $\mathbf{A}$ of Σ of lowest rank. According to Lemma 5.3, this is equivalent to Definition 5.1 when Σ is a differential ideal and $\mathcal{F}$ has characteristic zero, *provided we exclude the differential ideal* $\Sigma = \mathcal{R}$. In computations, we are typically given a finite subset Σ from which we hope to compute a characteristic set of the differential ideal generated by Σ. The process usually involves an ascending sequence of finite subsets

$$\Sigma_0 = \Sigma \subset \Sigma_1 \subset \cdots \subset \Sigma_k \subset \cdots$$

and their characteristic sets (also referred to as *basic sets*). The Ritt definition thus facilitates the development of theoretical properties of the process as long as the differential ideals generated by Σ_k $(k = 0, 1, \ldots)$ remain proper. However, allowing autoreduced sets to contain elements of $\mathcal{F}$, as discussed in Remark 4.16, would require either accompanying conventions on the meaning of leader, initial, separant, partial reduction, reduction, autoreduction and/or a meticulous and tedious re-examination (and perhaps re-statement) of many results proven in Kolchin [23].

Just as they speak of autoreduced sets in a polynomial subring $\mathcal{F}[V]$ where $V \subset \Theta Y$, some authors speak of characteristic sets of ideals in $\mathcal{F}[V]$ without clarifying their precise meaning. As discussed in Remark 4.21, this is potentially confusing. We remind readers the convention we adopt in Remark 4.21. In particular, an ideal J of $\mathcal{F}[V]$ is zero-reduced means *as a subset of* $\mathcal{R}$, it is zero-reduced, and this is the case if and only if the only differential polynomial $F \in J$ that is reduced (*in the differential sense* and *a fortiori* F is also partially reduced) with respect to $\mathbf{A}$ is the zero polynomial. Even if J is zero-reduced, there may be non-zero differential polynomials F, not partially reduced, such that for every $A \in \mathbf{A}$, the degree of F in u_A is less than the degree of A in u_A.

Example 5.6 Let $\mathcal{R} = \mathcal{F}\{\, y, z \,\}$ be an ordinary differential polynomial ring. Suppose the ranking is orderly and satisfies $z < y$. Let $A_1 = y^2 + z$, $A_2 = y' + y$. Then $\boldsymbol{\Gamma} \colon A_1 < A_2$ is "autoreduced" as a subset of polynomials in $\mathcal{S} = \mathcal{F}[z, y, y']$ and is a "characteristic set" of the ideal $J = (A_1, A_2)$ of $\mathcal{S}$, which is prime. However, $\boldsymbol{\Gamma}$ is not autoreduced, because A_2 is not partially reduced with respect to A_1. The differential ideal $\mathfrak{a} = [A_1, A_2]$, while also prime, has a characteristic set $\mathbf{A} \colon A_1 < A_3$ lower than $\boldsymbol{\Gamma}$, where $A_3 = A_1' + 2A_1 - 2yA_2 = z' + 2z$. We have $\mathfrak{a} = [\mathbf{A}] \colon 2y$ where $2y$ is the product of initials and separants of $\mathbf{A}$. Thus a "characteristic set" of a prime ideal need not be a characteristic set of the differential ideal it generates. The ideal J, being a subset of $\mathfrak{a}$, is zero-reduced with respect to $\mathbf{A}$, but contains A_2 which is "reduced" with respect to $\mathbf{A}$ as a polynomial in $\mathcal{F}[z, y, z', y']$. Also, A_3 is a non-zero differential polynomial which is reduced with respect to $\mathbf{\Gamma}$.

Exercise 5.7 Discuss the possible meanings and corresponding validity of Lemma 5.3 when $\mathfrak{a}$ is replaced by an ideal J, or even just a subset Σ, of $\mathcal{F}[V]$.

Example 5.8 Let $m = 2, n = 2$, let $h > 0$, and let $\mathfrak{a}_1$ be the differential ideal generated by $P_{1,1} = \delta_1^h y_1$, $P_{1,2} = \delta_1 \delta_2^{h-1} y_2 + y_1$. Then $\mathbf{A}_1 : P_{1,1}, P_{1,2}$ is a characteristic set of $\mathfrak{a}_1$. Similarly, we can define $\mathfrak{a}_2$ and its characteristic set $\mathbf{A}_2$. By Exercise 1.11, the differential ideal $\mathfrak{a}_0$ generated by $P_{1,1}, P_{1,2}, P_{2,2}, P_{2,1}$ is $[y_1, y_2]$ and hence it has a characteristic set $\mathbf{A}_0 : y_1, y_2$.

While there may be more than one characteristic set of a differential ideal $\mathfrak{a}$, the lengths of any two characteristic sets of $\mathfrak{a}$ and the corresponding ranks of their elements are always the same. We note that a characteristic set of $\mathfrak{a}$ usually does not generate $\mathfrak{a}$ as a differential ideal. However, when a differential ideal $\mathfrak{a}$ is generated by linear homogeneous differential polynomials, it is prime, and it is generated as a *differential* ideal by a canonical characteristic set that has a very nice property. We refer the readers to Chapter IV, Section 5 of Kolchin [23] for basics on linear differential ideals, and to Chapter 2, Section 3 of Sit [38] for the properties of canonical characteristic sets of linear differential ideals.

6   Reduction Algorithms

In this section, we introduce computational concepts in differential algebra using results from Section 2. Given a ranking $\leq$, a differential polynomial F, and an autoreduced set $\mathbf{A}: A_1 < A_2 < \cdots < A_p$, Ritt and Kolchin defined a *partial remainder* $\widetilde{F}$ and a *remainder* F_0 of F with respect to $\mathbf{A}$ as the results of two *specific* reduction procedures. Loosely speaking, the first uses a pseudo-remainder sequence using repeated pseudo-divisions by elements from the set of proper derivatives of elements of $\mathbf{A}$, and the second from the set $\mathbf{A}$ itself. Each procedure specifies an exact sequence of pseudo-divisions. However, in order to accommodate other reduction procedures, such as discussed in Remark 2.7, we give a more general definition below.

Definition 6.1 By a *partial remainder* of a differential polynomial F with respect to a non-empty autoreduced set $\mathbf{A}$, we mean any differential polynomial $\widetilde{F}$, partially reduced with respect to $\mathbf{A}$ such that there exist natural numbers $s_A, A \in \mathbf{A}$ satisfying the property that

$$\prod_{A \in \mathbf{A}} S_A^{s_A} \cdot F - \widetilde{F} \in [\mathbf{A}]. \tag{6.2}$$

Ritt and Kolchin gave a constructive proof that partial remainders exist.

Procedure 6.3 Ritt-Kolchin's Partial Remainder Algorithm

Input: *A ranking $\leq$, a non-empty autoreduced set $\mathbf{A}$, $F \in \mathcal{R}$*

Output: *$\widetilde{F}$, a partial remainder of F*

begin

$\quad \widetilde{F} := F$

$\%\quad s_A := 0$ *for all $A \in \mathbf{A}$*

$\%\quad Q_w := 0$ *for all $w \in \Theta Y, w < u_F$*

$\quad$**While** $D(\widetilde{F}, \mathbf{A}) \neq \emptyset$ **repeat**

$\qquad v := v(\widetilde{F}, \mathbf{A})$

$\qquad C := C(\widetilde{F}, \mathbf{A})$

$\qquad \theta := \theta(\widetilde{F}, \mathbf{A})$

$\qquad \mathbf{S} :=$ *largest polynomial subring of $\mathcal{R}_{[u_{\widetilde{F}}]}$ not containing v*

$\%\qquad Q_v := Q_v + Q(\widetilde{F}, \theta C, \mathbf{S}, v)$

$\%\qquad s_C := s_C + \deg_v(\widetilde{F})$

$\qquad \widetilde{F} := R(\widetilde{F}, \theta C, \mathbf{S}, v)$

$\quad$**return** $\widetilde{F}$

end

To describe this Ritt-Kolchin procedure, for any $F \in \mathcal{R}$, let $D(F, \mathbf{A})$ be the set of all proper derivatives $w = \theta u_A$ (of some leader u_A) that appear in F. By definition, F is partially reduced with respect to $\mathbf{A}$ if and only if $D(F, \mathbf{A}) = \emptyset$. If $D(F, \mathbf{A})$ is non-empty, it has a (unique) highest ranked element $v = v(F, \mathbf{A}) \in D(F, \mathbf{A})$, and there is a (unique) highest ranked element $C = C(F, \mathbf{A}) \in \mathbf{A}$ and a (unique) $\theta = \theta(F, \mathbf{A}) \in \Theta$ with $\theta \neq 1$ such that $v = \theta u_C$. Finally, given any $v \in \Theta Y$, and differential polynomials F, C, we may view both F and C as polynomials in the univariate polynomial ring $\mathcal{F}[\Theta Y \setminus \{v\}][v]$. If $v = v(F, \mathbf{A})$, $\theta = \theta(F, \mathbf{A})$, and $C = C(F, \mathbf{A})$, then we may even view F and θC in $\mathbf{S}[v]$, where $\mathbf{S}$ is the largest polynomial subring of $\mathcal{R}_{[u_F]}$ not containing v. Thus we may perform a pseudo-division of F

by θC over $\mathbf{S}$ with respect to the variable v and obtain the pseudo-quotient $Q = Q(F, \theta C, \mathbf{S}, v)$ and the pseudo-remainder $R = R(F, \theta C, \mathbf{S}, v)$. Note that in our case, $v = \theta u_C$, $\theta C = S_C v - T$ is linear in v, and both S_C and T belong to $\mathcal{R}_{(v)}$. If the degree of F in v is e, we have $S_C^e \cdot F = Q \cdot \theta C + R$, both $Q, R \in \mathbf{S}$, and either $D(R, \mathbf{A}) = \emptyset$ or $v(R, \mathbf{A}) < v(F, \mathbf{A})$. This last property of R guarantees that Procedure 6.3 terminates with $D(\widetilde{F}, \mathbf{A}) = \emptyset$. Its correctness is now clear.

Definition 6.4 The partial remainder computed by Procedure 6.3 will be referred to as the *Ritt-Kolchin partial remainder*.

Remark 6.5 A differential polynomial F is partially reduced with respect to $\mathbf{A}$ if and only if its Ritt-Kolchin partial remainder is itself. By including the commented lines (those beginning with a % sign) to keep track of the pseudo-quotients and exponents and thus output the corresponding $s_A, A \in \mathbf{A}$ and coefficients $Q_{\theta u_A}$, Procedure 6.3 returns a representation of $\prod_{A \in \mathbf{A}} S_A^{s_A} \cdot F - \widetilde{F}$ as a linear combination over $\mathcal{R}_{[u_F]}$ of proper derivatives θA, where $A \in \mathbf{A}$, $\theta u_A \leq u_F$, and $\theta \neq 1$. In particular, $\widetilde{F} \leq F$ and if $A \in \mathbf{A}$ and $u_A \geq u_F$, then $s_A = 0$ and no proper derivative of A appears in the linear combination. Kolchin's presentation of Ritt's construction appeared in Chapter I, Section 9, Lemma 6 of Kolchin [23]. If you are reading his proof, please note that on p. 78, line 3, "are lower than v" should be read as "are free of v".

To obtain a reduced polynomial corresponding to F, Ritt and Kolchin applied the successive pseudo-division algorithm to their partial remainder $\widetilde{F}$ and $\mathbf{A}$. Specifically, let $\mathbf{A}: A_1, \ldots, A_p$ and $v_k = u_{A_k}$, with $v_1 < \cdots < v_p$. Let $v = \max(v_p, u_{\widetilde{F}})$ and $V = \{v_1, \cdots, v_p\}$. There is a smallest polynomial subring $\mathbf{S}_0 = \mathcal{F}[U]$ of $R_{[v]}$, U being a subset of $\Theta Y \backslash V$ with every $u \in U$ partially reduced with respect to $\mathbf{A}$, such that $\mathbf{S}_0[V]$ contains $\widetilde{F}$ and $\mathbf{A}$. Since $\mathbf{A}$ is autoreduced, it is a triangular set in $\mathbf{S}_0[V]$ with respect to $v_1, \ldots, v_p$. Let the initial of A_k be denoted by I_k. Thus we obtain, using Proposition 2.9, $F_0 = R_1(\widetilde{F}, \mathbf{A}, \mathbf{S}_0, V)$. Let $i_k = E_k(\widetilde{F}, \mathbf{A}, \mathbf{S}_0, V)$ and $Q_k = Q_k(\widetilde{F}, \mathbf{A}, \mathbf{S}_0, V)$. Then we have

$$I_1^{i_1} \cdots I_p^{i_p} \widetilde{F} = Q_1 \cdot A_1 + \cdots + Q_p \cdot A_p + F_0. \tag{6.6}$$

Remark 6.7 It is clear that for each k $(1 \leq k \leq p)$, Q_k has rank lower than or equal to that of $\widetilde{F}$, that it is partially reduced with respect to $\mathbf{A}$, and that F_0 is reduced with respect to $\mathbf{A}$. Moreover, an examination of the proof of Proposition 2.9 shows that if k is the highest index such that v_k appears in

$\widetilde{F}$, then $Q_j = 0$, $R_j = \widetilde{F}$ and $i_j = 0$ for all $k < j \le p$. If F is reduced, then $F = F_0$, and $i_j = 0, Q_j = 0, R_j = F$ for all j.

Combining the partial remainder algorithm with these successive pseudo-divisions, we obtain

$$S_1^{s_1} \cdots S_p^{s_p} I_1^{i_1} \cdots I_p^{i_p} F \equiv F_0 \mod ([\mathbf{A}]). \tag{6.8}$$

Definition 6.9 By a *remainder* of given a differential polynomial $F \in \mathcal{R}$ with respect to a non-empty autoreduced set $\mathbf{A} : A_1, \ldots, A_p$, we mean a differential polynomial F_0 which is reduced with respect to $\mathbf{A}$ such that there exist non-negative integers $i_1, \ldots, i_p, s_1, \ldots, s_p$ such that Equation (6.8) holds. The remainder obtained by the method above will be referred as the *Ritt-Kolchin remainder*.

Remark 6.10 Definitions 6.1 and 6.9 are very liberal: for example, 0 is always a partial remainder for any $F \in [\mathbf{A}] : S_{\mathbf{A}}^{\infty}$ and a remainder of any $F \in [\mathbf{A}] : H_{\mathbf{A}}^{\infty}$. Thus it should be emphasized that the Ritt-Kolchin partial remainder $\widetilde{F}$ and remainder F_0 of F are obtained by carrying out a sequence of pseudo-divisions in a very specific manner. By Remark 6.7, a differential polynomial F is reduced if and only if F coincides with its Ritt-Kolchin remainder. This is *not* the case for remainders in general. However, some results may follow from the properties (6.2) (resp. (6.8)) of $\widetilde{F}$ (resp. F_0) rather than the particular integers $i_1, \ldots, i_p$ and $s_1, \ldots, s_p$ or corresponding partial remainder and remainder. Rosenfeld [34] recognized this in his paper. On the other hand, some results such as Procedure 10.1 will rely on the Ritt-Kolchin process. Recent works by Mansfield and Fackerell [26], Mansfield and Clarkson [27], Boulier *et al.* [5], Morrison [29], Hubert [17], and Sadik [36] either allow other sequences of pseudo-divisions, or do not necessarily require $\mathbf{A}$ to be autoreduced or even triangular. In this introductory article, we restrict these notions to autoreduced sets only. Observe that the Ritt-Kolchin partial remainder $\widetilde{F}$ (and hence also the Ritt-Kolchin remainder F_0) and corresponding exponents $i_1, \ldots, i_p$ and $s_1, \ldots, s_p$ have the additional property that when we represent the differential polynomials $\prod S_k^{s_k} F - \widetilde{F}$ and $\prod I_k^{i_k} S_k^{s_k} F - F_0$ in the differential ideal $[\mathbf{A}]$, we only need derivatives θA of $A \in \mathbf{A}$ which are lower than or equal to F, and their coefficients in the linear combination are also lower than or equal to F. Such a representation avoids cancellations of any derivative v higher than u_F in the entire reduction process and is similar to the property of standard representation using a Gröbner basis (see p. 218 of Becker and Weispfenning [3]). In Section 12, we shall discuss a more elaborate reduction method called the preparation process that also computes a partial remainder and a remainder.

Exercise 6.11 Give an example of an autoreduced set **A** and a non-zero differential polynomial F which is reduced but has a remainder 0.

Example 6.12 This example shows the difference between Gröbner reduction and Ritt-Kolchin reduction. Consider an ordinary differential polynomial ring, where we denote derivatives by the prime notation. Suppose the ranking is orderly with $y_1 < y_2$ and suppose the term ordering is pure lexicographic with respect to the ranking. Let $f_1 = y_2 y_2' + y_1'$, $f_2 = y_1 y_1'$. Then f_1 is partially reduced with respect to f_2 but not reduced. To reduce f_1 in the Ritt's sense, we multiply f_1 by y_1 and substract f_2. This is pseudo-reduction and results in $f_3 = y_1 y_2 y_2'$ which is in the ideal $J = (f_1, f_2)$ but f_2, f_3 do not generate J since f_1 only belongs to the saturation ideal $(f_2, f_3): y_1$. On the other hand, f_1 is reduced with respect to f_2 in the Gróbner sense since no term in f_1 is a multiple of the leading monomial of f_2. Indeed, $\{ f_1, f_2 \}$ is a Gröbner basis.

Corollary 6.13 Let $\mathfrak{a}$ be a proper differential ideal in $\mathcal{R} = \mathcal{F}\{y_1, \cdots, y_n\}$ and let $F \in \mathfrak{a}$. Then for every ranking and every characteristic set $\mathbf{A}: A_1, \ldots, A_p$ of $\mathfrak{a}$ relative to the ranking, every remainder F_0 of $F \in \mathfrak{a}$ with respect to $\mathbf{A}$ is zero. In particular, if $H = \prod_{k=1}^{p} I_k S_k$ is the product of initials and separants of $\mathbf{A}$, and if $\mathfrak{a}_H = [\mathbf{A}]: H^\infty$, then $\mathfrak{a} \subseteq \mathfrak{a}_H$. If $\mathfrak{a}$ is prime[8], then $\mathfrak{a} = \mathfrak{a}_H$. If $\mathfrak{a}$ is radical, then $\mathfrak{a}: H = \mathfrak{a}_H$, which is radical.

Proof. Since $F_0 \in \mathfrak{a}$ and is reduced, it is zero by Lemma 5.3. By Definition 6.9, there exist natural numbers i_k, s_k for $1 \leq k \leq p$ such that (6.8) holds. Hence $\mathfrak{a} \subseteq \mathfrak{a}_H$ and $\mathfrak{a}: H \subseteq \mathfrak{a}_H$. Moreover, letting $P \in \mathfrak{a}_H$, say $H^t P \in [\mathbf{A}]$ for some natural number t, we have $H^t P \in \mathfrak{a}$, $H \notin \mathfrak{a}$ (by Corollary 5.4). If $\mathfrak{a}$ is prime, then $P \in \mathfrak{a}$. If $\mathfrak{a}$ is radical, then $\mathfrak{a}: H$ is radical, and $HP \in \mathfrak{a}$ and $P \in \mathfrak{a}: H$. $\square$

Remark 6.14 Note that if we can compute a characteristic set **A** of a prime differential ideal $\mathfrak{p}$, Corollary 6.13 will provide a membership test for $\mathfrak{p}$, namely, $F \in \mathfrak{p}$ if and only if some remainder F_0 of F (for example, the Ritt-Kolchin remainder) with respect to **A** is zero.

7 Rosenfeld Properties of an Autoreduced Set

The link between algebra and differential algebra is made possible by a property of certain autoreduced sets first introduced in a lemma of Rosenfeld [34].

---

[8]From the proof, we only need that the initials and separants are *regular with respect to* $\mathfrak{a}$, that is, are not zero divisors of $\mathcal{R}/\mathfrak{a}$.

Broadly speaking, this property is what permits reducing computational problems in differential polynomial algebra to those of polynomial algebra. In this paper, we shall refer to this property as the Rosenfeld property, and introduce a slightly stronger variation.

Given an autoreduced set $\mathbf{A}$ in $\mathcal{R} = \mathcal{F}\{y_1, \cdots, y_n\}$, we shall investigate the relations between certain differential ideals and their polynomial counterparts.

Definition 7.1 Let V be a subset of ΘY such that $\mathbf{A} \subset \mathcal{F}[V]$. Let $\mathcal{S}_0 = \mathcal{F}[V \backslash \{v_1, \ldots, v_p\}]$. If we consider $\mathbf{A}$ as a triangular set, we shall always mean that, as a set of polynomials in $\mathcal{S}_0[v_1, \ldots, v_p]$, $\mathbf{A}$ is triangular over $\mathcal{S}_0$ with respect to $v_1, \ldots, v_p$. If $G = I$ (resp. $G = S$, resp. $G = H$), the ideal $(\mathbf{A}) : G^\infty$ of $\mathcal{S}_0[v_1, \ldots, v_p] = \mathcal{F}[V]$ will be denoted by J_G^V, the ideal $(\mathbf{A}) : G^\infty$ of $\mathcal{R}$ will be denoted by J_G, and the differential ideal $[\mathbf{A}] : G^\infty$ of $\mathcal{R}$ by $\mathfrak{a}_G$. We shall refer to it as the saturation (in case of $\mathfrak{a}_G$, differential) ideal of $\mathbf{A}$ with respect to initials (resp. separants, resp. initials and separants) in their (differential) rings.

Definition 7.2 We say that $\mathbf{A}$ has the *Rosenfeld property* (resp. *strong Rosenfeld property*) if every differential polynomial F partially reduced with respect to $\mathbf{A}$ belonging to the differential ideal $\mathfrak{a}_H$ (resp. $\mathfrak{a}_S$) already belongs to the ideal J_H (resp. J_S) in $\mathcal{R}$.

It is easy to see that if $\mathbf{A}$ has the strong Rosenfeld property, then it has the Rosenfeld property since the initials I_A are partially reduced with respect to $\mathbf{A}$. If $\mathbf{A}$ has the Rosenfeld property, then it is possible to answer certain questions about a differential ideal by answering similar questions about an ideal. For the differential ideals $\mathfrak{a}_H$ and $\mathfrak{a}_S$, we shall be interested in the properties of being prime, radical, and zero-reduced. The last property is an important one: recall that by Lemma 5.3, an autoreduced subset of a proper differential ideal $\mathfrak{a}$ is a characteristic set if $\mathfrak{a}$ is zero-reduced with respect to it. Which of these three properties may be deduced from corresponding properties of the ideals J_H, J_S? The answers to the first two are given below (the case for J_H is Lemma 6 of Kolchin [23], p. 137). The third is left as an exercise (Exercise 7.5).

Proposition 7.3 *Suppose* $\mathbf{A}$ *has the Rosenfeld property (resp. strong Rosenfeld property). Then* $\mathfrak{a}_H$ *(resp.* $\mathfrak{a}_S$*) is prime if and only if* J_H *(resp.* J_S*) is prime, and* $\mathfrak{a}_H$ *(resp.* $\mathfrak{a}_S$*) is radical if and only if* J_H *(resp.* J_S*) is radical.*

Proof. We will only prove the case when $\mathbf{A}$ has the strong Rosenfeld property. Suppose J_S is prime. Let $F, G \in \mathcal{R}$, $FG \in \mathfrak{a}_S$. Let $\widetilde{F}$ (resp. $\widetilde{G}$) be some partial remainder of F (resp. G) with respect to $\mathbf{A}$. By Equation (6.2), $\widetilde{F}\widetilde{G} \in \mathfrak{a}_S$ and is partially reduced and hence by the strong Rosenfeld property, $\widetilde{F}\widetilde{G} \in J_S$. So say $\widetilde{F} \in J_S \subseteq \mathfrak{a}_S$. Again by Equation (6.2), $F \in \mathfrak{a}_S \colon S^\infty = \mathfrak{a}_S$. Conversely, assume $\mathfrak{a}_S$ is prime, and let $F, G \in \mathfrak{R}$, $FG \in J_S$. Let $\widetilde{F}, \widetilde{G}$ be some partial remainders of F, G with respect to $\mathbf{A}$. Then $\widetilde{F}\widetilde{G} \in J_S \subseteq \mathfrak{a}_S$. Hence one, say $\widetilde{F}$, belongs to $\mathfrak{a}_S$. By the strong Rosenfeld property, $\widetilde{F} \in J_S$, and by (6.2), $F \in J_S$. So J_S is prime. The proof for the radical property is similar. $\qquad\square$

Remark 7.4 The ideal J_S (and therefore $J_H = J_S \colon I$ also) is always radical (as long as $\mathbf{A}$ is an autoreduced set, not necessarily having the Rosenfeld property). This fact is known as Lazard's Lemma (see Boulier *et al.* [5], and Hubert [17]; see also Proposition 3.4 of Morrison [29] for a generalization). Thus by the above, $\mathfrak{a}_H$ (resp. $\mathfrak{a}_S$) is radical if $\mathbf{A}$ has the Rosenfeld property (resp. strong Rosenfeld property). The readers may wonder if there is a similar result between J_I and $\mathfrak{a}_I$, and the answer is no: J_I may be prime but $\mathfrak{a}_I$ is not even radical! See Example 8.18 in the next section.

Exercise 7.5 Suppose $\mathbf{A}$ has the Rosenfeld property (resp. strong Rosenfeld property). Then $\mathfrak{a}_H$ (resp. $\mathfrak{a}_S$) is zero-reduced if and only if J_H (resp. J_S) is.

The ideals J_H and J_S belong to $\mathfrak{R}$ which is a polynomial ring in infinitely many indeterminates. In order to decide the properties above, we must work in a polynomial ring with finitely many indeterminates. Fortunately, these are properties of $\mathbf{A}$, and are independent of the ambient ring containing $\mathbf{A}$.

Exercise 7.6 Let $\mathbf{A}$ be a subset of $\mathfrak{R}$. Let V be a finite subset of ΘY such that $\mathbf{A} \subset \mathcal{F}[V]$. Let $G \in \mathcal{F}[V]$, $G \neq 0$. Let J_G be the ideal $(\mathbf{A}) \colon G^\infty$ in $\mathfrak{R}$ and let J_G^V be the ideal $(\mathbf{A}) \colon G^\infty$ in $\mathcal{F}[V]$. Then J_G is generated by J_G^V.

Exercise 7.7 Let V be any non-empty subset of ΘY and let J^V be an ideal in $\mathcal{F}[V]$. Let J be the ideal in $\mathfrak{R}$ generated by J^V. Show that J is prime (resp. radical, resp. zero-reduced with respect to an autoreduced set $\mathbf{A}$ of $\mathfrak{R}$ contained in $\mathcal{F}[V]$) if and only if J^V is.

Proposition 7.8 *Let J be an ideal generated by a given set Φ in a polynomial ring $\mathcal{F}[V]$ with V finite. Then the property of J being prime or radical is decidable. Moreover, if J is not radical, then one can find a polynomial $F \notin J$, and a natural number e such that $F^e \in J$; if J is not prime, then one can find a pair of polynomials $F, F' \notin J$ such that $FF' \in J$.*

Proof. Using the Gröbner basis method or the characteristic set method of Ritt-Wu for the algebraic case[9], we may assume that J is proper, compute

---

[9]For a recent review and comparison of algorithms to compute the radical of an ideal, the primary decomposition of an ideal, and other related algorithms see Decker *et al.* [15].

the radical $\sqrt{J}$ of J, and test the inclusion $\sqrt{J} \subseteq J$, thus deciding whether J is a radical ideal or not. If it is not, the Gröbner basis of $\sqrt{J}$ will include a polynomial F which is not in J. Now if J is not radical, then it cannot be prime. In that case, we can compute an exponent e such that $F^e \in J$ by using intermediate results during the usual radical ideal membership test to express 1 as a linear combination of generators of J and $tF - 1$, where t is a new indeterminate over $\mathcal{F}[V]$; an exponent is obtained by substituting $1/F$ for t and then clearing denominators. Then $e \geq 2$ and we can take $F' = F^{e-1}$. This proves the case for testing if J is radical. If J is radical, we can compute its prime decomposition. If there is only one prime component, then J is prime. Otherwise, suppose $J = J_1 \cap \cdots \cap J_r$ ($r \geq 2$) is the (irredundant) prime decomposition. For each prime component, we can find an $F_i \in J_i$ but $F_i \notin J$. Let k be the least integer i such that the product $F_1 F_2 \cdots F_i$ belongs to J (certainly one such product is when $i = r$). Then $k > 1$ and we can take $F = F_1 \cdots F_{k-1}$ and $F' = F_k$. □

We end this section with a sufficient condition for an autoreduced set $\mathbf{A}$ to be the characteristic set of a radical or prime differential ideal. Because of Corollary 3.20, this is basically a generalization of a result from Ritt [32], pp. 107–108, where he established that the Rosenfeld property holds for the ordinary case. However, his reasoning that $(\mathbf{A}): H^\infty$ (or, in his notation, the set Σ) is zero-reduced seems flawed. Our version will be modified to a necessary and sufficient condition and allows effective verification later (see Theorem 8.15).

Lemma 7.9 *Let* $\mathbf{A}: A_1 < \cdots < A_p$ *be an autoreduced set in* $\mathcal{R}$ *with the Rosenfeld property. Let* V *be a finite subset of* ΘY *such that* $\mathbf{A} \subset \mathcal{F}[V]$. *Suppose that* $\mathbf{A}$, *as a triangular set of polynomials in* $\mathcal{F}[V]$ *with respect to its leaders* $v_1, \ldots, v_p$ *has invertible initials, whose product is denoted by* I. *Suppose further that the ideal* J_I^V *of* $\mathcal{F}[V]$ *is radical and* $J_I^V = J_H^V$ (*resp.* J_I^V *is prime and does not contain any separants of* $\mathbf{A}$). *Then* $J_I = J_H$, *the differential ideal* $\mathfrak{a}_H$ *is radical (resp. prime) and* $\mathbf{A}$ *is its characteristic set.*

Proof. By Exercises 7.6 and 7.7, J_I is generated by J_I^V, and is radical (resp. prime). Observe that when J_I^V is prime and does not contain any separants of $\mathbf{A}$, $J_I^V = J_H^V$. Hence $J_I = J_H$. By Proposition 7.3, $\mathfrak{a}_H$ is radical (resp. prime). Now let $F \in \mathfrak{a}_H$ be reduced with respect to $\mathbf{A}$, that is, F is partially reduced with respect to $\mathbf{A}$, and F is its own pseudo-remainder with respect to $\mathbf{A}$. By the Rosenfeld property, $F \in J_H = J_I$. Expressing F as a sum $\sum C_M M$, where M is a monomial in $\overline{V} = \Theta Y \backslash V$ and $C_M \in \mathcal{F}[V]$, we see that each $C_M \in J_I^V$, and is reduced with respect to **A**. By Lemma 3.6, $C_M = 0$ for all M and $F = 0$. Hence $\mathfrak{a}_H$ is zero-reduced. Since the initials and separants of **A** are reduced, they are not in $\mathfrak{a}_H$, which is thus a proper ideal. By Lemma 5.3, **A** is a characteristic set of $\mathfrak{a}_H$. $\qquad\square$

Remark 7.10 We note that in the statement of Lemma 7.9 and later in Theorem 8.15, the condition $J_I^V = J_H^V$ is not really needed since it is equivalent to J_I^V being radical (see Proposition 3.3 of Hubert [17]). However, since the proof of that proposition is less elementary, we include the condition separately here. Algorithmically speaking, it may be easier to verify equality of two ideals than to test if an ideal is radical. We also note that the hypothesis J_I^V is radical is not necessary to conclude that $\mathfrak{a}_H$ is radical (Remark 7.4), but is needed to show that $\mathfrak{a}_H$ is zero-reduced.

Remark 7.11 As will be seen in Example 8.18 of the next section, it is possible that $\mathfrak{a}_H \neq \mathfrak{a}_I$ under the full hypothesis of Lemma 7.9 (in fact, not even under the additional assumption that **A** has invertible separants).

We leave the analogous result based on the strong Rosenfeld property as an exercise.

Exercise 7.12 Let $\mathbf{A}\colon A_1 < \cdots < A_p$ be an autoreduced set in $\mathfrak{R}$ with the strong Rosenfeld property. Let V be a finite subset of ΘY such that $\mathbf{A} \subset \mathcal{F}[V]$. Suppose that **A**, as a triangular set of polynomials in $\mathcal{F}[V]$ with respect to its leaders $v_1, \ldots, v_p$ has invertible initials. Suppose further that the ideal J_I^V of $\mathcal{F}[V]$ is radical (resp. prime) and $J_I^V = J_S^V$. Then $J_I = J_S = J_H$, the differential ideal $\mathfrak{a}_S$ is radical (resp. prime), $\mathfrak{a}_H = \mathfrak{a}_S = \{\,\mathbf{A}\,\}\colon S$, and **A** is a characteristic set of $\mathfrak{a}_S$.

8   Coherence and Rosenfeld's Lemma

To ensure that the Rosenfeld property holds and verifiable in a finite number of steps, Rosenfeld [34] introduced a sufficient condition he called *coherence* [10]. His lemma then states that a coherent autoreduced set has the Rosenfeld property. Kolchin [23] (Chapter III, Section 8, p. 135) generalized this to L-coherence relative to an ideal L (not necessarily differential, but having a set of generators partially reduced with respect to **A**) and over differential domains of arbitrary characteristic. Morrison [29] recently introduced new

---

[10]This notion is closely related to, and probably implies, the involutiveness, formal integrability, or other completeness properties of the system defined by **A**. However, the author is not aware of any research on the relationships among these various notions.

notions of semi-reduction, relative coherence, and Δ-completeness in her generalization of Rosenfeld's lemma. Her paper contains a careful study of the coherence property and its generalization. In this tutorial, we further study Rosenfeld's original version, and introduce a stronger version, and then two more, one equivalent to the Rosenfeld property, and the other to the strong Rosenfeld property. Both observations we believe are new. To avoid confusion with the literature and future extensions, it is suggested that each subsequent occurrence of the word "coherent" be replaced by "Rosenfeld-coherent" (but not recursively)!

Definition 8.1 For any $v \in \Theta Y$, let $\mathbf{A}_{(v)}$ be the set of all differential polynomials θA with $\theta \in \Theta$, $A \in \mathbf{A}$ and $\theta u_A < v$. We say an autoreduced set $\mathbf{A}$ is *coherent* (resp. *subcoherent*[11], resp. *strongly coherent*, resp. *strongly subcoherent*) if for all pairs $A, A' \in \mathbf{A}$ whose leaders $u_A, u_{A'}$ have a common derivative, say $v = \theta u_A = \theta' u_{A'}$, the differential polynomial $\Delta(A, A', v) = S_{A'} \theta A - S_A \theta' A'$, which has lower rank than v, belongs to the *ideal* $(\mathbf{A}_{(v)}) : H^\infty$ (resp. $(\mathbf{A} \cup \mathbf{A}_{(v)}) : H^\infty$, resp. $(\mathbf{A}_{(v)}) : S^\infty$, resp. $(\mathbf{A} \cup \mathbf{A}_{(v)}) : S^\infty$) in the polynomial ring $\mathcal{R}$.

Remark 8.2 Clearly, if $\mathbf{A}$ is strongly coherent, it is coherent and strongly subcoherent, and if it is coherent or strongly subcoherent, it is subcoherent. If the leaders of $\mathbf{A}$ have no common derivatives (for example, if $\mathcal{R}$ is a differential polynomial ring over an ordinary differential field, or if $\mathbf{A}$ consists of a singleton), then it is strongly coherent. The coherence or strong coherence property of an autoreduced set is decidable since it is enough to verify the condition on $\Delta(A, A', v)$ when v is a least common derivative of $u_A, u_{A'}$ (see Exercise 8.4 below). In recent literature, such differential polynomials $\Delta(A, A', v)$, where v is a least common derivative, are sometimes referred to as Δ-S-polynomials or differential S-polynomials to emphasize their similarity with S-polynomials[12] in Gröbner basis computations. Morrison [29] introduced a generalization called *pseudo-S-polynomial* and a differential version. Another variation is what Hubert [17] called the *cross-derivative* $X(A, A', v)$, defined as $\Delta(A, A', v) / \gcd(S_A, S_{A'})$. The cross-derivative involves an extra gcd computation and division, but may help to control intermediate expression swell in Gröbner basis computations. However, from a theoretical point of view, the cross-derivative has no advantage over $\Delta(A, A', v)$ since it is easy to see that for any $A, A' \in \mathbf{A}$, $\Delta(A, A', v)$ belongs to $(\mathbf{A}_{(v)}) : H^\infty$ (or any one of the other ideals in Definition 8.1) if and only if $X(A, A', v)$ does.

---

[11]Subcoherence is a special case of L-coherence with $L = (\mathbf{A})$.

[12]The S here stands for "syzygy", and is not related to the product of separants.

Exercise 8.3 Show that for any differential ring $\mathcal{R}$ and elements $a, b \in \mathcal{R}$, $\theta \in \Theta$ and $\operatorname{ord} \theta = s$, we can write $a^{s+1}\theta b$ as a linear combination (with coefficients in $\mathcal{R}$) of derivatives $\theta'(ab)$ where $\theta' \mid \theta$ (exact division in the monoid Θ). In particular, $a^{s+1}\theta b \in [ab]$.

Exercise 8.4 Apply Exercise 8.3 to show that if $\mathbf{A}$ is an autoreduced subset of $\mathcal{R} = \mathcal{F}\{y_1, \cdots, y_n\}$, and if for all $A, A' \in \mathbf{A}$ whose leaders have a common derivative, the differential S-polynomial $\Delta(A, A', v)$ with respect to lowest common derivatives v of $u_A, u_{A'}$ belongs to $(\mathbf{A}_{(v)}) : H^\infty$ (resp. $(\mathbf{A}_{(v)}) : S^\infty$), then $\mathbf{A}$ is coherent (resp. strongly coherent).

Corollary 8.5 *Given an autoreduced subset $\mathbf{A}$ of $\mathcal{R}$ relative to some ranking, the properties of coherence and strong coherence are decidable.*

Proof. We only need to consider the partial case. Observe, for any $A, A' \in \mathbf{A}$ whose leaders have a (least) common derivative $v = \theta u_A = \theta' u_{A'}$, that $\theta \neq 1$, $\theta' \neq 1$ and that $\theta A = S_A \theta u_A - T_\theta$, $\theta' A' = S_{A'} \theta' u_{A'} - T_{\theta'}$, where T_θ and $T_{\theta'}$ are of lower rank than v (see Exercise 4.14) and hence v does not appear in $F = \Delta(A, A', v)$. Now let G be S or H. By Exercise 7.6, $F \in (\mathbf{A}_{(v)}) : G^\infty$ if and only if F belongs to the ideal $(\mathbf{A}_{(v)}) : G^\infty$ of $\mathcal{R}_{[w]}$, where $w = \max\{v, (u_A)_{A \in \mathbf{A}}\}$. Thus, while not an efficient method, it is possible in principle to use the method of Gröbner basis to compute, for each of the finitely many pairs (A, A') that has a least common derivative v, the ideal $(\mathbf{A}_{(v)}) : G^\infty$ in the polynomial subring $\mathcal{R}_{[w]}$ of $\mathcal{R}$ and test membership of $\Delta(A, A', v)$. $\square$

Remark 8.6 A more practical method is to apply the following test: For $\mathbf{A}$ to be coherent (resp. strongly coherent), it is *sufficient* that the Ritt-Kolchin remainder (resp. partial remainder) of every $\Delta(A, A', v)$ be zero, since for any $F \in \mathcal{R}$ with rank $< v$ and whose Ritt-Kolchin remainder (resp. partial remainder) is zero, the Ritt-Kolchin reduction process (see Remark 6.10) shows that for some $s \in \mathbb{N}$, $H^s F \in (\mathbf{A}_{(v)})$ (resp. $S^s F \in (\mathbf{A}_{(v)})$) and hence $F \in (\mathbf{A}_{(v)}) : H^\infty$ (resp. $F \in (\mathbf{A}_{(v)}) : S^\infty$). From this, it also follows that any characteristic set of a differential ideal is coherent.

Remark 8.7 It is not known whether the properties of subcoherence and strongly subcoherence are decidable. However, they are respectively equivalent to the Rosenfeld property and strong Rosenfeld property, as we now show.

Lemma 8.8 (Rosenfeld) *An autoreduced set $\mathbf{A}$ is subcoherent (resp. strongly subcoherent) if and only if it has the Rosenfeld property (resp. strong Rosenfeld property).*

Proof. Suppose **A** is subcoherent and let $G = H$ (resp. suppose **A** is strongly subcoherent and let $G = S$). Let $\mathbf{A} : A_1 < \cdots < A_p$ with corresponding leaders $v_1 < \cdots < v_p$. Using notations as in Definition 7.1, let $F \in \mathfrak{a}_G$ be partially reduced with respect to **A**. For some $s \in \mathbb{N}$, $G^s F \in [\mathbf{A}]$. Let

$$G^s F = \sum_{j, \theta_j \neq 1,} C_j \theta_j A_{k_j} + \sum_i B_i A_i. \tag{8.9}$$

If possible, let v be the highest leader among all $\theta_j v_{k_j}$, and separate the terms in the first sum into two sums: one sum involving those summands with $\theta_j v_{k_j}$ having leaders $< v$ and the other involving those having leaders v. Thus

$$G^s F = \sum_{\substack{j, \theta_j \neq 1, \\ \theta_j v_{k_j} < v}} C_j \theta_j A_{k_j} + \sum_{\substack{\ell, \theta_\ell \neq 1, \\ \theta_\ell v_{k_\ell} = v}} D_\ell \theta_\ell A_{k_\ell} + \sum_i B_i A_i.$$

In the middle sum, choose an index h. Multiply the above by S_{k_h} (the separant of A_{k_h}) and rewrite the equation as

$$S_{k_h} G^s F = \sum_{\substack{j, \theta_j \neq 1, \\ \theta_j v_{k_j} < v}} C'_j \theta_j A_{k_j} + \sum_{\substack{\ell \neq h, \theta_\ell \neq 1, \\ \theta_\ell v_{k_\ell} = v}} D_\ell (S_{k_h} \theta_\ell A_{k_\ell} - S_{k_\ell} \theta_h A_{k_h})$$

$$+ D'_h \theta_h A_{k_h} + \sum_i B'_i A_i. \tag{8.10}$$

Now v, being a proper derivative of a leader v_{k_h}, does not appear in $S_{k_h} G^s F$ which is partially reduced with respect to **A**, and v does not appear in A_i (last sum), $\theta_j A_{k_j}$ (first sum). By the subcoherence (resp. strong subcoherence) of **A**, $S_{k_h} \theta_\ell A_{k_\ell} - S_{k_\ell} \theta_h A_{k_h} \in (\mathbf{A} \cup \mathbf{A}_{(v)}) : G^\infty$ and does not involve v. After multiplying (8.10) by a suitable product of initials and separants (resp. separants), we may distribute the resulting second sum among the first and last sums, obtaining an equation

$$G^{s'} F = \sum_{\substack{j, \theta_j \neq 1, \\ \theta_j v_{k_j} < v}} C''_j \theta_j A_{k_j} + + D''_h \theta_h A_{k_h} + \sum_i B''_i A_i. \tag{8.11}$$

Using $\theta_h A_{k_h} = S_{k_h} v - T_{k_h}$, where $T_{k_h} < v$, and treating v as an algebraic indeterminate, we may substitute $v = T_{k_h}/S_{k_h}$ into (8.11). After clearing denominators, multiplying by possibly other initials and separants (resp. separants) of **A**, and regrouping, we obtain, for some s'',

$$G^{s''} F = \sum_{j, \theta'_j \neq 1,} C'''_j \theta'_j A_{k'_j} + \sum_i B'''_i A_i.$$

This equation has the same form as (8.9), except that the highest derivative among $\theta'_j A_{k'_j}$ is $< v$. We now repeat the argument until we obtain an equation like (8.9) where the first sum is an empty summation, proving that[13] $F \in J_G$.

We now prove that the Rosenfeld property (resp. strong Rosenfeld property) implies subcoherence (resp. strong subcoherence). Let $F = \Delta(A, A', v)$, where v is some common derivative of $u_A, u_{A'}$. Let $\widetilde{F}$ be the Ritt-Kolchin partial remainder of F. Then by Equation (6.6) and Remark 6.10, we have $S^e F \equiv \widetilde{F} \bmod (\mathbf{A}_{(v)})$ for some suitable power of S. Thus $\widetilde{F} \in [\mathbf{A}]$ and by the Rosenfeld property (resp. strong Rosenfeld property), $\widetilde{F} \in J_G$. By multiplying by initials and separants, we may replace S^e by $G^{e'}$ so that $G^{e'} F \in (\mathbf{A} \cup \mathbf{A}_{(v)})$ and this proves that $\mathbf{A}$ is subcoherent (resp. strongly subcoherent). $\qquad\square$

When the subcoherence (resp. strong subcoherence) of an autoreduced set has been constructively verified, the proof above provides an algorithm to re-express any partially reduced element $F \in \mathfrak{a}_G$ as an element of J_G.

Remark 8.12 Subcoherence is probably what Rosenfeld [34] had in mind when he claimed (on page 397), wrongly, that coherence is equivalent to the Rosenfeld property. Hubert [17] (page 654) remarked that Rosenfeld's lemma "could be stated for" J_S and $\mathfrak{a}_S$, presumably under the coherence hypothesis. We are not able to verify that. We shall construct an example showing that subcoherence (resp. strong subcoherence) and coherence (resp. strong coherence) are not equivalent. But first, let us give a *failed attempt*[14] to prove that a subcoherent autoreduced set $\mathbf{A}$ is coherent.

Attempted Proof Suppose $\mathbf{A}: A_1 < \cdots < A_p$ is subcoherent. Let v_k, I_k, S_k denote respectively the leader, initial and separant of A_k and H the product of initials and separants. It suffices to show that for any $v \in \Theta Y$ which is some common derivative of leaders of $A, A' \in \mathbf{A}$, and $F = \Delta(A, A', v)$ that $F \in (\mathbf{A} \cup \mathbf{A}_{(v)}): H^\infty$ implies $F \in (\mathbf{A}_{(v)}): H^\infty$. Suppose for some $i_k, s_k \in \mathbb{N}$, $1 \le k \le h$, where $i_h + s_h \ne 0$, we have

$$I_1^{i_1} \cdots I_h^{i_h} S_1^{s_1} \cdots S_h^{s_h} F = \sum_{\substack{j, \theta_j \in \Theta, \\ \theta_j v_{k_j} < v}} C_j \theta_j A_{k_j} + \sum_{1 \le \ell \le p, v_\ell \ge v} B_\ell A_\ell. \qquad (8.13)$$

---

[13] If V is any finite subset of ΘY containing V_{pr} where V_{pr} is the set of $u \in \Theta Y$ that appears in F and A_i, and if J_G is replaced by J_G^V and $\mathcal{R}$ by $\mathcal{F}[V]$, we can prove that $F \in J_G^V$ by simply substituting $u = 0$ for any derivative $u \in \Theta Y, u \notin V_{pr}$ that appears in the final coefficient of every A_i.

[14] It is unusual to publish an attempted proof, especially a failed one, but this is a tutorial!

Let q be the highest subscript ℓ such that $B_\ell \neq 0$, and suppose among all possible equations (8.13) for F, ours is one for which q is highest. Let (e, Q_j, R_j) be a pseudo-remainder triple for C_j by A_q, and let (e, Q'_ℓ, R'_ℓ) be a pseudo-remainder triple for B_ℓ, $\ell \neq q$. Then we have

$$I_1^{i_1} \cdots I_h^{i_h} S_1^{s_1} \cdots S_h^{s_h} I_q^e F = \sum_{\substack{j, \theta_j \in \Theta, \\ \theta_j v_{k_j} < v}} R_j \theta_j A_{k_j} + \sum_{1 \leq \ell \leq p, v_\ell \geq v, \ell \neq q} R'_\ell A_\ell + B'_q A_q.$$

If $h < q$, the degree of v_q would be strictly higher in $B'_q A_q$ than any other term in the equation since $\mathbf{A}$ is autoreduced. Hence we must have $q \leq h$ and v_h is higher than v. Suppose $q < h$. Let (e', P_j, T_j) be a pseudo-remainder triple for C_j by A_h, and (e', P'_ℓ, T'_ℓ) be a pseudo-remainder triple for B_ℓ by A_h. Then we have

$$I_1^{i_1} \cdots I_h^{i_h} S_1^{s_1} \cdots S_h^{s_h} I_h^{e'} F = \sum_{\substack{j, \theta_j \in \Theta, \\ \theta_j v_{k_j} < v}} T_j \theta_j A_{k_j} + \sum_{1 \leq \ell \leq p, v_\ell \geq v} T'_\ell A_\ell + B'_h A_h.$$

By our choice of q, we must have $B'_h = 0$. This suggests the following construction of a counterexample when $q = h$. $\qquad\square$

Example 8.14 Let the ranking in $\mathcal{F}\{z, y, t\}$ be unmixed such that $z < y < t$ and orderly in each of the differential indeterminates. Let $A_1 = \delta_2^4 z + \delta_2^2 z$, $A_2 = \delta_1 y + z$, $A_3 = \delta_2^2 y$, and $A_4 = t^2$. $\mathbf{A}$ be the autoreduced set given by $A_1, \ldots, A_4$. Then $S_4 = 2t$. Let $v = \delta_1 \delta_2^2 y$. For any $\theta \in \Theta$, let $F_\theta = \Delta(A_2, A_3, \theta v) = \theta \delta_2^2 z$. Then $S_4^2 F_\theta = 4t^2 \theta \delta_2^2 z = 4\theta \delta_2^2 z A_4$. Thus $F_\theta \in (\mathbf{A} \cup \mathbf{A}_{(v)}) : S^\infty$ (which is the unit ideal) and hence $\mathbf{A}$ is strongly subcoherent (hence subcoherent). However, $\mathbf{A}$ is not coherent (and *a fortiori* not strongly coherent) since $F_1 = \Delta(A_2, A_3, v) \notin (\mathbf{A}_{(v)}) : H^\infty$. Observe that $(\mathbf{A}) : S^\infty$ is also the unit ideal and hence $\mathbf{A}$ has the strong Rosenfeld Property, showing that the strong Rosenfeld property is not equivalent to strong coherence.

Theorem 8.15 *Let* $\mathbf{A} : A_1 < \cdots < A_p$ *be an autoreduced set of* $\mathcal{R}$. *Let* V *be a finite subset of* ΘY *such that* $\mathbf{A} \subset \mathcal{F}[V]$. *Suppose the following three conditions hold:*

(a) *The Ritt-Kolchin remainder of* $\Delta(A, A', v)$ *with respect to* $\mathbf{A}$ *is zero for every pair* $A, A' \in \mathbf{A}$ *for which the leaders* $u_A, u_{A'}$ *have a least common derivative* v.

(b) $\mathbf{A}$, *as a triangular set of polynomials in* $\mathcal{F}[V]$ *with respect to its leaders* $v_1, \ldots, v_p$, *has invertible initials over* $\mathbf{S}_0 = \mathcal{F}[V \setminus \{v_1, \ldots, v_p\}]$.

(c) *The saturation ideal $J_I^V = (\mathbf{A}): I^\infty$ in $\mathcal{F}[V]$ with respect to initials of $\mathbf{A}$ is radical and $J_I^V = J_H^V$ (resp. prime and does not contain any separants of $\mathbf{A}$).*

Then $J_H = J_I$, and $\mathbf{A}$ is a characteristic set of the radical (resp. prime) differential ideal $\mathfrak{a}_H$. Conversely, if $\mathbf{A}$ is a characteristic set of a radical (resp. prime) differential ideal $\mathfrak{p}$ and if $J_H = J_I$, then $\mathfrak{p}: H = \mathfrak{a}_H$ (resp. $\mathfrak{p} = \mathfrak{a}_H$) and the three conditions hold.

Proof. Condition (a) implies that $\mathbf{A}$ is coherent (Remark 8.6), which implies that $\mathbf{A}$ is subcoherent (Remark 8.2), which implies that $\mathbf{A}$ has the Rosenfeld property (Lemma 8.8). The sufficiency of the conditions and further conclusions then follow from Lemma 7.9. For the necessity, suppose $\mathbf{A}$ is a characteristic set of a prime differential ideal $\mathfrak{p}$. Then it is coherent (Remark 8.6), $\mathfrak{p} = \mathfrak{a}_H$ (resp. $\mathfrak{p}: H = \mathfrak{a}_H$) (Corollary 6.13) and $\mathfrak{p}$ is zero-reduced with respect to $\mathbf{A}$ (Lemma 5.3). In particular, the Ritt-Kolchin remainder of every $\Delta(A, A', v)$, which belongs to $\mathfrak{p}$, is zero. By Proposition 7.3, J_I (which is also J_H by hypothesis) is radical (resp. prime), and hence by Exercises 7.6 and 7.7, J_I^V is radical (resp. prime). Since the initials and separants of $\mathbf{A}$ are not in $\mathfrak{p}$, they are not in J_I^V. In fact, we may replace V by the subset V' consisting of all $\theta y_j \in V$ that appears in some $A \in \mathbf{A}$, and $J_I^{V'}$ is radical (resp. prime).

Let $G \in J_I^{V'}$. Then G is partially reduced with respect to $\mathbf{A}$. The Ritt-Kolchin remainder of G with respect to $\mathbf{A}$ is simply the pseudo-remainder of G with respect to $\mathbf{A}$, and since $G \in \mathfrak{p}$, this remainder is zero by Corollary 6.13. By Corollary 3.20, $\mathbf{A}$ has invertible initials over $\mathcal{S}_0' = \mathcal{F}[V' \backslash \{v_1, \ldots, v_p\}]$ and hence also over $\mathcal{S}_0$. This completes the proof. $\square$

The three conditions on $\mathbf{A}$ in Theorem 8.15 are verifiable algorithmically. These conditions differ from and generalize the classical ones given by Lemma 2 on page 167 of Kolchin [23]. For comparison, we pose them here as exercises, and we add also the cases when ideals are radical.

Exercise 8.16 Suppose $\mathbf{A}$ is a characteristic set of a prime (resp. radical) differential ideal $\mathfrak{p}$. Prove that
(a) $\mathfrak{p} = \mathfrak{a}_H$ (resp. $\mathfrak{p}: H = \mathfrak{a}_H$),
(b) $\mathbf{A}$ is coherent,
(c) J_H is prime (resp. radical), and
(d) J_H is zero-reduced.

Exercise 8.17 Let $\mathbf{A}$ be an autoreduced set. Suppose
(a) $\mathbf{A}$ has the Rosenfeld property (resp. the strong Rosenfeld property),
(b) J_H (resp. J_S) is prime (or radical), and
(c) J_H (resp. J_S) is zero-reduced.

Prove that $\mathfrak{a}_H$ (resp. $\mathfrak{a}_S$) is a prime (or radical) differential ideal with $\mathbf{A}$ as a characteristic set; in particular, $\mathbf{A}$ is coherent.

Example 8.18 Consider again Example 5.6. Recall that we are working in an ordinary differential polynomial ring $\mathcal{R} = \mathcal{F}\{z, y\}$ under an orderly ranking such that $z < y$. For any $F \in \mathcal{R}$, let $F' = \delta F$. Let $A_1 = y^2 + z$, $A_2 = y' + y$, $A_3 = z' + 2z$. The set $\mathbf{A}: A_1, A_3$ is autoreduced, strongly coherent, and has invertible initials and invertible separants. A simple Gröbner basis computation verifies that $J_I = J_S = J_H = (A_1, A_3)$, which is prime and does not contain any separants. By Theorem 8.15 (and also Exercise 7.12), $\mathbf{A}$ is a characteristic set of the prime differential ideal $\mathfrak{a}_H = \mathfrak{a}_S = [A_1, A_3]:(2y)^\infty$, which is $[A_1, A_2]$. However, $\mathfrak{a}_I = [A_1, A_3] \neq \mathfrak{a}_H$ since $A_2 \notin \mathfrak{a}_I$ (Exercise 8.19). In fact, $\mathfrak{a}_I$ is not even a radical differential ideal since $A_2 \in \sqrt{\mathfrak{a}_I}$.

Exercise 8.19 In Example 8.18, show that $A_2 \notin \mathfrak{a}_I$ but $A_2 \in \sqrt{\mathfrak{a}_I}$.

9   Ritt-Raudenbush Basis Theorem

The direct analog of the Hilbert basis theorem is false, even in the simplest case when $\mathcal{F}$ is an ordinary differential field of characteristic zero and the differential polynomial ring $\mathcal{R} = \mathcal{F}\{y\}$ has only one differential indeterminate.

Exercise 9.1 Show that the following ascending sequence of differential ideals is infinite:

$$[y^2] \subset [y^2, (\delta y)^2] \subset \cdots \subset [y^2, (\delta y)^2, \cdots, (\delta^k y)^2] \subset \cdots$$

and hence the differential ideal $[y^2, (\delta y)^2, \cdots, (\delta^k y)^2, \cdots]$ is not finitely generated as a differential ideal.

There is, however, a weakened analog, the Ritt-Raudenbush Basis Theorem. Ritt proved the basis theorem in 1930 for radical differential ideals in a differential polynomial ring over a differential field of functions meromorphic in a region, and his result was generalized to abstract differential fields of characteristic zero as the coefficient field by Raudenbush in 1934. Kolchin further generalized the result to a wider class of coefficient domains. The version for differential fields of arbitrary characteristics was given by Seidenberg (see Chapter III, Section 5, p. 129 of Kolchin [23]). To present this result for the characteristic zero case, we begin with some general theory for an arbitrary differential ring $\mathcal{R}$. The proofs are very similar to the algebraic case and are left to the reader.

The set M of radical differential ideals of $\mathcal{R}$ (see Proposition 1.5) is an example of a *perfect differential conservative system* as defined in Chapter 0,

Sections 7, 8 of Kolchin [23]. The basis theorem holds for any Noetherian perfect differential conservative system with an additional hypothesis involving separability.

By Proposition 1.5, for any subset U of $\mathcal{R}$, there is a smallest radical differential ideal containing U, which we denote[15] by $\{U\}$. For any sets U, W of $\mathcal{R}$, let UW be the set of all products $u \in U$, $w \in W$. We note that for any subset U, if $F \in \{U\}$, then there is a finite subset Φ of U such that $F \in \{\Phi\}$.

Exercise 9.2 For characteristic $p \neq 0$, give an example showing that the radical of a differential ideal need not be differential. For characteristic zero, prove that $\{U\} = \sqrt{[U]}$.

Exercise 9.3 Prove that for any two subsets U, W of $\mathcal{R}$, we have

$$\{U\}\{W\} \subseteq \{UW\} = \{U\} \cap \{W\}.$$

Exercise 9.4 Prove that for any subset W of $\mathcal{R}$ and any $T \in \mathcal{R}$, we have $\{W\} = \{W\}:T \cap \{W, T\}$.

Exercise 9.5 Prove that the following are equivalent for $\mathcal{R}$:

(a) Every radical differential ideal is finitely generated as a radical differential ideal.

(b) For any subset U of $\mathcal{R}$, there exists a finite subset $\Phi \subseteq U$ such that $\{U\} = \{\Phi\}$.

(c) Every strictly increasing sequence of radical differential ideals is finite.

(d) Every non-empty set of radical differential ideal has a maximal element.

Definition 9.6 A differential ring $\mathcal{R}$ for which the conditions (a)–(d) of Exercise 9.5 hold is said to be *Noetherian with respect to radical differential ideals*.

Lemma 9.7 (Zorn) *Let S be a partially ordered set and suppose every totally ordered subset of S has an upper bound in S. Then S has a maximal element.*

Theorem 9.8 (Ritt-Raudenbush) *The differential polynomial ring $\mathcal{R} = \mathcal{F}\{y_1, \cdots, y_n\}$ is Noetherian with respect to radical differential ideals.*

---

[15]Occasionally, we still use $\{F_1, \ldots, F_r\}$ to indicate a *set* of differential polynomials $F_1, \ldots, F_r$ and not the radical differential ideal. When this is not clear from the context, we will always qualify the notation precisely.

Proof. Suppose the set N of radical differential ideals that are not finitely generated is not empty. By Proposition 1.5 (b) and Zorn's lemma, there is a maximal element $\mathfrak{p}$ in N. Clearly, $\mathfrak{p} \neq \mathcal{R}$, which is generated by 1. We claim $\mathfrak{p}$ is prime. Let $A, B \in \mathcal{R}$ and $A, B \notin \mathfrak{p}$. Then $\{\mathfrak{p}, A\} \supset \mathfrak{p}$ and by maximality, there exists a finite subset $\Phi \subset \mathfrak{p}$ such that $\{\mathfrak{p}, A\} = \{\Phi, A\}$. Similarly, there is a finite subset Ψ of $\mathfrak{p}$ such that $\{\mathfrak{p}, B\} = \{\Psi, B\}$. Now $\{\mathfrak{p}, AB\} = \{\mathfrak{p}, A\} \cap \{\mathfrak{p}, B\} = \{\Phi, A\} \cap \{\Psi, B\} = \{\Phi\Psi \cup \Phi B \cup A\Psi, AB\}$. If AB were in $\mathfrak{p}$, $\mathfrak{p}$ would be finitely generated. Whence $AB \notin \mathfrak{p}$ and $\mathfrak{p}$ is prime.

Let $\mathbf{A}$ be a characteristic set of $\mathfrak{p}$ and let $H = \prod_{A \in \mathbf{A}} I_A S_A$. Then $H \notin \mathfrak{p}$ by Corollary 5.4 and $H \cdot \mathfrak{p} \subseteq \{\mathbf{A}\}$. Also $\{\mathfrak{p}, H\} \supset \mathfrak{p}$, so that $\{\mathfrak{p}, H\} = \{\Sigma, H\}$ for some finite set $\Sigma \subset \mathfrak{p}$. Thus

$$\mathfrak{p} = \{\mathfrak{p}, H\} \cap \mathfrak{p} = \{\Sigma, H\} \cap \mathfrak{p} = \{\Sigma\mathfrak{p} \cup H\mathfrak{p}\} \subseteq \{(\Sigma) \cup \mathbf{A}\} \subseteq \mathfrak{p}$$

which shows that $\mathfrak{p}$ is finitely generated, a contradiction. $\square$

Definition 9.9 Let $\mathcal{R}$ be a differential ring and let $\mathfrak{a}$ be a radical differential ideal in $\mathcal{R}$. By a *component* (more precisely, an *$\mathcal{F}$-component*) of $\mathfrak{a}$ we mean any minimal prime differential ideal containing $\mathfrak{a}$. By Proposition 1.5(1) and Zorn's lemma, every prime differential ideal containing $\mathfrak{a}$ also contains a component of $\mathfrak{a}$.

Exercise 9.10 (Prime Decomposition) Let $\mathcal{R}$ be a differential ring, Noetherian with respect to radical differential ideals. Then every radical differential ideal $\mathfrak{a}$ is an intersection of a finite number of prime differential ideals.

As in the commutative ring case, we call a representation of a radical differential ideal $\mathfrak{a}$ as the finite intersection of prime differential ideal *irredundant* if none of the prime differential ideals can be omitted. Any irredundant decomposition is unique up to a permutation and the prime differential ideals in such a decomposition are the components.

10 Decomposition Problems

Just as for polynomial ideals, it is natural to compute the prime components of a radical differential ideal. More precisely, the problem is:

Decomposition Problem: Given a finite subset W of $\mathcal{R}$, decompose the radical differential ideal $\mathfrak{a} = \{W\}$ as an irredundant intersection of prime differential ideals: $\{W\} = \mathfrak{p}_1 \cap \mathfrak{p}_2 \cap \cdots \cap \mathfrak{p}_r$.

To this end, Ritt [32] (on page 109) gave an algorithm which depends on factorization over a tower of extensions. The algorithm is presented in Chapter

IV, Section 9, p. 166 of Kolchin [23], where he separated the decomposition problem into the following subproblems:

Problem 1: Find a finite set $\mathcal{A}$ of autoreduced sets of $\mathcal{R}$, each of which is a characteristic set of a prime differential ideal containing W, and for $k = 1, \ldots, r$, $\mathcal{A}$ contains a characteristic set $\mathbf{A}_k$ of the component $\mathfrak{p}_k$.

Problem 2: Given an autoreduced subset $\mathbf{A}$ of $\mathcal{R}$, decide whether it is a characteristic set of some component $\mathfrak{p}_k$.

Problem 3 (resp. **3'**): Given that $\mathbf{A}$ and $\mathbf{B}$ are respectively characteristic sets of prime differential ideals $\mathfrak{p}$ and $\mathfrak{q}$, decide whether $\mathfrak{p} \supseteq \mathfrak{q}$ (resp. $\mathfrak{p} = \mathfrak{q}$).

Clearly, solving Problems 1 and 2 will solve the decomposition problem. Alternatively, solving Problems 1 and 3 will also solve the decomposition problem. Assuming that Problems 1 and 3' are solvable (they are), Problems 2 and 3 are equivalent, that is, solving one solves the other. However, both Problems 2 and 3 are unsolved except for very special cases.

In this section, we shall show that Problem 1, the decomposition into prime differential ideals (not necessarily irredundant), is algorithmic.

Seidenberg [37] showed that the membership problem is solvable for radical differential ideals. Our presentation of Procedure 10.1: $\mathcal{A}(W)$, is built on the works of Ritt, Kolchin, and recent authors. This procedure may be combined with Remark 6.14 to solve the membership problem for radical differential ideals. In his book, Kolchin did not elaborate on the fine details of the constructive aspects; in particular, the test for a polynomial ideal to be zero-reduced with respect to $\mathbf{A}$ is buried in one single 5-line sentence (pp. 169–170 of Kolchin [23], which *also* includes a test for primeness). The algorithm requires factorization over algebraic extensions of $\mathcal{F}$. This test is now simplified by using the notion of invertibility of initials. We also show how to modify the procedure to give a decomposition into radical differential ideals satisfying the conditions of Theorem 8.15. Since it is possible to test membership for these radical differential ideals, such a decomposition will allow membership testing for any radical differential ideal $\{W\}$.

In Procedure 10.1, we adopt the following conventions: The intersection of an empty set of ideals is the unit ideal. All computations are performed in a finitely generated polynomial subring $\mathcal{R}_{(w)}$ of $\mathcal{R}$ for some $w = w(W) \in \Theta Y$ that is of higher rank than elements in the input set W and in the set of differential polynomials $\Delta(A, A', v)$ where $A, A' \in W$ and v is the least common derivative of $u_A, u_{A'}$. Following Ritt, the notation $W + T$ is used to mean the set obtained by adjoining a differential polynomial T to W.

Procedure 10.1 Algorithm for Problem 1: $\mathcal{A}(W)$

Input: A differential polynomial ring $\mathcal{R} = \mathcal{F}\{y_1, \cdots, y_n\}$ with a ranking on ΘY and a non-empty finite set $W \subset \mathcal{R}$

Output: A finite, possibly empty, set $\mathcal{A} = \mathcal{A}(W)$ of autoreduced sets of $\mathcal{R}$ such that for each $\mathbf{A} \in \mathcal{A}$, $\mathfrak{q_A} = [\mathbf{A}]: H_{\mathbf{A}}^{\infty}$ is prime with characteristic set $\mathbf{A}$, and $\{W\} = \cap_{\mathbf{A} \in \mathcal{A}} \mathfrak{q_A}$.

Step 1:   If W contains a non-zero element of $\mathcal{F}$, Return $\emptyset$.

Step 2: $\mathbf{A} :=$ autoreduced subset of W of lowest rank

Step 3:   For all $F \in W, F \notin \mathbf{A}$ do

Step 3a: $F_0 :=$ Ritt-Kolchin remainder of F with respect to $\mathbf{A}$

Step 3b: If $F_0 \neq 0$, Return $\mathcal{A}(W + F_0)$.

Step 4:   For all $A, A' \in \mathbf{A}$ do

Step 4a: If $u_A, u_{A'}$ has a least common derivative v

Step 4a.1: $F := \Delta(A, A', v)$

Step 4a.2: $F_0 :=$ Ritt-Kolchin remainder of F

Step 4a.3:   If $F_0 \neq 0$, Return $\mathcal{A}(W + F_0)$.

Step 5:   Sort $\mathbf{A}: A_1 < \ldots < A_p$ by rank.
$V :=$ the set of derivatives θy_j appearing in $\mathbf{A}$
$\mathcal{S}_0 := \mathcal{F}[V \backslash \{v_1, \ldots, v_p\}]$, where v_k is leader of A_k

Step 5a: For $k = 2, \ldots, p$ do

Step 5a.1:   If the initial I_k of A_k is not invertible with respect to $\mathbf{A}_{k-1}$,

Step 5a.1.1:   find a non-zero $G_k \in \mathcal{S}_0[v_1, \ldots, v_{k-1}]$, $G_k \notin (\mathbf{A}_{k-1})$,
G_k reduced with respect to $\mathbf{A}_{k-1}$ and $G_k I_k \in (\mathbf{A}_{k-1})$

Step 5a.1.2:   Return $\mathcal{A}(W + I_k) \cup \mathcal{A}(W + G_k)$.

Step 6: $I := \prod_{k=1}^{p} I_k$; $J_I^V := (\mathbf{A}): I^{\infty}$ (compute a Gröbner basis of J_I^V)

Step 6a: If $J_I^V = (1)$, Return the union of $\mathcal{A}(W + I_k)$ for all $1 \leq k \leq p$.

Step 6b: If J_I^V is not prime

Step 6b.1:   Find non-zero $F, F' \in \mathcal{S}_0[v_1, \ldots, v_p]$ such that
$FF' \in J_I^V$, $F, F' \notin J_I^V$

Step 6b.2: $F_0 :=$ Ritt-Kolchin remainder of F

Step 6b.3: $F_0' :=$ Ritt-Kolchin remainder of F'

Step 6b.4:   Return the union of $\mathcal{A}(W + I_k)$ for all $1 \leq k \leq p$,
and $\mathcal{A}(W + F_0)$, $\mathcal{A}(W + F_0')$.

Step 7:   If $S := \prod_{k=1}^{p} S_k \in J_I^V$

Step 7a: Return the union of $\mathcal{A}(W + I_k)$ and $\mathcal{A}(W + S_k)$ for all $1 \leq k \leq p$.

Step 8:   Return the union of $\mathcal{A}(W + I_k)$ and $\mathcal{A}(W + S_k)$ for all $1 \leq k \leq p$,
and the set with the singleton $\mathbf{A}$.

Theorem 10.2 *Procedure 10.1 solves Problem 1. Furthermore, if $\mathbf{A} \in \mathcal{A}(W)$, then $\mathbf{A}$ satisfies the three conditions in Theorem 8.15.*

Proof. Procedure 10.1 is a recursive routine and its termination is guaranteed by the well-founded property (Proposition 4.24) because every recursive call involves an enlarged input set $W + T$, where T is a non-zero differential polynomial that is reduced with respect to the current autoreduced subset $\mathbf{A} = \mathbf{A}(W)$ of lowest rank in W. Every such T (in particular, the initial and separant of any $A \in \mathbf{A}$) cannot be in W, making $\mathbf{A}(W + T)$ an autoreduced set of strictly lower rank than $\mathbf{A}(W)$. The procedure ends without making recursive calls when and only when W contains a non-zero element of $\mathcal{F}$ (in Step 1). In every other situation (Steps 3b, 4a.3, 5a.1.2, 6a, 6b.4, 7a, 8), correctness depends on the fact that if $\{W\}$ contains (for example, if $[\mathbf{A}]$ does) a product $T = T_1 \cdots T_r$, where each factor T_i is non-zero, is reduced, and is not in W, then $\{W\} = \{W + T\} = \cap_{i=1}^{r}\{W + T_i\}$. This latter equation guarantees the validity of the returned output $\mathcal{A}(W) = \cup_{i=1}^{r}\mathcal{A}(W + T_i)$.

The cases in Steps 3b, 4a.3, 5a.1.2 are obvious since $F_0 \in \{W\}$ and $G_k I_k \in \{W\}$. At Step 4, W is zero-reduced with respect to $\mathbf{A}$ and $W \subset \mathfrak{a}_H = [\mathbf{A}] : H^\infty$. At Step 5, $\mathbf{A}$ satisfies condition (a) of Theorem 8.15. Steps 5a.1 and 5a.1.1 are both effective by Theorem 3.22. At Step 6, $\mathbf{A}$ has invertible initials (condition (b) of Theorem 8.15). Steps 6, 6a are both effective, and when $J_I^V = (1)$, some power of I belongs to $(\mathbf{A})$ and hence $I \in \{W\}$, so we may apply recursion with $T = I$. Steps 6b and 6b.1 are effective by Proposition 7.8, and Steps 6b.2, 6b.3 only involves pseudo-division since F, F' are already partially-reduced with respect to $\mathbf{A}$ by the choice of V in Step 5. Note that F_0, F_0' are non-zero since $F, F' \notin J_I^V$, and cannot be in W, which is zero-reduced after Step 4. At Step 6b.4, $T = IF_0 F_0' \in \{W\}$ and the recursion is valid. At Step 7, J_I^V is a prime ideal. If $S \in J_I^V$, then $T = IS \in \{W\}$. Finally, at Step 8, J_I^V satisfies condition (c) of Theorem 8.15, hence $\mathfrak{a}_H$ is a prime differential ideal, $J_I = J_H$, and $\mathbf{A}$ is a characteristic set of $\mathfrak{a}_H$. Since $\{W\} \subset \mathfrak{a}_H$, $\{W\}$ is zero-reduced with respect to $\mathbf{A}$ and hence $\mathbf{A}$ is a characteristic set of $\{W\}$ by Lemma 5.3. Moreover, by Corollary 6.13, $\{W\}: H = \mathfrak{a}_H$. We have, by Exercise 9.4,

$$\{W\} = \{W\}: H \cap \{W + H\}. \tag{10.3}$$

The returned result at Step 8 is correct. □

Remark 10.4 In Kolchin's book [23] (see Problem(a), p. 169), the algorithm calls for a test that J_I^V is zero-reduced with respect to $\mathbf{A}$, and if the test fails, it proceeds to find a non-zero $G \in J_I^V$ which is reduced. By using a V consisting of only partially reduced derivatives θy_j, this becomes equivalent, by Corollary 3.20 and Remark 3.21, to testing whether $\mathbf{A}$ has invertible initials (Steps 5a through 5a.1.2). This improvement is due to the works of Kandri Rody $et\,al.$ [18], Bouziane $et\,al.$ [10], and Sadik [36].

Procedure 10.1 provides the model for many variations that yield decompositions for which the autoreduced sets $\mathbf{A}$ in the output list enjoy specific properties. These specific properties are described using some rather confusing terminologies by recent authors. We refrain from coining a new combined name for the three conditions of Theorem 8.15 and let each property speak for itself. We explore some of these variations.

Variation 10.5 Recent interest in the decomposition problem was due mainly to a desire to solve the membership problem for a radical differential ideal $\{W\}$, after the analogous problem was solved in the polynomial case by Buchberger's algorithm [9]. For a differential ideal of the form[16] $\mathfrak{a}_H = [\mathbf{A}]\!:\! H^\infty$ where $\mathbf{A}$ is a characteristic set of $\mathfrak{a}_H$, a differential polynomial F belongs to $\mathfrak{a}_H$ if and only if the Ritt-Kolchin remainder of F with respect to $\mathbf{A}$ is zero. It is thus natural to solve the membership problem of $\{W\}$ by expressing $\{W\}$ as an intersection of such differential ideals. To modify Procedure 10.1 for this purpose, we can replace Steps 6b through 6b.4 by either testing that J_I^V is radical (Proposition 7.8), or testing that $J_I^V = J_H^V$ (Remark 7.10). If the radical test returns false, we can then find an $F \notin J_I^V$ but $F^e \in J_I^V$ for some $e \in \mathbb{N}$. Or if $J_H^V \not\subseteq J_I^V$, we can find an $F \notin J_I^V$ but $F \in J_H^V$. In either case, the Ritt-Kolchin remainder F_0 of F cannot be zero (because F is partially reduced, and $F \notin J_I^V$), and cannot be in W (because after Step 4, W is zero-reduced). Thus $LF_0 \in \{W\}$ for some product L of initials and separants, and we can apply recursion with $T = IF_0$ (if the radical test fails) or with $T = ISF_0$ (if the equality test fails). If either test returns true, we can continue with Steps 7, 7a, and 8. Correctness of this variation follows from the radical ideal version of Theorem 8.15.

Variation 10.6 Instead of Theorem 8.15, we may apply the strong version: Exercise 7.12 (see also Remark after the solution in the Appendix). We can replace Steps 4 through 4a.3 by applying the algorithm in Corollary 8.5 to make $\mathbf{A}$ strongly coherent. As with Variation 10.5, we can replace Steps 6b through 6b.4 by testing that J_I^V is radical (Proposition 7.8). We will make a recursive call if J_I^V is not radical. If J_I^V is radical, we have $J_S^V \subseteq J_H^V = J_I^V$ (either by Remark 7.10 or ensuring $J_H^V = J_I^V$ as in Variation 10.5). and

---

[16]According to Hubert [17], such a differential ideal $\mathfrak{a}$ is called $characterizable$ (with respect to the given ranking), $\mathbf{A}$ is said to be $characteristic$ and to $characterizes$ $\mathfrak{a}$, and a decomposition of $\{W\}$ into characterizable differential ideals is a $characteristic$ $decomposition$.

further test whether $J_I^V \subseteq J_S^V$. If $J_I^V = J_S^V$, we can continue as before with Steps 7, 7a, 8; by Exercise 7.12, every $\mathbf{A} \in \mathcal{A}(W)$ is the characteristic set of the radical differential ideal $\mathfrak{a}_S = [\mathbf{A}]:S^\infty$. If $J_I^V \not\subseteq J_S^V$, by a result (Corollary 2.1.1 of Sadik [36]), $\mathbf{A}$ does not have invertible separants. Thus we can find a separant S_k that is not invertible with respect to $\mathcal{A}_k$ and find a non-zero F, reduced with respect to $\mathbf{A}$ such that $FS_k \in (\mathbf{A})$. We can apply a recursive call with $T = S_k F$.

Exercise 10.7 Explain why, in Variation 10.6, we cannot simply make a recursion call after finding an $F \in J_I^V$ which is not in J_S^V.

Variation 10.8 At Step 6, since $\mathbf{A}$ has invertible initials, we may improve the efficiency of the algorithm and Variations 10.5 and 10.6 as follows. By Theorem 3.22, for each $k, 1 \le k \le p$, we can find a non-zero, differential polynomial $L_k \in (\mathbf{A}_{k-1}, I_k) \cap \mathbf{S}_0$ that is reduced with respect to $\mathbf{A}$ (this adds a Step 5a.2 at very little extra cost). Suppose $L_k = M_k I_k + N_k$ as in (b) of Theorem 3.22. If $T = I_k^e F \in \{W\}$, then $T' = L_k^e F \in \{W\}$. Thus in the Return statements at Steps 6a, 6b.4, 7a, and 8, we may replace I_k by L_k. This will be more efficient because L_k does not involve $v_1, \dots, v_p$ (and may perhaps even belong to $\mathcal{F}$, effectively making the recursive call trivial). Note that we are *not*[17] replacing I_k by L_k in A_k! Similarly, by adding another step to test the invertibility of separants, we may replace separants by non-zero, reduced, differential polynomials of lower ranks in the recursive calls in Steps 7a and 8.

Variation 10.9 Instead of using J_I^V, one can use J_S^V (resp. J_H^V) in Steps 6 through 7. If we only want an intersection of radical differential ideals, then after making sure that $\mathbf{A}$ is strongly coherent in case of $\mathfrak{a}_S$, we know that $\mathfrak{a}_S$ (resp. $\mathfrak{a}_H$) is radical (Remark 7.4). Hence it suffices to test equality of ideals $J_I^V = J_S^V$ (resp. $J_I^V = J_H^V$) in order to apply Exercise 7.12 (resp. Theorem 8.15). In case the test fails, a recursive call can be made in a manner similar to Variation 10.6.

Variation 10.10 When the input W is linear, Steps 6b through 6b.4 are not necessary. Several variations in this special case are used to compute integrability conditions of differential equations, with applications to determining Lie symmetries (see, for example, Carminati and Vu [12] and references there).

---

[17]That this *may* be done requires delving into the form of Gröbner basis for $(\mathbf{A})$ over the quotient field of $\mathbf{S}_0$. See Sadik [36]. However, we do not think the extra knowledge is necessary to improve the algorithm. Besides, our argument works also for the separants.

Of course, these variations may be applied to the prime decomposition. Improvements in implementation are still hot topics of current research. Boulier *et al.* [5,6,7,8] generalized the decomposition for $\{W\}$ into a decomposition of *regular differential systems*, which include not only the systems of equations $W = 0$, but also inequations $U \neq 0$, and these must satisfy certain more relaxed conditions compared to those on coherence and autoreduction. By studying more carefully the dimension relationships among the radical differential ideals and their polynomial counterparts, a decomposition into unmixed dimension radical differential ideals is possible (see Hubert [17], Kandri Rody *et al.* [18], Bouziane *et al.* [10], and Sadik [36]). We leave the readers to pursue this rich set of results.

11   Component Theorems

In this section, we continue to explore the decomposition problem. The solution to Problem 3' is quite easy, and is left to the reader.

Exercise 11.1 The equality problem between prime differential ideals is decidable: Given two prime differential ideals $\mathfrak{p}$ and $\mathfrak{q}$ and corresponding characteristic sets $\mathbf{A}$ and $\mathbf{B}$ with respect to the same ranking,[18] we have $\mathfrak{p} = \mathfrak{q} \iff \mathbf{B} \subset \mathfrak{p}$ and $\mathbf{A} \subset \mathfrak{q}$.

Despite the similarity between Problem 3 and Problem 3', the former is still open. Problem 3 is a notoriously difficult problem because even in very special cases, several of the deepest results in differential algebra are needed. To provide even a paraphrase, not to mention any detail or proof of these advanced topics would be beyond the scope of this introductory paper. Instead we shall be content with an exposition of the Component Theorem and the Low Power Theorem, referring readers to Kolchin's text[19] for more information. These advanced topics are:

- Leading Coefficient Theorem (Theorem 4, p. 172)
- Levi's Lemma (Lemma 3, p. 177)
- Shapiro's Lemma (Lemma 17, pp. 53–56)
- Domination Lemma (Lemma 6, p. 181)
- Preparation Lemmas (Lemma 7, p. 80; Chap. IV, Sect. 13, p. 183–185)
- Component Theorem (Theorem 5, p. 185)

---

[18]This is emphasized because of a recent paper of Boulier *et al.* [8], where two different characteristic sets with respect to different rankings are computed for the same prime differential ideal.

[19]All references cited with page number(s) are from Kolchin [23] unless otherwise stated.

- Low Power Theorem (Theorems 6, 7, pp. 187–188)
- The Ritt Problem (Chap. IV, Sect. 16, pp. 190–193)

We begin by describing the special case for the Decomposition Problem when the input W contains a single differential polynomial A. We are interested in determining the components of $\{A\}$. When A is irreducible, we have the following theorem (easy by now, if you have read this far) on the general component of A.

Theorem 11.2 (Ritt's General Component Theorem) *Let A be a differential polynomial in $\mathfrak{R} = \mathfrak{F}\{y_1, \cdots, y_n\}$ irreducible over $\mathfrak{F}$, and let $\{A\} = \mathfrak{p}_1 \cap \cdots \cap \mathfrak{p}_r$ be an irredundant decomposition of the radical differential ideal generated by A as the intersection of prime differential ideals $\mathfrak{p}_i$. Then among the components of $\{A\}$, there is one, denoted by $\mathfrak{p}_{\mathcal{F}}(A)$, that does not contain any separant of A. Each of the other components of $\{A\}$ contains every separant of A. Furthermore, for any separant S of A, $\mathfrak{p}_{\mathcal{F}}(A) = [A]:S^\infty = \{A\}:S$ and an element of $\mathfrak{p}_{\mathcal{F}}(A)$ that is partially reduced with respect to A must be divisible by A.*

Proof. Fix a ranking $\leq$ and let v be the leader and S be the separant of A. Let $\mathfrak{p} = \mathfrak{a}_S = [A]:S^\infty$ and let $\mathbf{A}$ be the singleton set consisting of A alone. Clearly, $\mathbf{A}$ is autoreduced and strongly coherent. If $F \in \mathfrak{p}$ is partially reduced with respect to $\mathbf{A}$, then by Lemma 8.8, $F \in (A):S^\infty$. Thus for some $e \in \mathbb{N}$, $S^e F \in (A)$. Since A is irreducible and does not divide S, A must divide F. Hence $\mathfrak{p}$ is a proper differential ideal (S is not divisible by A) and is zero-reduced with respect to $\mathbf{A}$. Moreover if $G = G_1 G_2 \in \mathfrak{p}$, and if $\widetilde{G_1}, \widetilde{G_2}$ are the corresponding Ritt-Kolchin partial remainders of G_1, G_2, then A divides $\widetilde{G_1}$ or $\widetilde{G_2}$ so that G_1 or $G_2 \in \mathfrak{p}$. Thus $\mathfrak{p}$ is prime, $\mathfrak{p} = \mathfrak{a}_H$ (Corollary 6.13), and $\mathbf{A}$ is a characteristic set of $\mathfrak{p}$. Now by Exercise 9.4, $\{A\} = \mathfrak{p} \cap \{A, S\}$. Let $\mathfrak{q}_1, \ldots, \mathfrak{q}_\ell$ be the prime components of $\{A, S\}$. Then every $\mathfrak{p}_j$ must be either $\mathfrak{p}$ or some $\mathfrak{q}_i$. Now $\mathfrak{p} \supseteq \mathfrak{p}_j$ for some j, $S \notin \mathfrak{p}$, and $S \in \mathfrak{q}_i$. Hence $\mathfrak{p}_j \neq \mathfrak{q}_i$ for any i and $\mathfrak{p}_j = \mathfrak{p}$ and $\mathfrak{p}$ is a component of $\{A\}$. Every other component must be some $\mathfrak{q}_i$ and hence contains S. If v' is the leader and S' the corresponding separant of A relative to some other ranking $\leq'$, then no proper derivative of v can appear in S', S' is partially reduced with respect to A relative to $\leq$, and is not divisible by A. Hence $S' \notin \mathfrak{p}$. By what we just proved, every component of $\{A\}$ other than $\mathfrak{a}_{S'}$ must contain S'. So $\mathfrak{a}_{S'} = \mathfrak{p}$ and this completes the proof. □

Remark 11.3 An examination of the proof shows that we used four properties of $\mathbf{A}$: (1) $\mathbf{A}$ is autoreduced with respect to every ranking, (2) $\mathbf{A}$ has

the strong Rosenfeld property with respect to every ranking, (3) the ideal
(**A**) is prime, and (4) the product S of separants of **A** with respect to every
ranking does not lie in (**A**). There are sets satisfying all these properties
without reducing to a single irreducible differential polynomial, for example,
when **A** consists of ordinary linear homogeneous differential polynomials, each
involving only one, distinct differential indeterminate. Of course in that case,
[**A**] is prime. Presumably it is possible to have non-linear systems with these
properties satisfying a similar theorem.

Definition 11.4 For any irreducible differential polynomial A, the component $\mathfrak{p}_{\mathcal{F}}(A)$ is called the *general component* (more correctly, *general
$\mathcal{F}$-component*) of A. All other components are called *singular components*.
Geometrically, an element $\eta = (\eta_1, \ldots, \eta_n)$ where each η_i belongs to some differential extension field of $\mathcal{F}$ is a *solution* or *zero* of $A \in \mathcal{R} = \mathcal{F}\{y_1, \cdots, y_n\}$
if $A(\eta) = 0$. We say η is a *non-singular* solution if some separant (based on
some ranking) of A does not vanish at η. Otherwise, we say η is a *singular*
solution if every separant of A vanishes at η.

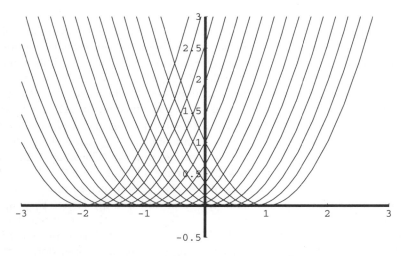

Figure 1. Example 11.5: $y = (x + c)^2$

Example 11.5 The ordinary differential polynomial $A = (y')^2 - 4y$ is
irreducible (here $y' = dy/dx$). Its only separant is $S_A = 2y'$. Differentiating A, we have $A' = 2y'y'' - 4y' = 2y'(y'' - 2)$. It is easy to see that
the functions $\eta_c = (x + c)^2$, for an arbitrary constant c, give the one parameter family of non-singular solutions (commonly known as the "general

solution"), that $\eta = 0$ is the only singular solution, and that η is neither in the family nor a limit of the family (see accompanying figure). Algebraically, the prime decomposition is given by $\{A\} = \{A, y'' - 2\} \cap \{A, y'\}$. Thus $\mathfrak{p}_1 = \{A, y'' - 2\} = \{A\} : y'$ is the general component $\mathfrak{p}_{\mathcal{F}}(A)$. The other (singular) component is $\mathfrak{p}_2 = \{A, y'\} = [y]$ which is the general component of $B = y$. It is clear that $\mathfrak{p}_{\mathcal{F}}(A) \not\subseteq [y]$ and thus the decomposition is irredundant.

While the radical differential ideal generated by every differential polynomial A admits a decomposition into prime differential ideals, we emphasize that the concepts of general and singular components as defined in Kolchin [23] apply only to an *irreducible* differential polynomial and are *independent of the ranking* and thus intrinsic to A. For example, if $F = yz \in \mathcal{F}\{y, z\}$, then $\{F\} = \{y\} \cap \{z\}$ but neither $[y]$ nor $[z]$ can be the general component or a singular component of F since each contains only one of the two possible separants. Readers should be aware that Definition 4.2 of Hubert [16] used a more liberal definition for a singular component of a differential polynomial, not necessarily irreducible, and that the definition depends on the ranking. According to that definition, $[y]$ would be a singular component of $\{F\}$ when the ranking is such that $y < z$.

For an arbitrary differential polynomial, the next theorem (Theorem 5, p. 185) gives the structure of the decomposition, relating the components to the general components of irreducible differential polynomials. The theorem is well illustrated by Example 11.5 above.

Theorem 11.6 (Ritt's Component Theorem) *Let F be a differential polynomial in $\mathcal{R} = \mathcal{F}\{y_1, \cdots, y_n\}$, $F \notin \mathcal{F}$ and not necessarily irreducible. Then for any component $\mathfrak{p}$ of $\{F\}$ in $\mathcal{R}$, there exists an irreducible differential polynomial $A \in \mathcal{R}$ such that $\mathfrak{p} = \mathfrak{p}_{\mathcal{F}}(A)$. If moreover F is irreducible, and if $\mathfrak{p} = \mathfrak{p}_{\mathcal{F}}(A)$ is a singular component of $\{F\}$, then F involves a proper derivative of the leader of A relative to any ranking, and $\operatorname{ord} A < \operatorname{ord} F$.*

Proof. We omit the proof of the main claim which requires deep results not covered here. When F is irreducible, F cannot be divisible by A since $F \neq A$ when $\mathfrak{p}_{\mathcal{F}}(A)$ is a singular component of $\{F\}$. By Theorem 11.2, F is not partially reduced with respect to A for any ranking; in particular, for an orderly ranking. Thus the second claim follows. $\square$

With these two theorems, the decomposition problem for a single differential polynomial F may be attacked as follows. We can use Procedure 10.1 to compute $\mathcal{A}(W)$, where W consists of the singleton F (for the first pass before any recursive calls, Steps 1 through 5a.1.2, and 6a are trivial). For any

$\mathbf{B} \in \mathcal{A}(W)$, the associated prime differential ideal $\mathfrak{q_B} = [\mathbf{B}] : H_\mathbf{B}^\infty$ may be a component only if $\mathfrak{q_B} = \mathfrak{p}_{\mathcal{F}}(A)$, where A is an irreducible differential polynomial. Since $\mathbf{A} : A$ is a characteristic set of $\mathfrak{p}_{\mathcal{F}}(A)$, it follows by the proof of Exercise 11.1 that $\mathbf{B}$ consists of a singleton B and A and B have the same rank. Since B is then partially reduced with respect to A, B must be divisible by A by Theorem 11.2. If $B = AB'$, then B' must be non-zero, of lower rank than A, and does not involve the leader of A or B. Thus to find A, we first discard every $\mathbf{B} \in \mathcal{A}(W)$ for which $\mathbf{B}$ is not a singleton set. Then we factor each singleton B as a univariate polynomial in its leader v and discard those with more than one irreducible factor involving v. For those B that has only one irreducible factor A involving v, we have to decide whether $\mathfrak{p}_{\mathcal{F}}(A)$ is a component of $\{F\}$. We end this section with another example, quite similar to Example 11.5, but different outcome.

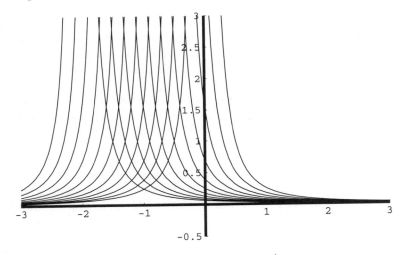

Figure 2. Example 11.7: $y = \dfrac{1}{4(x+c)^2}$

Example 11.7 The ordinary differential polynomial $A = (y')^2 - y^3$ is irreducible. The functions $\eta_c = (x+c)^{-2}/4$, for an arbitrary constant c, give a one-parameter family of non-singular solutions. Since the only separant is $S_A = 2y'$, the only singular solution is $\eta = 0$, which is also $\lim_{c \to \infty}(x+c)^{-2}/4$. Thus the singular solution is geometrically considered as "embedded in the general solution" (see accompanying figure). Algebraically, we have $A' = 2y'y'' - 3y^2y' = 2(y'' - \frac{3}{2}y^2)y'$ and $\{A\} = \{A, y'' - \frac{3}{2}y^2\} \cap [y] = \{A, y'' - \frac{3}{2}y^2\}$. Hence $\{A\} : y'$ is the general and only component and $\mathfrak{p}_{\mathcal{F}}(A) = \{A\} : y' \subset [y]$.

12 The Low Power Theorem

The key result that helps one to distinguish between the two algebraically similar examples in Section 11 is the Low Power Theorem (see Chapter III of Ritt [32]). We will explain the version (Theorem 6, p. 187) by Levi [25] for partial differential polynomials and give Ritt's version as a corollary. Kolchin has an even more general version (Theorem 7, p. 188). The Low Power Theorem is stated in terms of a *preparation congruence*.

We begin by describing and proving the construction of a *preparation equation*. In the algebraic case, this is an equation involving a polynomial F, a subset $\mathbf{G}: G_1, \ldots, G_r$ of polynomials triangular with respect to $u_1, \ldots, u_r$ over some integral domain $\mathbf{S}_0$, and a product L of corresponding initials. A preparation equation is very similar to the pseudo-remainder equation (Equation (2.10) in Section 2, with a change in notations), except that, instead of expressing LF as a linear combination of $G_1, \ldots, G_r$ with coefficients $Q_1, \ldots, Q_r$, and a remainder R, in a preparation equation, we now express LF as a polynomial P in $G_1, \ldots, G_r$. By carefully controlling the process to compute a preparation equation, we can ensure that the non-zero coefficients of the polynomial P have degrees in u_i strictly lower than $g_i = \deg_{u_i} G_i$. One may naively think this can be done by first applying the pseudo-division algorithm to obtain $Q_1, \ldots, Q_r$ and R, and apply induction to the coefficients $Q_1, \ldots, Q_r$. This idea does not work without modification, since in order to obtain the final relation, we must multiply, for example, R, by some product of initials, which may increase the degrees in the variables $u_1, \ldots, u_r$ of the coefficients of the monomials in $G_1, \ldots, G_r$ appearing so far.

We shall prove the algorithm for polynomials in case $r = 1$, which is the most important case. We then generalize the algorithm for polynomials to arbitrary r as a corollary, which may be useful to applications of polynomial algebra.

In the differential case, a preparation equation for a differential polynomial F will involve an autoreduced set $\mathbf{A}$ and possibly proper derivatives of its elements, the collection of which forms $G_1, \ldots, G_r$. Unfortunately, the corollary cannot be applied to the preparation process for differential polynomials, for the simple reason that we do not have *a priori* a set of inputs $G_1, \ldots, G_r$ since the choice of the derivatives of $\mathbf{A}$ depends on F and $\mathbf{A}$. The process is similar to the Ritt-Kolchin algorithm to obtain a remainder of F (Equation (6.8)). Whereas the Ritt-Kolchin algorithm completes all partial reductions involving proper derivatives of elements of $\mathbf{A}$ (differential reduction) before pseudo-reduction involving only elements of $\mathbf{A}$ (algebraic reduction), we now mix the two so that an algebraic reduction may be performed under certain circumstances even though F may not be partially reduced yet. In subsequent reductions, we have to be careful not to increase the degrees of $v_1, \ldots, v_k$ in the partial results. We continue this process to obtain a polynomial P in the elements of $A \in \mathbf{A}$ and their derivatives, stopping only when all the coefficients in P are reduced. The following is a rigorous treatment that may be skipped if the reader finds the above paraphrase sufficiently clear. We think however, for any one interested in implementing the algorithm, the detailed proofs are crucial to a thorough understanding.

Proposition 12.1 *Let V be a family of indeterminates over an integral domain $\mathbf{S}_0$, and let $\mathbf{S} = \mathbf{S}_0[V]$. Let u, X be two indeterminates over $\mathbf{S}$. Let F be a polynomial in $\mathbf{S}[u]$, and let G be a polynomial in $\mathbf{S}_0[u]$. Let $f = \deg_u F$, $g = \deg_u G$, and suppose $g \geq 1$. Let $I_G \in \mathbf{S}_0$ be the coefficient of u^g in G. Then we can find $e \in \mathbb{N}$ and $P(X) \in \mathbf{S}[u][X]$ such that*
(a) $\deg_X P \leq f$,
(b) $P(X) = 0$ if and only if $F = 0$,
(c) $\deg_X P = 0$ if and only if $f < g$,
(d) $P(G) = I_G^e F$,
(e) $(e, (P(G) - P(0))/G, P(0))$ is a pseudo-division triple of F by G over $\mathbf{S}$, and
(f) *Viewing $P(X) \in \mathbf{S}_0[V, u, X]$, we have $\deg_u P < g$, and $\deg_v P \leq \deg_v F$ for every $v \in V$.*

Proof. If $f < g$, we can take $e = 0, P(X) = F$. Suppose then $f \geq g$. Write $G = I_G u^g + B$, where $B \in \mathbf{S}_0[u]$ has degree less than g. Similarly, write $F = I_F u^f + C$, where $I_F \in \mathbf{S}$, and $C \in \mathbf{S}[u]$ has degree in u less than f. Clearly $I_G u^f = u^{f-g} G - u^{f-g} B$. Replacing $I_G u^f$ in $I_G F$ by $u^{f-g} X - u^{f-g} B$ yields a polynomial $Q(X)$ of the form $F_1 X + F_2$, where $F_1 = I_F u^{f-g}$, $F_2 = -I_F u^{f-g} B + I_G C$ and $Q(G) = I_G F$. Observe that $F_1, F_2 \in \mathbf{S}[u]$, have degrees in u less than f, and for every $v \in V$ and $j = 1, 2$, $\deg_v F_j \leq \deg_v F$. By induction, we can find $e_1, e_2 \in \mathbb{N}$, $Q_1(X), Q_2(X) \in \mathbf{S}[u][X]$ such that for $j = 1, 2$, $Q_j(G) = I_G^{e_j} F_j$, $\deg_u Q_j < g$, $\deg_X Q_j < f$, and for every $v \in V$, $\deg_v Q_j \leq \deg_v F$. Let $P(X) = I_G^{e-e_1} Q_1(X) X + I_G^{e-e_2} Q_2(X)$, where $e = \max(e_1, e_2)$. Then $P(G) = I_G^e F_1 G + I_G^e F_2 = I_G^{e+1} F$. Clearly, $\deg_X P \leq f$, $\deg_u P < g$ and $\deg_v P \leq \deg_v F$ for $v \in V$ since I_G involves neither u nor v. The pseudo-division property comes just from Definition 2.6. Finally, if $P = 0$, then clearly $F = 0$, and if $f \geq g$ and $\deg_X P = 0$, then $Q_1 = 0$, which is not possible since $F_1 \neq 0$. $\square$

Remark 12.2 When $g = 1$, there is a more efficient method than the inductive one: simply let $P(X) = I_G^f F((X-B)/I_G)$, that is, substitute $(X-B)/I_G$ for u in F and clear denominators. The next corollary may be handy in a purely polynomial situation.

Corollary 12.3 *Let V be a family of indeterminates over an integral domain $\mathcal{S}_0$ and let $\mathcal{S} = \mathcal{S}_0[V]$. Let $u = (u_r, \ldots, u_1)$, $X = (X_r, \ldots, X_1)$ be two families of indeterminates over $\mathcal{S}$. Let $\mathbf{G} \colon G_r, \ldots, G_1$ be a subset in $\mathcal{S}_0[u]$, triangular over $\mathcal{S}_0$ with respect to $u_r, \ldots, u_1$. Let $g_i = \deg_{u_i} G_i$, and let $I_{G_i} \in \mathcal{S}_0$ be the initial of G_i. Let $F \in \mathcal{S}[u]$ and let $f_i = \deg_{u_i} F$. Then we can find $e_r, \ldots, e_1 \in \mathbb{N}$ and $P(X) \in \mathcal{S}[u][X]$ such that for $1 \le i \le r$, $\deg_{u_i} P < g_i$, $\deg_v P \le \deg_v F$ for every $v \in V$, and*

$$P(\mathbf{G}) = P(G_r, \ldots, G_1) = I_{G_r}^{e_r} \cdots I_{G_1}^{e_1} F. \tag{12.4}$$

Moreover, $P(0, \ldots, 0)$ is a pseudo-remainder of F with respect to $\mathbf{G}$ and $P(X) = 0$ if and only if $F = 0$.

Proof. We prove this corollary by induction on r. The case $r = 1$ is given in Proposition 12.1. Let $\mathcal{S}_0' = \mathcal{S}_0[u_r]$ and $\mathcal{S}' = \mathcal{S}_0'[V]$. Applying the induction assumption to $F \in \mathcal{S}'[u_{r-1}, \ldots, u_1]$ and $\mathbf{G}' \colon G_{r-1}, \ldots, G_1$, there exist $Q(X_{r-1}, \ldots, X_1) \in \mathcal{S}'[u_{r-1}, \ldots, u_1][X_{r-1}, \ldots, X_1]$ and $e_{r-1}, \ldots, e_1 \in \mathbb{N}$ such that $Q(G_{r-1}, \ldots, G_1) = I_{G_{r-1}}^{e_{r-1}} \cdots I_{G_1}^{e_1} F$, with the properties that $\deg_{u_i} Q < g_i$ for $1 \le i \le r - 1$ and $\deg_v Q \le \deg_v F$ for every $v \in V$. Write $Q = \sum F_M M$ where the summation runs through a finite set Φ of monomials in $X_{r-1}, \ldots, X_1$ and $F_M \in \mathcal{S}'[u_{r-1}, \ldots, u_1]$ with $\deg_{u_i} F_M < g_i$ for $1 \le i \le r - 1$, and $\deg_v Q \le \deg_v F$ for $v \in V$. Now let $V'' = V \cup \{u_{r-1}, \ldots, u_1\}$, $\mathcal{S}'' = \mathcal{S}_0[V'']$. By Proposition 12.1 applied to each $F_M \in \mathcal{S}''[u_r]$ and $G_r \in \mathcal{S}_0[u_r]$, there exist $e_M \in \mathbb{N}$, $Q_M(X_r) \in \mathcal{S}''[u_r][X_r]$ such that $Q_M(G_r) = I_{G_r}^{e_M} F_M$, and $\deg_{u_r} Q_M < g_r$, $\deg_{u_i} Q_M \le \deg_{u_i} F_M < g_i$ for $1 \le i \le r - 1$, and $\deg_v Q_M \le \deg_v F_M \le \deg_v F$ for every $v \in V''$. Let e_r be the maximum of e_M for $M \in \Phi$. Then we have,

$$\begin{aligned} I_{G_r}^{e_r} I_{G_{r-1}}^{e_{r-1}} \cdots I_{G_1}^{e_1} F &= I_{G_r}^{e_r} Q(G_{r-1}, \ldots, G_1) \\ &= \sum_{M \in \Phi} I_{G_r}^{e_r - e_M} Q_M(G_r) M(G_{r-1}, \ldots, G_1). \end{aligned}$$

Let $P(X) = \sum_{M \in \Phi} I_{G_r}^{e_r - e_M} Q_M M$. Then P clearly satisfies Equation (12.4), $\deg_{u_i} P < g_i$ for $1 \le i \le r$ and $\deg_v P \le \deg_v F$ for every $v \in V$ since $I_{G_r} \in \mathcal{S}_0$. If $P(X) = 0$, then $Q_M = 0$ for all $M \in \Phi$, and by Proposition 12.1, $F_M = 0$ for all $M \in \Phi$ and hence $Q = 0$. By induction, $F = 0$. Finally, $P(0)$ is a pseudo-remainder by Definition 2.6. $\qquad\square$

Remark 12.5 At this point, we remind the reader that when F, G are differential polynomials and the variables u and those in V are derivatives in ΘY, in both Proposition 12.1 and Corollary 12.3, despite the suggestive notation, the initials I_G, I_F of G, F with respect to u are not necessarily the initials of G, F with respect to their leaders. This will be especially important since even Kolchin [23] seemed to assume that u is the leader of F in his proof (where u is denoted by v) of Lemma 7 on p. 80. Even though it may be "obvious" to some that his proof is still valid, making it rigorous is an interesting exercise on induction.

We now apply the previous results in polynomial algebra to differential algebra. We need to introduce the notion of a choice function.

Let $\mathbf{A}: A_1 < \cdots < A_p$ be an autoreduced subset of $\mathcal{R} = \mathcal{F}\{y_1, \cdots, y_n\}$ with leaders $v_1, \ldots, v_p$. Let $Z = (z_1, \ldots, z_p)$ be p differential indeterminates. Let ΘZ be the set of derivatives in Z. Let $E(\mathbf{A})$ be the set of derivatives u of the form θv_k with $\theta \in \Theta$, $1 \leq k \leq p$.

Definition 12.6 A *choice function* for $\mathbf{A}$ is a map $c: E(\mathbf{A}) \longrightarrow \Theta Z$ such that if we denote $c(u)$ by $\theta_u z_{k(u)}$, then $\theta_u v_{k(u)}$, the leader of $\theta_u A_{k(u)}$, is u.

A choice function preselects a $\theta_u \in \Theta$ and an $A_{k(u)} \in \mathbf{A}$ to represent a derivative u which is a derivative of the leader of some $A \in \mathbf{A}$. Our notation differs slightly from Kolchin [23], who uses the pair $(\theta_u, k(u))$ to denote $c(u)$. When $m = 1$ (ordinary case) or when $\mathbf{A}$ is a singleton, there exists only one choice function since the leaders of elements of $\mathbf{A}$ can have no common derivatives. When $m > 1$ and $\mathbf{A}$ consists of more than one element, the representation is quite arbitrary since there is no compatibility condition required between the choice function and differentiation.

For any $u \in E(\mathbf{A})$, let $X_u = c(u) = \theta_u z_{k(u)}$ and $G_u = \theta_u A_{k(u)}$. If $u_1, \ldots, u_r \in E(\mathbf{A})$ are given, we shall write simply X_i for X_{u_i} and G_i for G_{u_i}. The set $\mathbf{X} = c(E(\mathbf{A}))$ will be treated as a family $(X_u)_{u \in E(\mathbf{A})}$ of algebraic indeterminates over $\mathcal{R}$. There is a substitution map $\varphi_c: \mathcal{R}[\mathbf{X}] \longrightarrow \mathcal{R}$ such that $\varphi_c(X_u) = G_u$ for $u \in E(\mathbf{A})$ and $\varphi_c(F) = F$ for $F \in \mathcal{R}$. Kolchin viewed any $M \in \mathcal{R}[\mathbf{X}]$ as $M \in \mathcal{R}\{z_1, \ldots, z_p\}$ and thus for $M(z_1, \ldots, z_p) \in \mathcal{R}[\mathbf{X}]$, its image $\varphi_c(M)$ would be written as $M(A_1, \ldots, A_p) \in \mathcal{R}$. The main property of the choice function is that any two distinct derivatives $X_1, X_2 \in \mathbf{X}$ the corresponding leaders u_1, u_2 of G_1, G_2 are distinct.

Definition 12.7 Let $Q \in \mathcal{R}[\mathbf{X}]$ and write $Q = \sum_M Q_M M$ where M runs through a finite set Φ of monomials in the variables $\mathbf{X}$, and $Q_M \in \mathcal{R}$ for each M. The *leader* of Q, denoted by u_Q, is the highest ranked derivative among the leaders u_{Q_M} of Q_M for $M \in \Phi$. We say Q is *reduced* (with respect to $\mathbf{A}$) if every Q_M is reduced. If Q is not reduced, let u be the highest derivative $v \in E(\mathbf{A})$ that appears in some Q_M and $\deg_v Q_M \geq \deg_v G_v$. We call u the *head reduction derivative* of Q with respect to $\mathbf{A}$ *and* c, and we denote it by $\text{head}(Q, \mathbf{A}, c)$ or simply $\text{head}(Q)$ if $\mathbf{A}$ and c are clear and fixed.

Remark 12.8 Whenever Q is not reduced, the head reduction derivative is well defined since Q involves either the leader $v = v_i$ of some A_i to a degree higher than $\deg_{v_i} A_i$ (in which case, since $\mathbf{A}$ is autoreduced, $c(v)$ must be z_i, that is, $\theta_v = 1, k(v) = i$, and $G_v = A_i$), or a proper derivative v of the leader v_i of some A_i (in which case, $\deg_v Q \geq \deg_v G_v = 1$ always holds, since $G_v = \theta_v A_{k(v)}$ has leader v, no matter the choice for $c(v) = \theta_v z_{k(v)}$). Clearly, $\text{head}(Q) \leq u_Q$.

Lemma 12.9 Let $\mathbf{A}$ be an autoreduced set of $\mathcal{R}$. Let $c \colon E(\mathbf{A}) \longrightarrow \Theta Z$ be a choice function. Let $u_1, \ldots, u_j \in E(\mathbf{A})$. Suppose $Q \in \mathcal{R}[X_1, \ldots, X_j]$ is not reduced with respect to $\mathbf{A}$ and let $u = \text{head}(Q)$. Then we can compute a $P \in \mathcal{R}[X_1, \ldots, X_j, X_u]$, and $e \in \mathbb{N}$ such that $I_G^e \varphi_c(Q) = \varphi_c(P)$. Furthermore, if P is not reduced with respect to $\mathbf{A}$, then $\text{head}(P) < \text{head}(Q)$.

Proof. Let $F = Q$, $X = X_u$, $G = G_u$. Let V_0 be the set of derivatives appearing in G that is not u. Let V' be any set of derivatives $v \in \Theta Y$ containing those that appear in F but not in G. Let $V = V' \cup \{X_1, \ldots, X_j\}$. Let $\mathcal{S}_0 = \mathcal{F}[V_0]$ and $\mathcal{S} = \mathcal{S}_0[V]$. Apply Proposition 12.1 with this setup and obtain a polynomial $P(X) \in \mathcal{S}[u][X]$ and $e \in \mathbb{N}$ such that $I_G^e Q = P(G)$ and satisfying the other properties stated there. We may view $P \in \mathcal{F}[V_0][V'][u][X_1, \ldots, X_j, X] \subseteq \mathcal{R}[\mathbf{X}]$. Clearly, $I_G^e \varphi_c(Q) = \varphi_c(P)$. Now suppose P is not reduced with respect to $\mathbf{A}$ and let $w = \text{head}(P)$. Firstly, $\deg_u P < \deg_u G_u$, so $w \neq u$. Secondly, u is the leader of G and if $w \in V_0$, then $w < u$. Lastly, if $w \in V'$ and if $w > u$, we would have $\deg_w Q = \deg_w F \geq \deg_w P \geq \deg_w G_w$, which would contradict the definition of u. So $w < u$ and we are done. $\qquad \square$

Proposition 12.10 Let $\mathbf{A} \colon A_1 < \ldots < A_p$ be an autoreduced set in $\mathcal{R}$. Let $c \colon E(\mathbf{A}) \longrightarrow \Theta Z$ be a choice function for $\mathbf{A}$, and let $F \in \mathcal{R}$. If $F \notin \mathcal{F}$, let u_F be its leader. Then we can compute a product L of initials and separants of elements of $\mathbf{A}$, $r \in \mathbb{N}$, $u_1, \ldots, u_r \in E(\mathbf{A})$ and a polynomial $P(X_1, \ldots, X_r)$ in $\mathcal{R}[X_1, \ldots, X_r]$, where $X_j = c(u_j) = \theta_{u_j} z_{k(u_j)}$, such that either $r = 0$ (in which case $L = 1, P = F$ is reduced), or $u_r < u_{r-1} < \cdots < u_1 \leq u_F$, the non-zero coefficients of P are reduced with respect to $\mathbf{A}$, and if $G_i = \theta_{u_i} A_{k(u_i)}$, then

$$LF = P(G_1, \ldots, G_r) = P(\theta_{u_1} A_{k(u_1)}, \ldots, \theta_{u_r} A_{k(u_r)}). \qquad (12.11)$$

Proof. We begin by setting $L_0 = 1$, and $P_0 = F$. Suppose by induction, we have constructed for $j \geq 0$, $u_1, \ldots, u_j \in E(\mathbf{A})$ such that $u_j < \cdots < u_1 \leq u_F$, a non-zero polynomial $P_j(X_1, \ldots, X_j)$ in $\mathcal{R}[X_1, \ldots, X_j]$ such that P_j is not reduced with respect to $\mathbf{A}$, and a product L_j of initials and separants of elements of $\mathbf{A}$ such that $L_j F = P_j(G_1, \ldots, G_j)$. Let $Q = P_j$, $u_{j+1} = \text{head}(Q)$ and apply Lemma 12.9 to obtain $P \in \mathcal{R}[X_1, \ldots, X_{j+1}]$ and $e \in \mathbb{N}$ with properties stated there. If P is not reduced, we let $P_{j+1} = P$ and extend the sequence to $u_1 > \ldots > u_{j+1}$. This construction must stop since a ranking is a well-ordering. If this stops at $j = r-1 \geq 0$ when P_r is reduced with respect to $\mathbf{A}$, then we are done since $L_{r-1} I_{G_{r-1}}^e F = I_{G_{r-1}}^e P_{r-1}(G_1, \ldots, G_{r-1}) = P_r(G_1, \ldots, G_r)$ and we may take $L = L_{r-1} I_{G_{r-1}}^e$. $\qquad\square$

We now state the version as given in Lemma 7, p. 80 and pp. 183–184 of Kolchin [23].

Proposition 12.12 *Let $\mathbf{A}: A_1 < \ldots < A_p$ be a characteristic set of a prime differential ideal $\mathfrak{p}$ of $\mathcal{R} = \mathcal{F}\{y_1, \cdots, y_n\}$, let $v_1, \ldots, v_p$ be the leaders of $\mathbf{A}$, let $Z = (z_1, \ldots, z_p)$ be differential indeterminates, let $u \mapsto (\theta_u, k(u))$, $u \in E(\mathbf{A})$, be a choice function for $\mathbf{A}$, and let $F \in \mathcal{R}$. Then we can find an equation of the form*

$$LF = \sum_{j=1}^{s} C_j M_j(A_1, \cdots, A_p), \tag{12.13}$$

where L is a product of initials and separants, $C_1, \cdots, C_s \notin \mathfrak{p}$, and $M_1, \ldots, M_s$ are distinct differential monomials in Z with the property that every factor θz_k of an M_j has the form $\theta_u z_{k(u)}$ for some $u \in E(\mathbf{A})$.

Definition 12.14 When $\mathbf{A}$ is a characteristic set of a prime differential ideal $\mathfrak{p}$, an equation of the form Equation (12.13), where the coefficients $C_1, \ldots, C_s$ are not in $\mathfrak{p}$ (but not necessarily reduced) is called a *preparation equation* of F with respect to $\mathbf{A}$ (and the choice function c).

In Equation (12.11), let $M_1, \ldots, M_s$ be all the distinct monomials in $X_1, \ldots, X_r$ appearing with non-zero coefficients in $P(X_1, \ldots, X_r)$, and let their corresponding coefficients be $C_1, \ldots, C_s$. Then $C_1, \ldots, C_s$ are reduced with respect to $\mathbf{A}$. If $\mathbf{A}$ is the characteristic set of a prime differential ideal $\mathfrak{p}$, then $C_i \notin \mathfrak{p}$. Thus, when $M_1, \ldots, M_s$ are viewed as differential polynomials in $z_1, \ldots, z_p$ via $X_u = \theta_u z_{k(u)}$, Equation (12.11) is a preparation equation of F with respect to $\mathbf{A}$ and c. The proofs of Proposition 12.1, Lemma 12.9, and Proposition 12.10 give an algorithm which constructs a preparation equation.

We will refer to the algorithm as the *Ritt-Kolchin*[20] *preparation algorithm* and the preparation equation thus obtained as the *Ritt-Kolchin preparation equation*. We summarize the properties of the Ritt-Kolchin preparation equation below:

- $C_1, \ldots, C_s \in \mathfrak{R}$ are (non-zero and) reduced with respect to $\mathbf{A}$.
- If θz_k and $\theta' z_{k'}$ are two distinct factors present in at least one M_j, then $\theta v_k \neq \theta' v_{k'}$.
- If $F \notin \mathfrak{F}$, and u_F is its leader, then $\theta v_k \leq u_F$ for every factor θz_k of some M_j.
- The rank of C_j is less than or equal to the rank of F for $1 \leq j \leq s$, and
- If $M_i = 1$ is among the monomials $M_1, \ldots, M_s$, then C_i is a remainder of F.

If q is the lowest degree of the differential monomials $N_1, \ldots, N_\ell$ among $M_1, \ldots, M_s$, we may rewrite (12.13) in the form of a congruence (with a change of notation)

$$LF \equiv \sum_{\lambda=1}^{\ell} D_\lambda N_\lambda(A_1, \cdots, A_p) \qquad (\mathrm{mod}\, [A_1, \cdots, A_p]^{q+1}). \qquad (12.15)$$

Definition 12.16 We call (12.15) a *preparation congruence* of F with respect to $\mathbf{A}$ and c.

Give $F \in \mathfrak{R}$ and a prime differential ideal $\mathfrak{p} \subset \mathfrak{R}$, a preparation congruence, just like a preparation equation, is not unique and depends on the ranking, the characteristic set, and the choice function. By a result of Hillman (Lemma 7, p. 184), the lowest integer q and the set of differential monomials $N_1, \ldots, N_\ell$ of degree q, are unique and independent of the ranking, the characteristic set, the choice function, and the preparation congruence. The proof involves geometric notions that we have not covered: for example, the *differential algebraic variety* defined by $\mathfrak{p}$, the *generic zero* of $\mathfrak{p}$, and the *multiplicity* (which is q) of F at a generic zero of $\mathfrak{p}$.

In the case when $\mathbf{A}$ consists of a single irreducible differential polynomial A, the choice function is unique. We state below first this simplified version of the Low Power Theorem for general m (partial case) and n (arbitrary number of differential indeterminates), followed by the most commonly illustrated version when $n = 1$ and $A = y \in \mathfrak{F}\{y\}$.

---

[20]The author has not done a historical research to justify this credit. This attribution is only a matter of convenience.

Theorem 12.17 (Low Power Theorem) *Let A and F be differential polynomials in $\mathcal{R} = \mathcal{F}\{y_1, \cdots, y_n\}$, with A irreducible and $F \neq 0$. Let*

$$LF \equiv \sum_{\lambda=1}^{\ell} D_\lambda N_\lambda(A) \qquad (\mathrm{mod}\,[A]^{q+1})$$

be a preparation congruence of F with respect to A. A necessary and sufficient condition that $\mathfrak{p}_{\mathcal{F}}(A)$ be a component of $\{F\}$ is that $q \neq 0, \ell = 1$, and $N_1 = z^q$.

Corollary 12.18 *A necessary and sufficient condition for $[y]$ to be a component of a differential polynomial $F \in \mathcal{F}\{y\}$ is that F contains a term in y alone which is of lower degree than any other terms of F.*

Proof. Since the conditions in the Low Power Theorem are independent of the choice of a preparation congruence (by Hillman's Lemma), we may use the Ritt-Kolchin preparation congruence. When $A = y$, the Ritt-Kolchin preparation equation algorithm simply replaces every proper derivative θy of y (resp. y) occurring in F by θz (resp. z) (see proof of Proposition 12.1). Hence $F(y) = F(z)$ is the Ritt-Kolchin preparation equation. The conditions $q \neq 0$, $\ell = 1$ and $N_1 = z^q$ mean that $F(0) = 0$, the lowest degree term of F is of the form $D_1 y^q$, where $D_1 \in \mathcal{F}$ and is non-zero. $\qquad\square$

Example 12.19 By Corollary 12.18, $[y]$ is not a component of $yy''' - y''$, nor of $y''y''' - y^2$.

Exercise 12.20 Show that the differential ideal $[y]$ is a component of $F = y'y'' - y$ and find the general component of F (F being irreducible itself).

Example 12.21 (Ritt [32], page 133) Let $\mathcal{R} = \mathcal{F}\{x, y, w\}$ be an ordinary differential polynomial ring with differential indeterminates x, y, w with derivation δ. Assume an orderly ranking with $x > y > w$. Let

$$F = x^5 - y^5 + w(x\delta y - y\delta x)^2.$$

Let ζ be a fifth root of unity. For $1 \leq j \leq 5$, let $A_j = x - \zeta^{j-1}y$ and let B_j be defined by $A_j B_j = x^5 - y^5$. For every j, A_j is irreducible (as is F), the leader of A_j is x, and F is not partially reduced with respect to A_j. The rank of F is $(\delta x, 2)$. We differentiate A_j to get $\delta A_j = \delta x - \zeta^{j-1}\delta y$. It is easy to verify that $x\delta y - y\delta x = \delta y\, A_j - y\,\delta A_j$. Moreover, $F = B_j A_j + w(\delta y\, A_j - y\,\delta A_j)^2$ (resp. $F \equiv B_j A_j \pmod{[A_j]^2}$) is a preparation equation (resp. congruence) with respect to the autoreduced set consisting of A_j alone (the reader should verify that B_j, $w(\delta y)^2$, $wy\delta y$, and wy^2 do not belong to $[A_j]$). By the Low

Power Theorem, $[A_j]$ is a (singular) component of F. These together with the general component of F, shows that $\{F\}$ has 6 components.

The more general Low Power Theorem tries to solve Problem 3 (deciding if $\mathfrak{p} \supseteq \mathfrak{q}$) of Section 10 for a wider class of prime differential ideals $\mathfrak{p}, \mathfrak{q}$. When $\mathfrak{p}$ is given and $F \in \mathfrak{p}$, the question is to decide whether $\mathfrak{p}$ contains an irreducible differential polynomial G for which $\mathfrak{p}_{\mathcal{F}}(G)$ is a component of $\{F\}$ contained in $\mathfrak{p}$. Given a characteristic set $\mathbf{A}$: $A_1, \ldots, A_r$ of $\mathfrak{p}$, and a preparation equation like Equation (12.13) of F with respect to $\mathbf{A}$, the general Low Power Theorem provides a condition (based on the existence of one differential monomial M_j being *dominated* by the rest) under which two sets $\Lambda_0 \subseteq \Lambda_1$ of possible components (in the form $\mathfrak{p}_{\mathcal{F}}(G)$ where $G = \theta A_k$) of $\{F\}$ are obtained. Those from Λ_0 are components of $\{F\}$ contained in $\mathfrak{p}$, and every such component is among those in Λ_1. We refer interested readers to Theorem 7, p. 188 of Kolchin [23] and for recent research related to this, see Hubert [16].

We end this tutorial with a statement of the Ritt problem (see Section 16, Chapter IV), which is yet a more special case of Problem 3 of Section 10, but expressed more geometrically. Kolchin's text provides results on sufficiency and on necessity, but beyond that and the original papers, no progress has been made since.

The Ritt Problem: Given an irreducible differential polynomial A in $\mathcal{R}$ such that A vanishes at $(0, \ldots, 0)$, decide whether $(0, \ldots, 0)$ is a zero of the general component $\mathfrak{p}_F(A)$ of A. In other words, given $\mathfrak{p} = [y_1, \ldots, y_n]$ and $A \in \mathfrak{p}$, decide whether $\mathfrak{p} \supseteq \mathfrak{p}_F(A)$.

Appendix: Solutions and hints to selected exercises

Exercise 1.10 The number of derivatives of order $\leq s$ is $n\binom{m+s}{m}$, when $\mathrm{Card}(\Delta) = m$. There are infinitely many differential monomials of degree 2.

Exercise 1.11 We have $y_2 = P_{2,1} - \delta_1^{h-1}\delta_2 P_{1,2} + \delta_1^h P_{2,2}$, and a similar expression for y_1.

Exercise 4.4 See Kolchin [23], Lemma 15, p. 49–50.

Exercise 4.6 See Kolchin [23], Exercises 1, 2 on p. 53 and Robbiano [33].

Exercise 4.7 See Carrá-Ferro and Sit [13], Reid and Rust [35].

Exercise 4.13 If there is an element of $\mathcal{F}$ in the set, this is one with lowest rank. Otherwise, among differential polynomials in the set with leader a derivative of y_j, select those with the least j, and among those, select one with the least order h in y_j, and finally, among those, select one with the least degree in $\delta^h y_j$.

Exercise 4.14 By induction on the order of θ. For $\delta \in \Delta$, using the representation as in (4.9) of Definition 4.8, we have $\delta A = S_A \delta u_A + \sum_{i=0}^{d} \delta I_i u_A{}^i$.

Exercise 4.19 If an element $A \in \mathbf{A}$ has a leader $u_A = y_j$, $\mathbf{A}$ autoreduced implies y_j cannot appear in any other members of $\mathbf{A}$.

Exercise 4.20 The first part follows from definitions. The second part follows because every infinite sequence of derivatives has an infinite subsequence of derivatives of some y_j, which in turn has an infinite subsequence such that each one is a proper derivative of the previous one, which contradicts the property of autoreduction.

Exercise 4.22 In $\mathcal{F}\{w, z\} = \mathcal{F}[w, z]$ with an orderly ranking such that $w < z$, the set $\{w^2, wz, z^2\}$ is a Gröbner basis, but is not triangular. The set $\{w + z, z\}$ is triangular but not autoreduced.

Exercise 6.11 If $\mathbf{A} = \{\delta_1 y + z, \delta_2 y\}$, then $\delta_2 z \in [\mathbf{A}]$, is reduced, but 0 is a remainder according to Definition 6.9.

Exercise 7.5 Let $J = J_H$ or J_S as the case may be, and similarly define $\mathfrak{a}$. Clearly, if $\mathfrak{a}$ is zero-reduced, then J is. Conversely, suppose J is zero-reduced, and let $F \in \mathfrak{a}$ be reduced with respect to $\mathbf{A}$. Then in particular, F is partially reduced and hence belongs to J, and hence is zero.

Exercise 7.6 Let $\overline{V}$ be the complement of V in ΘY. Every differential polynomial $F \in \mathcal{R}$ may be written uniquely as a finite sum $\sum C_M M$, where M is a monomial in the $u \in \overline{V}$ and $C_M \in \mathcal{F}[V]$. It is easy to see that $F \in (\mathbf{A})$ if and only if $C_M \in \mathcal{F}[V] \cdot (\mathbf{A})$ for all M. Suppose $F \in J_G$. Write $F = \sum C_M M$ as above. For some $e \in \mathbb{N}$, $G^e F \in (\mathbf{A})$. Since $G^e \in \mathcal{F}[V]$, the uniqueness of the representation means that $G^e C_M \in \mathcal{F}[V] \cdot (\mathbf{A})$, and hence $C_M \in J_G^V$, for all M. Thus F belongs to the ideal of $\mathcal{R}$ generated by J_G^V.

Exercise 7.7 Let $\overline{V}$ be the complement of V in ΘY. Every differential polynomial $F \in \mathcal{R}$ may be written uniquely as a finite sum $\sum C_M M$, where M is a monomial in the $u \in \overline{V}$ and $C_M \in \mathcal{F}[V]$. It is easy to see that $F \in J$ if and only if $C_M \in J^V$ for all M. Hence $J^V = J \cap \mathcal{F}[V]$. From this, a Gauss lemma like argument shows that J is prime (resp. radical) if and only if J^V is prime (resp. radical). Moreover, if F is partially reduced with respect to $\mathbf{A}$, then any derivative $u \in \Theta Y$ that is a proper derivative of some leader u_A may be replaced by zero in the equation $F = \sum C_M M$, yielding a new sum $F = \sum C'_M M$ where M and C'_M are now partially reduced. The uniqueness implies then that $C_M = C'_M$ and must be partially reduced to begin with. Thus F is partially reduced if and only if all C_M and M are partially reduced. If F is reduced, then $C_M M$, and *a fortiori* C_M, must be also. This shows that J is zero-reduced if and only if J^V is.

Exercise 7.12 The proof that $\mathfrak{a}_S$ is radical (resp. prime) and $\mathbf{A}$ is its characteristic set is identical to that for Lemma 7.9, by replacing J_H by J_S, $\mathfrak{a}_H$ by $\mathfrak{a}_S$, and the Rosenfeld property by the strong Rosenfeld property. In particular, $J_I = J_S$. Since J_I is radical, by Proposition 3.3 of Hubert [17], $J_I = J_H$. Since $\mathbf{A}$ is a characteristic set of $\mathfrak{a}_S$, by Corollary 6.13, $\mathfrak{a}_S = \mathfrak{a}_H$ when $\mathfrak{a}_S$ is prime. When $\mathfrak{a}_S$ is radical, we have $\mathfrak{a}_S : H = \mathfrak{a}_H$. But if $F \in \mathfrak{a}_H$, and if $\tilde{F}$ is its Ritt-Kolchin partial remainder with respect to $\mathbf{A}$, then $\tilde{F} \in J_H = J_S$. So $F \in \mathfrak{a}_S$.

Remark (Exercise 7.12) By Exercise 3, p. 171 of Kolchin [23], when $\mathbf{A}$ is a characteristic set of a prime differential ideal $\mathfrak{p}$, then $\mathfrak{p} = \{\mathbf{A}\} : S$. The proof is very similar to the proof of Proposition 3.3 of Hubert [17]. Note that $\mathfrak{p} = \mathfrak{a}_H$ by Corollary 6.13. We have $\mathfrak{a}_S \subseteq \{\mathbf{A}\} : S$ in general, but not equality unless $\mathfrak{a}_S$ is radical. Thus we do not have $\mathfrak{a}_H = \mathfrak{a}_S$ unless $\mathfrak{a}_S$ is radical, and this can be obtained using the strong Rosenfeld property since J_S is radical by Lazard's Lemma. If we add $J_I^V = J_H^V$ to the hypothesis, then we need neither Hubert's result nor Lazard's lemma.

Exercise 8.3 See Lemma 1 on p. 62 of Kolchin [23] or use induction on s.

Exercise 8.4 Let $T = S$ or H (or any non-zero differential polynomial in $\mathcal{R}$). Suppose for all lowest common derivatives $v = \theta u_A = \theta' u_{A'}$, $F_v = \Delta(A, A', v) \in (\mathbf{A}_{(v)}) : T^\infty$. Suppose $v \in \Theta Y$ is such that $T^e F_v \in (\mathbf{A}_{(v)})$ for some $e \in \mathbb{N}$. Let $w = \delta v$ with $\delta \in \Delta$ and $F_w = \Delta(A, A', w)$. By Exercise 8.3,

$$(T^e)^2 \cdot \delta F_v \in \left(\delta(T^e F_v),\ T^e F_v\right) \subseteq \left(\mathbf{A}_{(w)}\right).$$

Now

$$\delta F_v = \delta(S_{A'} \cdot \theta A - S_A \cdot \theta' A')$$
$$= \delta(S_{A'}) \cdot \theta A + S_{A'} \cdot \delta\theta A - \delta(S_A) \cdot \theta' A' - S_A \cdot \delta\theta' A'$$

Hence $F_w = \delta F_v - \delta(S_{A'}) \cdot \theta A + \delta(S_A) \cdot \theta' A' \in \left(\mathbf{A}_{(w)}\right) : T^\infty$. The result now follows by induction.

Exercise 8.16 (a) and (d) were proved in Corollary 6.13 and (b) follows from Remark 8.2. (c) now follows from Proposition 7.3 since $\mathfrak{p} : H$ is radical if $\mathfrak{p}$ is.

Exercise 8.17 Let $\mathfrak{a} = \mathfrak{a}_H$ or $\mathfrak{a}_S$ and correspondingly let $J = J_H$ or J_S as the case may be. First, $\mathfrak{a}$ is prime (or radical) by Proposition 7.3. By (a), any element of $\mathfrak{a}$ reduced with respect to $\mathbf{A}$ lies in J, and hence is zero by (c). The element $1 \in \mathcal{F}$ is reduced with respect to $\mathbf{A}$ and hence is not in $\mathfrak{a}$. By Lemma 5.3, $\mathbf{A}$ is a characteristic set of $\mathfrak{a}$.

Exercise 8.19 Suppose $A_2 \in \mathfrak{a}_I$. We can then write

$$A_2 = Q_0 A_1 + R_0 A_3 + Q_1 A_1' + R_1 A_3' + \cdots + Q_k A_1^{(k)} + R_k A_3^{(k)}$$

for some $Q_i, R_i \in \mathcal{R}$, $0 \le i \le k$. Now $A_1' = 2yy' + z'$ and for $i > 1$,

$$A_1^{(i)} = (2yy' + z')^{(i-1)}$$
$$= 2\left(yy^{(i)} + \binom{i-1}{1} y' y^{(i-2)} + \cdots + \binom{i-1}{i-2} y^{(i-2)} y'' + y^{(i-1)} y' \right) + z^{(i)}.$$

Substituting this and $A_3^{(i)} = z^{(i+1)} + z^{(i)}$ into the first equation and then substituting 0 for all derivatives of y of order higher than 1 and 0 for all derivatives of z, we obtain for some $\overline{Q}_0, \overline{Q}_1, \overline{Q}_2 \in \mathcal{R}$ an equation

$$y' + y = \overline{Q}_0 y^2 + \overline{Q}_1 \cdot 2yy' + \overline{Q}_2 \cdot (2y'^2).$$

Rearranging the terms, we get

$$y'(1 - \overline{Q}_2 \cdot 2y') = y(\overline{Q}_0 y + \overline{Q}_1 \cdot 2y' - 1).$$

This is not possible since y does not divide the left hand side.

Now $A_3 = A_1' + 2A_1 - 2yA_2$. It is easy to verify that $A_2^3 = (yA_2)'A_2 + (yA_2)(A_2 - A_2')$ and hence $A_2 \in \sqrt{[A_1, A_3]}$. Alternatively, A_2 vanishes at every (differential) zero (α, β) of A_1, A_3. By the differential Nullstellensatz[21], $A_2 \in \sqrt{\mathfrak{a}_I}$.

Exercise 9.1 The proof is not trivial (see Ritt [30,32], pp. 11–13).

Exercise 9.2 (Kaplansky [20], p. 12) Let $\mathbb{F}[x]$ be the polynomial ring in x over the prime field of two elements. In $\mathcal{R} = \mathbb{F}_2[x]/(x^2)$, which is a differential ring under d/dx, $(x) = \sqrt{[0]}$ is not a differential ideal. For characteristic zero case, show by induction that if $a^n \in [U]$, then $(\delta a)^{2n-1} \in [U]$ and hence $\sqrt{[U]}$ is a *differential* ideal.

Exercise 9.3 Let $u \in \{U\}$, $w \in W$. Now $\{UW\} : w$ is a radical differential ideal containing U, and hence it contains $\{U\}$. In particular, $uw \in \{UW\}$

---

[21]Theorem 1, page 146 of Kolchin [23].

and $w \in \{UW\} : u$. Since the latter is a radical differential ideal, $\{W\} \subseteq \{UW\} : u$ and hence $\{U\}\{W\} \subseteq \{UW\}$.

Exercise 9.4 By Exercise 9.3, we have

$$\{W\} \subseteq (\{W\}:T) \cap \{W,T\} \subseteq \{(\{W\}:T)W, (\{W\}:T)T\} \subseteq \{W\}.$$

Exercise 9.10 By property in Exercise 9.5 (d), if the proposition were false, we could find a radical differential ideal I maximal with respect to the property that I is not the intersection of a finite number of prime differential ideals. Then I could not be prime, and there would exist $a, b \in \mathcal{R}$ such that $ab \in I$, but $a \notin I$ and $b \notin I$. Since $I = \{I, a\} \cap \{I, b\}$ and both $\{I, a\}$ and $\{I, b\}$ are finite intersections of prime differential ideals, we would have a contradiction.

Exercise 10.7 All we know is that $I^e F \in (\mathbf{A})$ for some $e \in \mathbb{N}$. Since F is already partially reduced by the choice of V, by Lemma 3.6, F has zero as its Ritt-Kolchin remainder. In order to make a recursion call, we need a product of non-zero, reduced elements.

Exercise 11.1 Clearly, if $\mathfrak{p} = \mathfrak{q}$, then the conditions are satisfied. Suppose that $\mathbf{B} \subset \mathfrak{p}$ and $\mathbf{A} \subset \mathfrak{q}$. Now, both $\mathbf{A}$ and $\mathbf{B}$ are autoreduced subsets of $\mathfrak{p}$ and of $\mathfrak{q}$. Hence, $\mathbf{A}$ and $\mathbf{B}$ must have the equal rank. In other words, both are characteristic sets of $\mathfrak{p}$ and of $\mathfrak{q}$. It follows from Corollary 6.13 that $\mathfrak{p} = \mathfrak{q}$.
Remark (Exercise 11.1) Kolchin left Problem 3' as an exercise (Exercise 1, p. 171) in which he included the conditions $H_{\mathbf{B}} \notin \mathfrak{p}$ and $H_{\mathbf{A}} \notin \mathfrak{q}$. These are superfluous in the characteristic zero case because of Corollary 5.4.

Acknowledgements

I would like to thank the other editors, Li Guo, William Keigher, and Phyllis J. Cassidy of these proceedings for their infinite patience and encouragement. My thanks go specially to Cassidy, who moved from Chicago to New York, spent days listening to my totally unprepared talks, and asked for examples after examples that substantially lengthen the paper (but hopefully make it more understandable). This paper originated from some very crude notes prepared from 1998 to 2000 at various workshops organized by B. Sturmfels, M. Singer, M. Roy, Ziming Li, Xiao-Shan Gao, Li Guo, and William Keigher. I thank these organizers for the opportunity to present the lectures.

References

1. Aubry, P., Lazard, D., Maza, M. M. *On the theories of triangular sets*, J. Symbolic Comput. **28** (1999), 105–124.
2. Aubry, P., Maza, M. M. *Triangular sets for solving polynomial systems: a comparative implementation of four methods*, J. Symbolic Comput. **28** (1999), 125–154.
3. Becker, T., Weispfenning, V. *Gröbner Bases: A Computational Approach to Commutative Algebra*, Graduate Texts in Mathematics **141**, Springer-Verlag, New York, 1993.
4. Boulier, F. *Étude et implantation de quelques algorithmes en algèbre différentielle*, Thése, L'Université des Sciences et Technologies de Lille, 1994.
5. Boulier, F., Lazard, D., Ollivier, F., Petitot, M. *Representation for the radical of a finitely generated differential ideal*, Technical Report IT-**267**, LIFL; Levelt, A., ed., ISSAC'95 Proceedings, ACM Press, 1995.
6. Boulier, F. *Some improvements of a lemma of Rosenfeld*, Technical Report IT-**268**, LIFL, 1996.
7. Boulier, F., Lazard, D., Ollivier, F., Petitot, M. *Computing representations for radicals of finitely generated differential ideals*, Technical Report IT-**306**, LIFL, 1997.
8. Boulier, F., Lemaire, F., Moreno Maza, M. *PARDI!*, Proc. ISSAC'01, ACM Press, New York, 2001.
9. Buchberger, B. (1985). *Gröbner Bases: An Algorithmic Method in Polynomial Ideal Theory*, Bose, N.K. ed., Recent Trends in Multidimensional System Theory, Reidel, Dordrecht, 1985, 184–232.
10. Bouziane, D., Rody, K. A., Maârouf, H. *Unmixed-dimensional decomposition of a finitely generated perfect differential ideal*, J. Symbolic Comput. **31** (2001), 631–649.
11. Buium, A., Cassidy, P. J. *Differential algebraic geometry and differential algebraic groups: from algebraic differential equations to Diophantine geometry*, Bass, H. *et al.* eds., *Selected Works of Ellis Kolchin, with Commentary*, A.M.S., Providence, 1999, 567–636.
12. Carminati, J., Vu, K. *Symbolic computation and differential equations: Lie symmetries*, J. Symbolic Comput. **29**(1) (2000), 95–116.
13. Carrà-Ferro, G., Sit, W. Y. *On term-orderings and rankings*, Fischer, K. G. *et al.* eds., *Computational Algebra*, LNPAM **151**, Marcel Dekker, 1993, 31–78.
14. Chou, S. C. *Mechanical Geometry Theorem Proving*, D. Reidel Pub. Co., Dordrecht, Holland, 1998.

15. Decker, W., Greuel, G., Pfister, G. *Primary decomposition: algorithms and comparisons*, Algorithmic Algebra and Number Theory (Heidelberg, 1997), Springer-Verlag, Berlin, 1999, 187–220.

16. Hubert, E. *Essential components of an algebraic differential equation*, J. Symbolic Comput. **28** (1999), 657–680.

17. Hubert, E. *Factorization-free decomposition algorithms in differential algebra*, J. Symbolic Comput. **29** (2000), 641–662.

18. Kandri Rody, A., Maârouf, H., Ssafini, M. *Triviality and dimension of a system of algebraic differential equations*, J. Automatic Reasoning **20** (1998), 365–385.

19. Kalkbrener, M. *A generalized Euclidean algorithm for computing triangular representations of algebraic varieties*, J. Symbolic Comput. **15** (1993), 143–167.

20. Kaplansky, I. *An Introduction to Differential Algebra*, Hermann, 1957.

21. Knuth, D. E. *The Art of Computing, Volume 2*, Addison-Wesley, Reading, Mass, 1969.

22. Kolchin, E. R. *Singular solutions of algebraic differential equations and a lemma of Arnold Shapiro*, Topology **3** (1965), Suppl. **2**, 309–318.

23. Kolchin, E. R. *Differential Algebra and Algebraic Groups*, Chapters 0–IV, Academic Press, 1973.

24. Lazard, D. *Solving zero-dimensional algebraic systems*, J. Symbolic Comput. **13** (1992), 117–131.

25. Levi, H. *The low power theorem for partial differential polynomials*, Ann. of Math. **46** (1945), 113–119.

26. Mansfield, E. L., Fackerell, E. D. *Differential Gröbner bases*, Macquarie Univ. Preprint **92/108**, 1992.

27. Mansfield, E. L., Clarkson, P. A. *Applications of the differential algebra package* diffgrob2 *to classical symmetries of differential equations*, J. Symbolic Comput. **23** (1997), 517–533.

28. Mishra, B. *Algorithmic Algebra*, Springer-Verlag, New York, 1993.

29. Morrison, S. *The differential ideal* $[P] : M^\infty$, J. Symbolic Comput. **28** (1999), 631–656.

30. Ritt, J. F. *Differential Equations from the Algebraic Standpoint*, A.M.S. Colloq. Pub. **14**, A.M.S., New York, 1932.

31. Ritt, J. F. *On the singular solutions of algebraic differential equations*, Ann. of Math. **37** (1936), 552–617.

32. Ritt, J. F. *Differential Algebra*, Dover, 1950.

33. Robbiano, L. *Term orderings on the polynomial ring*, Proc. EUROCAL 1985, LCNS **204**, Springer-Verlag, 1985, 513–517.

34. Rosenfeld, A. *Specializations in differential algebra*, Trans. A.M.S. **90** (1959), 394–407.

35. Rust, C. J., Reid, G. J. *Rankings of partial derivatives*, Proc. ISSAC'97, ACM Press, 1997, 9–16.

36. Sadik, B. *Study of some formal methods in the theory of differential ideals*, Private Communication, 2001.

37. Seidenberg, A. *An elimination theory for differential algebra*, Wolf *et al.* eds., Univ. Calif. Pub. Math., NS **3**(2) (1956), 31–66.

38. Sit, W. Y. *Differential algebraic Subgroups of SL(2) and strong normality in simple extensions*, Amer. J. Math. **97** (3) (1975), 627–698.

39. Wu, Wen Tsun. *A constructive theory of differential algebraic geometry based on works of J. F. Ritt with particular applications to mechanical theorem-proving of differential geometries*, Differential Geometry and Differential Equations (Shanghai, 1985), Lecture Notes in Math. **1255**, Springer-Verlag, Berlin-New York, 1987, 173–189.

Differential Algebra and Related Topics, pp. 71–94
Proceedings of the International Workshop
Eds. L. Guo, P. J. Cassidy, W. F. Keigher & W. Y. Sit

DIFFERENTIAL SCHEMES

JERALD J. KOVACIC

Department of Mathematics,
The City College of The City University of New York,
New York, NY 10031, USA
E-mail: jkovacic@member.ams.org
http://members.bellatlantic.net/~jkovacic

The language of schemes, which has proven to be of value to algebraic geometry, has not yet been widely accepted into differential algebraic geometry. One reason may be that there are "challenges"; some hoped-for properties of the ring of global sections are absent. In this paper we examine a class of differential rings (or modules) called AAD (Annihilators Are Differential) which meets this challenge. Any differential ring has a smallest ideal whose quotient is AAD. We use this to show that products exist in the category of AAD schemes.

1 Introduction

Differential algebraic geometry started with the work of Ritt, and his students Raudenbush and Levi, on the "manifolds" of solutions of algebraic differential equations. Kolchin, influenced by the work of Weil and Chevalley, modernized differential algebra and developed differential algebraic geometry and the theory of differential algebraic groups. For an excellent account of this, and an extensive bibliography, see Buium and Cassidy [4].

Most of the work so far has used the so-called "Weil" approach - solution sets of radical ideals in some affine or projective space over a given universe. This approach has worked very well and many important results have been obtained; see, for example, Chapter IV of Kolchin [18].

The theory of differential algebraic groups is also well developed. The papers of Cassidy (for example [8]) are ground-breaking. In [19], Kolchin broke from the Weil tradition by axiomatizing the notion of differential algebraic group. This approach is an elegant tour de force, but has not become widely accepted.

Algebraic geometry has benefited greatly from the language of schemes; it is now time for differential algebraic geometry to embrace that language.

The study of differential schemes began in the work of Keigher [13–17] and was continued by Carra' Ferro [5,6] and Buium [2]. In Carra' Ferro [7], a different approach is taken than what we take here; the definition of the structure sheaf is different. Buium [3] has yet another approach. In that work a differential scheme is a scheme whose structure sheaf consists of differential rings. One might call such an object a "scheme with differentiation".

We start with the elementary definitions and theorems of differential schemes. Very quickly, however, we find a "challenge": the ring of global sections is not isomorphic to the original ring. To meet this challenge we introduce the notion of AAD (Annihilators Are Differential). All rings with trivial derivations (i.e. non-differential rings) are AAD as well as all reduced differential rings. So our theory includes all of algebraic geometry as well as "classical" differential algebraic geometry.

In any differential ring there is a smallest differential ideal giving an AAD quotient. This is analogous to the nil radical, the quotient by which is reduced.

Using this notion we can develop some of the basic notions in the theory of differential schemes. We end by showing that products exist in the category of AAD differential schemes. However it is clear that one could go much further simply using analogies with the theory of (non-differential) schemes.

In this paper we often deal with modules rather than rings. We do not do this systematically, but only when the proof for modules is same as that for rings. The theory of quasi-coherent sheaves of modules awaits development.

We do not make any restrictive assumption about our differential rings (e.g. Ritt algebras or Keigher rings); instead, we state all hypotheses whenever they are needed.

2 Differential rings

In this section we record some facts for use in subsequent sections. References for this material are Kaplansky [12, Chapter I] and Kolchin [18, Chapters I-III].

By a ring we always mean a commutative ring with identity 1; the 0 ring being the only ring in which $1 = 0$. A differential ring is a ring together with a fixed set of commuting derivations denoted by $\Delta = \{\delta_1, \ldots, \delta_m\}$. A differential ring is *ordinary* if $m = 1$. For ordinary differential rings we denote the derivation by $'$. The free commutative semigroup generated by Δ is denoted by Θ. Every $\theta \in \Theta$ has the form $\theta = \delta_1^{e_1} \cdots \delta_m^{e_m}$, where $e_1, \ldots, e_m \in \mathbb{N}$. For an ordinary Δ-ring we write $a^{(n)}$ for $\delta_1^n a$.

We shall use the terms Δ-ring, Δ-ideal, etc., to mean differential ring, differential ideal, etc.

Let $\mathcal{R}$ be a Δ-ring. A Δ-$\mathcal{R}$-module is a $\mathcal{R}$-module $\mathcal{M}$ together with derivations $\delta : \mathcal{M} \to \mathcal{M}$, for $\delta \in \Delta$, such that $\delta(rm) = \delta r\, m + r\, \delta m$ for $r \in \mathcal{R}$ and $m \in \mathcal{M}$.

Notation. Throughout this paper $\mathcal{R}$ denotes a Δ-ring and $\mathcal{M}$ a Δ-$\mathcal{R}$-module.

If S is a subset of $\mathcal{R}$, we denote by $[S]$ the smallest Δ-ideal of $\mathcal{R}$ containing S and by $\{S\}$ the smallest radical Δ-ideal of $\mathcal{R}$ containing S. $\{S\}$ may be described recursively as follows (see Kolchin [18, p. 122]):

$$\{S\}_0 = S, \qquad \{S\}_1 = \sqrt{[S]}, \qquad \{S\}_{k+1} = \{\{S\}_k\}_1 = \sqrt{[\{S\}_k]},$$
$$\{S\} = \bigcup_{k=0}^{\infty} \{S\}_k.$$

Under favorable circumstances this can be simplified. Recall the following definition from Keigher [14, p. 239].

Definition 2.1 For any subset S of $\mathcal{R}$, $S_\# = \{a \in S \mid \theta a \in S, \text{ for all } \theta \in \Theta\}$.

Proposition 2.2 *The following are equivalent:*

(a) *If P is a prime ideal, then $P_\#$ is a prime Δ-ideal.*

(b) *If $\mathfrak{a}$ is a Δ-ideal and Σ a multiplicative set disjoint from $\mathfrak{a}$, then a Δ-ideal maximal among Δ-ideals that contain $\mathfrak{a}$ and are disjoint from Σ is prime.*

(c) *If $\mathfrak{a}$ is a Δ-ideal, then so is $\sqrt{\mathfrak{a}}$.*

(d) *If S is any set, then $\{S\} = \sqrt{[S]}$.*

Proof. $(a) \Leftrightarrow (b)$: Keigher [14, Proposition 1.11, p. 244]. $(b) \Leftrightarrow (c)$: Gorman [9, Lemma 2, p. 27]. $(c) \Leftrightarrow (d)$: Obvious. $\qquad\square$

Keigher [14, p. 242] uses condition (a) as the definition of a *special* Δ-ring. Gorman [9, p. 25] uses condition (b) as the definition of a *d-MP* Δ-ring.

Definition 2.3 A Δ-ring $\mathcal{R}$ that satisfies the conditions of the previous proposition is called a *Keigher* Δ-ring.

Any ring $\mathcal{R}$ with trivial derivation ($\delta r = 0$ for all r and all $\delta \in \Delta$) is a Keigher Δ-ring. In particular the 0 ring is a Keigher Δ-ring. If $\mathcal{R}$ is a Δ-ring that contains $\mathbb{Q}$ (often called a *Ritt algebra*) then $\mathcal{R}$ is a Keigher Δ-ring (Kaplansky [12, Lemma 1.8, p. 12] or Keigher [14, Proposition 1.5, p. 242]). Were we to use Hasse-Schmidt derivations, every Δ-ring would be a Keigher Δ-ring by Okugawa [21, Theorem 1, p. 63].

In this paper we do not assume that $\mathcal{R}$ is a Keigher Δ-ring. This generality is at the expense of slightly more complicated proofs; the following propositions assist in our efforts.

Proposition 2.4 *Suppose that S is a subset of $\mathcal{R}$ and $a \in \mathcal{R}$. If $a \in \{S\}$ then there is a finite subset S_0 of S such that $a \in \{S_0\}$.*

Proof. First observe that $a \in \{S\}_k = \sqrt{[\{S\}_{k-1}]}$ for some $k \in \mathbb{N}$, and that there is a finite subset T of $\{S\}_{k-1}$ with $a \in \sqrt{[T]}$, then use induction. $\quad\square$

The following proposition will be used frequently.

Proposition 2.5 *Let S and T be subsets of $\mathcal{R}$. Then $\{S\}\{T\} \subset \{ST\}$.*

Proof. Kaplansky [12, Lemma 1.6, p. 12]. $\quad\square$

Suppose Σ is a multiplicative set of $\mathcal{R}$. Then $\mathcal{M}\Sigma^{-1}$ has a natural structure of Δ-module, obtained by using the quotient rule. If P is a prime ideal of $\mathcal{R}$ and $\Sigma = \mathcal{R} \setminus P$, we write $\mathcal{M}_P$ instead of $\mathcal{M}\Sigma^{-1}$. Note that $\mathcal{R}_P$ is a local ring and a Δ-ring, but the maximal ideal is not necessarily a Δ-ideal. If $\mathfrak{p}$ is a prime Δ-ideal then the maximal ideal of $\mathcal{R}_\mathfrak{p}$ is indeed a Δ-ideal. In this case we say that $\mathcal{R}_\mathfrak{p}$ is a *local Δ-ring*. If $b \in \mathcal{R}$ and $\Sigma = \{1, b^e \,|\, e \in \mathbb{N}\}$ we write $\mathcal{M}_b$ instead of $\mathcal{M}\Sigma^{-1}$. This is the 0 module if b is nilpotent.

Proposition 2.6 *Let $\mathfrak{a}$ be a radical Δ-ideal of $\mathcal{R}$ and Σ a multiplicative set. Suppose that $\mathfrak{a} \cap \Sigma = \emptyset$. Then there exists a prime Δ-ideal $\mathfrak{p}$ containing $\mathfrak{a}$ such that $\mathfrak{p} \cap \Sigma = \emptyset$.*

Proof. By Zorn's Lemma there is a radical Δ-ideal maximal with respect to containing $\mathfrak{a}$ and not meeting Σ. By Kaplansky [12, Lemma, p. 13], this ideal is prime. $\quad\square$

We often use this proposition in the following form.

Corollary 2.7 *Let S a subset of $\mathcal{R}$ and $b \in \mathcal{R}$. Then $b \in \{S\}$ if and only if every prime Δ-ideal containing S also contains b. Hence $1 \in \{S\}$ if and only if S is not contained in any prime Δ-ideal of $\mathcal{R}$.*

Proof. For the first statement, take $\Sigma = \{1, b^e \,|\, e \in \mathbb{N}\}$; take $b = 1$ for the second. $\quad\square$

Corollary 2.8 *Let S be a subset of $\mathcal{R}$. Then $\{S\}$ is the intersection of all the prime Δ-ideals that contain S.*

Proof. If $b \notin \{S\}$ there is a prime Δ-ideal containing S not containing b. $\quad\square$

The use of the symbol $\{\ \}$ in the previous corollaries is essential; for example, it is not true that $1 \in [S]$ if and only if S is not contained in any prime Δ-ideal. Consider the ordinary Δ-ring $\mathcal{R} = \mathbb{Z}[x]$ where $x' = 1$. Then, for any $n \in \mathbb{N}$, (n, x^n) is a Δ-ideal since $(x^n)' = n\,x^{n-1}$, but any prime (even radical) Δ-ideal containing (n, x^n) would contain x and therefore 1.

Corollary 2.9 *Suppose that $\mathcal{R}$ is a Keigher Δ-ring, S a subset of $\mathcal{R}$ and $b \in \mathcal{R}$. Then $b^e \in [S]$, for some $e \in \mathbb{N}$, if and only if every prime Δ-ideal containing S also contains b. $1 \in [S]$ if and only if S is not contained in any prime Δ-ideal of $\mathcal{R}$.*

Proof. Use condition (d) of Proposition 2.2 and Corollary 2.7. $\qquad\square$

3 Differential spectrum

This material here can be found in various forms in the papers of Keigher and Buium. For the non-differential case see Grothendieck and Dieudonné [10, Chapitre I] or Hartshorne [11, Chapter II].

Definition 3.1 Denote by $X = \operatorname{diffspec}\mathcal{R}$ the set of all prime Δ-ideals of $\mathcal{R}$. For any set $S \subset \mathcal{R}$, denote by $V(S)$ the set of $\mathfrak{p} \in \operatorname{diffspec}\mathcal{R}$ with $\mathfrak{p} \supset S$. For $f \in \mathcal{R}$, denote by $D(f)$ the set of $\mathfrak{p} \in \operatorname{diffspec}\mathcal{R}$ with $f \notin \mathfrak{p}$.

Throughout this section we shall use X to denote $\operatorname{diffspec}\mathcal{R}$.

It is curious that a non-zero Δ-ring $\mathcal{R}$ can have empty diffspec. Consider the ordinary Δ-ring $\mathcal{R} = \mathbb{Z}[x]/(n, x^n)$ where $x' = 1$ and $n \in \mathbb{N}$. Any prime Δ-ideal would have to contain the image of x and therefore 1. This phenomenon cannot occur for a Keigher Δ-ring by Corollary 2.9 (take S to be the set consisting of 0 alone, and $b = 1$).

Proposition 3.2 *Let $\mathcal{R}$ be a Δ-ring.*

(a) $V((0)) = X$, $V((1)) = \emptyset$.

(b) *If $(S_i)_{i \in I}$ is a family of subsets of $\mathcal{R}$, then*

$$V\left(\bigcup_{i \in I} S_i\right) = V\left(\sum_{i \in I} S_i\right) = \bigcap_{i \in I} V(S_i).$$

(c) *If S and T are subsets of $\mathcal{R}$, then $V(S) \subset V(T)$ if and only if $\{S\} \supset \{T\}$.*

(d) *If S is a subset of $\mathcal{R}$, then $V(S) = V(\{S\})$.*

(e) *If $\mathfrak{a}$ and $\mathfrak{b}$ are Δ-ideals of $\mathcal{R}$, then $V(\mathfrak{a} \cap \mathfrak{b}) = V(\mathfrak{ab}) = V(\mathfrak{a}) \cup V(\mathfrak{b})$.*

Proof. The first two statements are immediate from the definitions. The third and fourth come from Corollary 2.8. For the last, observe that a prime Δ-ideal contains $\mathfrak{a}$ or $\mathfrak{b}$ if and only if contains $\mathfrak{ab}$, or if it contains $\mathfrak{a} \cap \mathfrak{b}$. $\qquad\square$

Definition 3.3 Let $\mathcal{R}$ be a Δ-ring. We define a topology on $X = \operatorname{diffspec}\mathcal{R}$, called the *Kolchin topology*, by decreeing $V(S)$ to be closed for every $S \subset \mathcal{R}$.

Evidently diffspec $\mathcal{R}$ is a subset of spec $\mathcal{R}$ and the Kolchin topology is the subspace topology of the Zariski topology on spec $\mathcal{R}$.

Proposition 3.4 *Let $\mathcal{R}$ be a Δ-ring.*

(a) $D(0) = \emptyset$, $D(1) = X$.

(b) *If $f, g \in \mathcal{R}$, then $D(fg) = D(f) \cap D(g)$.*

(c) $D(f) \subset D(g)$ *if and only if $\{f\} \subset \{g\}$.*

Proof. The first and second statements are obvious. For the third, observe that $D(f) = X \setminus V(\{f\})$. $\qquad\square$

Note that the set $D(f)$ is open, being the complement of $V(\{f\})$. The following proposition states that these sets form a basis of the topology.

Proposition 3.5 *Let $\mathfrak{p} \in X$ and U be an open neighborhood of $\mathfrak{p}$. Then there exists $b \in \mathcal{R}$ with $\mathfrak{p} \in D(b) \subset U$.*

Proof. If $U = X \setminus V(\mathfrak{a})$, choose any $b \in \mathfrak{a} \setminus \mathfrak{p}$. $\qquad\square$

Proposition 3.6 *X is quasi-compact and T_0.*

Proof. Let U_i, $i \in I$, be an open cover of X. By the previous proposition we may assume that $U_i = D(b_i)$ for some $b_i \in \mathcal{R}$. For any $\mathfrak{p} \in X$ there exists $i \in I$ such that $b_i \notin \mathfrak{p}$, so, by Corollary 2.7, $1 \in \{(b_i)_{i \in I}\}$. Then, by Proposition 2.4, there is a finite subset $b_1, \ldots, b_n$ with $1 \in \{b_1, \ldots, b_n\}$.

To see that X is T_0, observe that given any two distinct Δ-prime ideals $\mathfrak{p}$ and $\mathfrak{q}$ of $\mathcal{R}$, one of them does not contain the other. Thus $\mathfrak{p} \notin V(\mathfrak{q})$ or $\mathfrak{q} \notin V(\mathfrak{p})$. $\qquad\square$

Keigher [14, Theorem 2.4, p.248] proves that X is a spectral space (which implies the above) if $\mathcal{R}$ is a Keigher ring. Carra' Ferro [5, Teorema 1.3, p.5] extends this to arbitrary $\mathcal{R}$.

Let S be a Δ-ring and set $Y = \text{diffspec}\, S$. Suppose that $\phi : \mathcal{R} \to S$ is a Δ-homomorphism. If $\mathfrak{b}$ is any Δ-ideal of S then $\phi^{-1}(\mathfrak{b})$ is a Δ-ideal of $\mathcal{R}$. If $\mathfrak{b}$ is prime or radical then so is $\phi^{-1}(\mathfrak{b})$. Thus we have a mapping $^a\phi : Y \to X$, given by $^a\phi(\mathfrak{p}) = \phi^{-1}(\mathfrak{p})$. It is called the "adjoint" of ϕ. The corresponding mapping from radical Δ-ideals of S to radical Δ-ideals of $\mathcal{R}$ is denoted by $^r\phi$.

For a subset Z of X define, following Bourbaki [1, Chapter II, §4.3, p. 99],

$$\mathfrak{J}(Z) = \bigcap_{\mathfrak{p} \in Z} \mathfrak{p}.$$

Then, by Corollary 2.8, $\mathfrak{J}(V(S)) = \{S\}$ for any subset S of $\mathcal{R}$.

Proposition 3.7 Let $\phi : \mathcal{R} \to \mathcal{S}$ be a Δ-homomorphism of Δ-rings. If S is a subset of $\mathcal{R}$, then

$$^a\phi^{-1}(V(S)) = V(\phi(S)).$$

In particular, for $b \in \mathcal{R}$,

$$^a\phi^{-1}(D(b)) = D(\phi(b)).$$

If T is a subset of $\mathcal{S}$, then

$$\overline{^a\phi(V(T))} = V(\phi^{-1}(T)).$$

Proof. This is similar to Grothendieck and Dieudonné [10, Proposition 1.2.2, p.196]. However their notation is different so we sketch the argument here.

Now $\mathfrak{p} \in {}^a\phi^{-1}(V(S))$ if and only if $S \subset {}^a\phi(\mathfrak{p}) = \phi^{-1}(\mathfrak{p})$, which is true if and only if $\phi(S) \subset \mathfrak{p}$, that is $\mathfrak{p} \in V(\phi(S))$. For the second statement write $D(b) = X \setminus V(\{b\})$.

Let $S \subset \mathcal{R}$ be such that

$$V(S) = \overline{^a\phi(V(T))}.$$

If $\mathfrak{p} \in {}^a\phi(V(T))$, then $\phi^{-1}T \subset \mathfrak{p}$ so $\mathfrak{p} \in V(\phi^{-1}(T))$. Therefore

$$V(S) = \overline{^a\phi(V(T))} \subset V(\phi^{-1}(T)).$$

But we also have

$$\{S\} = \mathfrak{J}(V(S)) \subset \mathfrak{J}(\overline{^a\phi(V(T))}) \subset \mathfrak{J}(V(\phi^{-1}(T))) = \{\phi^{-1}(T)\}.$$

Hence, by Proposition 3.2, $V(\phi^{-1}(T)) \subset V(S)$. $\quad\square$

Corollary 3.8 Let $\phi : \mathcal{R} \to \mathcal{S}$ be a Δ-homomorphism of Δ-rings. Then $^a\phi :$ diffspec $\mathcal{S} \to$ diffspec $\mathcal{R}$ is continuous.

Proof. This comes from the first assertion of the preceding proposition. $\quad\square$

Proposition 3.9 Let $\phi : \mathcal{R} \to \mathcal{S}$ be a Δ-homomorphism. Then the following conditions are equivalent.

(a) $^a\phi$ is a homeomorphism of $Y =$ diffspec $\mathcal{S}$ onto its image.

(b) $^r\phi$ is injective.

(c) If $\mathfrak{b}$ and $\mathfrak{q}$ are Δ-ideals of $\mathcal{S}$ with $\mathfrak{q}$ prime, $\mathfrak{b}$ radical and $\phi^{-1}(\mathfrak{b}) \subset \phi^{-1}(\mathfrak{q})$, then $\mathfrak{b} \subset \mathfrak{q}$.

Proof. $(a) \Rightarrow (b)$: Let $\mathfrak{b}_1, \mathfrak{b}_2$ be radical Δ-ideals of $\mathcal{S}$ with $^r\phi(\mathfrak{b}_1) = {}^r\phi(\mathfrak{b}_2)$. For every $\mathfrak{q} \in V(\mathfrak{b}_1)$, $^a\phi(\mathfrak{q}) \in V(\phi^{-1}(\mathfrak{b}_1)) \cap {}^a\phi(Y)$. By the last condition of previous proposition, and the assumption that $^a\phi$ is a closed mapping,

$$^a\phi(\mathfrak{q}) \in V(\phi^{-1}(\mathfrak{b}_1)) \cap {}^a\phi(Y) = \overline{{}^a\phi(V(\mathfrak{b}_1))} = {}^a\phi(V(\mathfrak{b}_1)) = {}^a\phi(V(\mathfrak{b}_2)).$$

Since $^a\phi$ is injective, $\mathfrak{q} \in V(\mathfrak{b}_2)$. Therefore

$$\mathfrak{b}_1 = \mathfrak{J}(V(\mathfrak{b}_1)) \subset \mathfrak{J}(V(\mathfrak{b}_2)) = \mathfrak{b}_2.$$

By symmetry, $\mathfrak{b}_1 = \mathfrak{b}_2$.

$(b) \Rightarrow (c)$: Let $\mathfrak{q}$ and $\mathfrak{b}$ be as in the statement of condition 3. Then

$$^r\phi(\mathfrak{b}) = \phi^{-1}(\mathfrak{b}) = \phi^{-1}(\mathfrak{b}) \cap \phi^{-1}(\mathfrak{q}) = \phi^{-1}(\mathfrak{b} \cap \mathfrak{q}) = {}^r\phi(\mathfrak{b} \cap \mathfrak{q}).$$

Since $^r\phi$ is injective, $\mathfrak{b} = \mathfrak{b} \cap \mathfrak{q}$.

$(c) \Rightarrow (a)$: If $\mathfrak{q}_1, \mathfrak{q}_2 \in Y$ with $^a\phi(\mathfrak{q}_1) = {}^a\phi(\mathfrak{q}_2)$, then $\mathfrak{q}_1 \subset \mathfrak{q}_2$ (taking $\mathfrak{b} = \mathfrak{q}_1$ and $\mathfrak{q} = \mathfrak{q}_2$). It follows that $^a\phi$ is injective. We need to show, for any radical Δ-ideal $\mathfrak{b} \subset \mathcal{S}$, that $^a\phi(V(\mathfrak{b}))$ is closed. But $^a\phi(\mathfrak{q}) \in V(\phi^{-1}(\mathfrak{b}))$ implies that $\phi^{-1}(\mathfrak{q}) \supset \phi^{-1}(\mathfrak{b})$, hence $\mathfrak{q} \supset \mathfrak{b}$. Therefore

$$V(\phi^{-1}(\mathfrak{b})) \cap {}^a\phi(Y) \subset {}^a\phi(V(\mathfrak{b})) \subset \overline{{}^a\phi(V(\mathfrak{b}))} = V(\phi^{-1}(\mathfrak{b})). \qquad \square$$

Corollary 3.10 *Let $\phi : \mathcal{R} \to \mathcal{S}$ be a surjective Δ-homomorphism with kernel $\mathfrak{a}$. Then $^a\phi$ is a homeomorphism of diffspec $\mathcal{S}$ onto $V(\mathfrak{a}) \subset X$.*

Proof. This is because ϕ satisfies the third condition and the image of $^a\phi$ is evidently $V(\mathfrak{a})$. $\qquad \square$

Corollary 3.11 *Let S a multiplicative subset of $\mathcal{R}$, and let $\phi : \mathcal{R} \to \mathcal{R}S^{-1}$ be the canonical mapping. Then $^a\phi$ is a homeomorphism from diffspec $\mathcal{R}S^{-1}$ onto the set of prime Δ-ideals of $\mathcal{R}$ not meeting S. If $b \in \mathcal{R}$, the canonical mapping $\mathcal{R} \to \mathcal{R}_b$ induces a homeomorphism of diffspec $(\mathcal{R}_b)$ onto $D(b)$.*

Proof. This is because $^r\phi$ is injective. $\qquad \square$

4 Structure sheaf

We define the structure sheaf exactly as in Hartshorne [11, p. 70].

Definition 4.1 Let $\mathcal{R}$ be a Δ-ring, $X = $ diffspec $\mathcal{R}$ and $\mathcal{M}$ a Δ-$\mathcal{R}$-module. For each open set U of X, let $\mathcal{O}_X(U)$ be the set of functions

$$s : U \to \coprod_{\mathfrak{p} \in U} \mathcal{R}_\mathfrak{p}$$

and $\widetilde{\mathcal{M}}(U)$ the set of functions

$$t : U \to \coprod_{\mathfrak{p} \in U} \mathcal{M}_{\mathfrak{p}}$$

satisfying the following:

(a) For each $\mathfrak{p} \in U$, $s(\mathfrak{p}) \in \mathcal{R}_{\mathfrak{p}}$, $t(\mathfrak{p}) \in \mathcal{M}_{\mathfrak{p}}$, and

(b) there is an open cover U_i of U, and $a_i, b_i \in \mathcal{R}$, $m_i \in \mathcal{M}$, such that, for each $\mathfrak{q} \in U_i$, $b_i \notin \mathfrak{q}$ and $s(\mathfrak{q}) = a_i/b_i \in \mathcal{R}_q$, $t(\mathfrak{q}) = m_i/b_i \in \mathcal{M}_q$.

The set $\mathcal{O}_X(U)$ inherits the structure of Δ-ring from $\mathcal{R}_{\mathfrak{p}}$ by the formula $\delta(s)(\mathfrak{p}) = \delta(s(\mathfrak{p}))$ for $\delta \in \Delta$. If $V \subset U$ are two open sets we define the "restriction" $\mathcal{O}_X(U) \to \mathcal{O}_X(V)$ by restricting the functions. One easily sees that $\mathcal{O}_X$ is a sheaf of Δ-rings. Also $\widetilde{\mathcal{M}}$ is a Δ-$\widetilde{\mathcal{R}}$-module; i.e. for each open $U \subset X$, $\widetilde{\mathcal{M}}(U)$ is a differential $\widetilde{\mathcal{R}}(U) = \mathcal{O}_X(U)$-module.

Proposition 4.2 For $\mathfrak{p} \in X$, the stalk of $\mathcal{O}_X$ at $\mathfrak{p}$, denoted by $\mathcal{O}_{X,\mathfrak{p}}$, is Δ-isomorphic to $\mathcal{R}_{\mathfrak{p}}$. The stalk $\widetilde{\mathcal{M}}_{\mathfrak{p}}$ is Δ-isomorphic to $\mathcal{M}_{\mathfrak{p}}$.

Proof. The proofs are the same as Proposition 2.2(a), p. 71 and Proposition 5.1(b), p. 110 in Hartshorne [11] . □

Unfortunately Hartshorne's Proposition 2.2(b,c) and Proposition 5.1(c,d) are, in general, false; it is not necessarily true that $\mathcal{O}_X(X)$ is isomorphic to $\mathcal{R}$, nor that $\widetilde{\mathcal{M}}(X)$ is isomorphic to $\mathcal{M}$. We give the relevant definitions here, but leave the discussion to later sections (and even more so to Kovacic [20]).

Definition 4.3 $\widehat{\mathcal{R}}$ denotes the ring of global sections $\mathcal{O}_X(X) = \Gamma(X, \mathcal{O}_X)$. $\widehat{\mathcal{M}}$ denotes the $\widehat{\mathcal{R}}$-module $\widetilde{\mathcal{M}}(X) = \Gamma(X, \widetilde{\mathcal{M}})$.

Definition 4.4 Denote by $\iota_{\mathcal{R}}$ the Δ-homomorphism $\mathcal{R} \to \widehat{\mathcal{R}}$ with

$$\iota_{\mathcal{R}}(r)(\mathfrak{p}) = \frac{r}{1} \in \mathcal{R}_{\mathfrak{p}} \qquad (\mathfrak{p} \in X).$$

If $\mathcal{M}$ is a Δ-$\mathcal{R}$-module, then $\iota_{\mathcal{M}} : \mathcal{M} \to \widehat{\mathcal{M}}$ is the Δ-homomorphism with

$$\iota_{\mathcal{M}}(m)(\mathfrak{p}) = \frac{m}{1} \in \mathcal{M}_{\mathfrak{p}} \qquad (\mathfrak{p} \in X).$$

We call these the *canonical homomorphisms*.

Although the proof of the next proposition is not difficult, it is surprisingly less trivial than the corresponding proof for spec.

Proposition 4.5 *Let $\mathcal{R}$ be a Δ-ring, $X = \text{diffspec}\,\mathcal{R}$ and $\mathcal{M}$ a Δ-$\mathcal{R}$-module. Let $s \in \widehat{\mathcal{M}}$. Then, for some $n \in \mathbb{N}$, there exist $m_1, \ldots, m_n \in \mathcal{M}$ and $b_1, \ldots, b_n \in \mathcal{R}$ such that $1 \in \{b_1, \ldots, b_n\}$ and $s(\mathfrak{p}) = m_i/b_i \in \mathcal{M}_\mathfrak{p}$ whenever $\mathfrak{p} \in D(b_i)$.*

Proof. By definition, there exists an open cover U_i of X, $m_i \in \mathcal{M}$, $b_i \in \mathcal{R}$ such that for every $\mathfrak{q} \in U_i$, $b_i \notin \mathfrak{q}$ and $s(\mathfrak{q}) = m_i/b_i$. By Proposition 3.5 we may assume that $U_i = D(f_i)$ and by Proposition 3.6 a finite number of these, say for $i = 1, \ldots, n$, cover X.

Since $\mathfrak{q} \in D(f_i)$ implies $b_i \notin \mathfrak{q}$, $D(f_i) \subset D(b_i)$. In the non-differential case this implies that $f_i \in \sqrt{(b_i)}$, i.e. that $f_i^e \in (b_i)$, from which the result easily follows. In our case we merely have that $f_i \in \{b_i\}$. However, using Proposition 3.4, we have

$$D(f_i b_i) = D(f_i) \cap D(b_i) = D(f_i) = U_i.$$

Now replace m_i by $f_i m_i$ and b_i by $f_i b_i$.

It remains to show that $1 \in \{f_1 b_1, \ldots, f_n b_n\}$. For any $\mathfrak{p} \in X$ there exists i with $\mathfrak{p} \in D(f_i b_i)$, therefore $\{f_1 b_1, \ldots, f_n b_n\}$ is not contained in any prime Δ-ideal. The result follows from Corollary 2.7. $\qquad\square$

5 Morphisms

The following definition is from Keigher [13, p.110]; see also Hartshorne [11, p.72].

Definition 5.1 By an *LDR* (Local Differential Ringed) space we mean a pair $(X, \mathcal{O}_X)$ where X is a topological space and $\mathcal{O}_X$ is a sheaf of Δ-rings whose stalks are local Δ-rings (which means that the maximal ideal is a Δ-ideal).

Proposition 5.2 *Let $X = \text{diffspec}\,\mathcal{R}$. Then $(X, \mathcal{O}_X)$ is an LDR space.*

Proof. This follows immediately from Proposition 4.2. $\qquad\square$

Definition 5.3 By a morphism of LDR spaces we mean a morphism of local ringed spaces whose sheaf morphism is a morphism of sheaves of Δ-rings.

Thus a morphism $(Y, \mathcal{O}_Y) \to (X, \mathcal{O}_X)$ is a pair $(f, f^\#)$ where $f : Y \to X$ is a continuous mapping and $f^\# : \mathcal{O}_X \to f_* \mathcal{O}_Y$ is a mapping of sheaves of Δ-rings. In addition the induced mapping on stalks, $f_y^\# : \mathcal{O}_{X,f(y)} \to \mathcal{O}_{Y,y}$, is a local homomorphism; it carries the maximal ideal into the maximal ideal.

Proposition 5.4 *Let* $\phi : \mathcal{R} \to \mathcal{S}$ *be* Δ-*homomorphism. Let* $X = \text{diffspec}\,\mathcal{R}$ *and* $Y = \text{diffspec}\,\mathcal{S}$. *Then* ϕ *induces a morphism of LDR spaces*

$$({}^a\phi, \phi^{\#}) : (Y, \mathcal{O}_Y) \to (X, \mathcal{O}_X).$$

Proof. We have already seen that ${}^a\phi : Y \to X$ is a continuous mapping (Proposition 3.8). See Hartshorne [11, Proposition 2.3(b), p. 73], for the definition of $\phi^{\#}$, and remainder of the proof. □

The converse, Hartshorne [11, Proposition 2.3(c)], is not true in general: a morphism of differential spectra does not necessarily come from a homomorphism of Δ-rings. For complete details see Kovacic [20].

Proposition 5.5 *Let* $b \in \mathcal{R}$. *Then the canonical mapping* $\phi : \mathcal{R} \to \mathcal{R}_b$ *induces an isomorphism of* $D(b) \to \text{diffspec}\,(\mathcal{R}_b)$.

Proof. By Corollary 3.11, ${}^a\phi$ is a homeomorphism onto $D(b)$. To show that the sheaves are isomorphic we must show that ϕ induces an isomorphism $\mathcal{R}_{\mathfrak{p}} \to (\mathcal{R}_b)_{\mathfrak{p}\mathcal{R}_b}$ whenever $\mathfrak{p} \in D(b)$, which is straightforward. □

6 Δ-Schemes

Definition 6.1 An *affine* Δ-*scheme* is an LDR space which is isomorphic to $(X, \mathcal{O}_X)$ where $X = \text{diffspec}\,\mathcal{R}$ for some Δ-ring $\mathcal{R}$.

Definition 6.2 A Δ-*scheme* is an LDR space in which every point has an open neighborhood that is an affine Δ-scheme.

By Proposition 5.5, $D(f)$ is an affine Δ-scheme for any $f \in \mathcal{R}$.

Proposition 6.3 *If* U *is an open subset of* X, *then* U *is a* Δ-*scheme.*

Proof. Here we understand that the sheaf on U is the restriction of the sheaf on X. This follows from Proposition 3.5 and the above remark. □

7 Δ-Zeros

Observe that the zero of a ring or module is characterized by the condition that $1 \in \text{ann}(0)$ or $1 \in \sqrt{\text{ann}(0)}$. The following definition generalizes that condition.

Definition 7.1 We say that $m \in \mathcal{M}$ is a Δ-*zero* of $\mathcal{M}$ if $1 \in \{\text{ann}(m)\} \subset \mathcal{R}$. The set of Δ-zeros of $\mathcal{M}$ is denoted by $\mathfrak{Z}(\mathcal{M})$.

For a Keigher Δ-ring, this is equivalent to the condition that $1 \in [\mathrm{ann}(m)]$ by Proposition 2.2. This definition indicates that annihilators are not necessarily Δ-ideals. Indeed, if $am = 0$ we can only conclude that $\delta a\, m + a\, \delta m = 0$, but not that $\delta a\, m = 0$. The importance of Δ-zeros is given by the following proposition.

Proposition 7.2 *The kernel of the canonical homomorphism* $\iota_{\mathcal{M}} : \mathcal{M} \to \widehat{\mathcal{M}}$ *is* $3(\mathcal{M})$.

Proof. We have $m \in \ker \iota_{\mathcal{M}}$ if and only if $m = 0/1 \in \mathcal{M}_{\mathfrak{p}}$ for every $\mathfrak{p} \in X$, which happens if and only if there exists $a \in \mathcal{R}$, $a \notin \mathfrak{p}$ with $am = 0$. By Proposition 2.7 this is equivalent to $1 \in \{\mathrm{ann}(m)\}$. $\qquad\square$

Example 7.3 Consider the ordinary Δ-ring $\mathcal{R} = \mathbb{Q}[x]\{\eta\} = \mathbb{Q}[x]\{y\}/[xy]$, where y is a Δ-indeterminate over $\mathbb{Q}[x]$ and $x' = 1$. Since $x\eta = 0$, $x \in \mathrm{ann}(\eta)$, and $1 = x' \in [\mathrm{ann}(\eta)]$, therefore $\eta \in 3(\mathcal{R})$.

Proposition 7.4 *Let* $\mathfrak{n}$ *be the nil radical of* $\mathcal{R}$. *Then* $3(\mathcal{R}) \subset \{\mathfrak{n}\}$. *If* $\mathcal{R}$ *is a Keigher ring, then* $3(\mathcal{R}) \subset \mathfrak{n}$. *If* $\mathcal{R}$ *is reduced then* $3(\mathcal{R}) = 0$.

Proof. Let $r \in 3(\mathcal{R})$. Then $1 \in \{\mathrm{ann}(r)\}$ so, by Proposition 2.5,

$$r \in \{r\}\{\mathrm{ann}(r)\} \subset \{\mathrm{rann}(r)\} = \{0\} = \{\sqrt{(0)}\} = \{\mathfrak{n}\}.$$

If $\mathcal{R}$ is Keigher, then $\{\sqrt{(0)}\} = \sqrt{(0)} = \mathfrak{n}$. If $\mathcal{R}$ is reduced, then $\mathfrak{n} = \sqrt{(0)} = (0)$, and we can use the formulae preceding Definition 2.1 to conclude that $\{0\} = (0)$. $\qquad\square$

Proposition 7.5 *The quotient homomorphism* $\pi : \mathcal{R} \to \mathcal{R}/3(\mathcal{R})$ *induces an isomorphism* $\mathrm{diffspec}\,(\mathcal{R}/3(\mathcal{R})) \to X$.

Proof. By Proposition 3.8, $^a\pi$ is a homeomorphism of Y onto $V(3(\mathcal{R}))$. Since $3(\mathcal{R}) \subset \{0\}$, $3(\mathcal{R})$ is contained in every element of X. Therefore $^a\pi : Y \to X$ is bijective.

To complete the proof we need to show that, for every $\mathfrak{p} \in X$,

$$\pi_{\mathfrak{p}} : \mathcal{R}_{\mathfrak{p}} \to (\mathcal{R}/3(\mathcal{R}))_{\pi\mathfrak{p}}, \qquad \pi_{\mathfrak{p}}\left(\frac{a}{b}\right) = \frac{\pi a}{\pi b},$$

is an isomorphism. It is clearly surjective. If $\pi_{\mathfrak{p}}(a/b) = 0$, then there exists $r \in \mathcal{R}$, $r \notin \mathfrak{p}$ such that $ra \in 3(\mathcal{R})$. Since $1 \in \{\mathrm{ann}(ra)\}$, $\mathrm{ann}(ra)$ is not contained in $\mathfrak{p}$. Hence there exists $s \in \mathrm{ann}(ra)$, $s \notin \mathfrak{p}$. Hence $sra = 0$ and $a/b = 0 \in \mathcal{R}_{\mathfrak{p}}$. $\qquad\square$

8 Differential spectrum of $\widehat{\mathcal{R}}$

This material generalizes Carra' Ferro [6].

Lemma 8.1 *For any* $\mathfrak{p} \in X$ *define* $\widehat{\mathfrak{p}} = \{s \in \widehat{\mathcal{R}} \,|\, s(\mathfrak{p}) \in \mathfrak{p}\mathcal{R}_\mathfrak{p}\}$. *Then* $\widehat{\mathfrak{p}}$ *is a prime* Δ-*ideal of* $\widehat{\mathcal{R}}$ *and* $\iota_{\widehat{\mathcal{R}}}^{-1}(\widehat{\mathfrak{p}}) = \mathfrak{p}$.

Proof. $\widehat{\mathfrak{p}}$ is the inverse image of the maximal ideal of the stalk at $\mathfrak{p}$, so is a prime Δ-ideal. Also $a \in \iota_{\mathcal{R}}^{-1}(\widehat{\mathfrak{p}})$ if and only if $\iota_{\mathcal{R}}(a)(\mathfrak{p}) = a/1 \in \mathfrak{p}\mathcal{R}_\mathfrak{p}$, i.e. $a \in \mathfrak{p}$. $\qquad\square$

Proposition 8.2 $3(\widehat{\mathcal{M}}) = (0)$.

Proof. Let $s \in 3(\widehat{\mathcal{M}})$. Then $1 \in \{\mathrm{ann}(s)\} \subset \widehat{\mathcal{R}}$ so, for any $\mathfrak{p} \in X$, there exists $t \in \mathrm{ann}(s)$ with $t \notin \widehat{\mathfrak{p}}$. Thus $t(\mathfrak{p}) = a/b$, where both $a, b \notin \mathfrak{p}$. If $s(\mathfrak{p}) = c/d$, with $d \notin \mathfrak{p}$, then

$$0 = t(\mathfrak{p})s(\mathfrak{p}) = \frac{a\,c}{b\,d} \in \mathcal{R}_\mathfrak{p}.$$

There exists $e \notin \mathfrak{p}$ such that $eac = 0 \in \mathcal{R}$, hence $c/d = 0 \in \mathcal{R}_\mathfrak{p}$. This being so for every $\mathfrak{p} \in X$, we have $s = 0$. $\qquad\square$

The following proposition shows that $\widehat{\mathcal{R}}$ and $\widehat{\mathcal{M}}$ may be thought of as "closures" of $\mathcal{R}$ and $\mathcal{M}$.

Theorem 8.3 *The canonical mappings* $\iota_{\widehat{\mathcal{R}}} : \widehat{\mathcal{R}} \to \widehat{\widehat{\mathcal{R}}}$ *and* $\iota_{\widehat{\mathcal{M}}} : \widehat{\mathcal{M}} \to \widehat{\widehat{\mathcal{M}}}$ *are isomorphisms. Hence* diffspec $\widehat{\mathcal{R}}$ *and* diffspec$\widehat{\widehat{\mathcal{R}}}$ *are isomorphic.*

Proof. The previous proposition and Proposition 7.2 shows that $\iota_{\widehat{\mathcal{M}}}$ is injective. For $\widehat{s} \in \widehat{\widehat{\mathcal{M}}}$ we need to find $t \in \widehat{\mathcal{M}}$ such that $\iota_{\widehat{\mathcal{M}}}(t) = \widehat{s}$.

Let $\mathfrak{p} \in X$. There exist $f \in \widehat{\mathcal{M}}$ and $g \in \widehat{\mathcal{R}}$ such that $\widehat{s}(\widehat{\mathfrak{p}}) = f/g \in \widehat{\mathcal{M}}_{\widehat{\mathfrak{p}}}$. There also exist $m \in \mathcal{M}$, $b, c, d \in \mathcal{R}$ such that $f(\mathfrak{p}) = m/b \in \mathcal{M}_\mathfrak{p}$ and $g(\mathfrak{p}) = c/d \in \mathcal{R}_\mathfrak{p}$. Observe that $g \notin \widehat{\mathfrak{p}}$ implies that $c \notin \mathfrak{p}$. We then define

$$t(\mathfrak{p}) = \frac{d\,m}{b\,c} \in \mathcal{M}_\mathfrak{p}.$$

It is straightforward to verify that this formula is independent of the choice of f, g, m, b, c, d. Using Proposition 4.5, we may ensure that $\widehat{s}(\mathfrak{q}) = f/g$ for every $\mathfrak{q} \in D(g) \subset \widehat{X}$ and for all $\mathfrak{p} \in D(bc)$. Hence $t \in \widehat{\mathcal{M}}$.

For any $\mathfrak{q} \in D(g) \subset \widehat{X}$ and any $\mathfrak{p} \in X$,

$$\iota_{\widehat{\mathcal{M}}}(t)(\mathfrak{q})(\mathfrak{p}) = t(\mathfrak{p}) = \frac{d\,m}{b\,c} = \frac{m/b}{c/d} = \widehat{s}(\mathfrak{q})(\mathfrak{p}) \in \mathcal{M}_\mathfrak{p}. \qquad\square$$

In Proposition 5.4 we saw that a Δ-ring homomorphism induces a morphism of schemes. For $\widehat{\mathcal{R}}$ the converse is also true.

Proposition 8.4 *Let* $\widehat{X} = \text{diffspec } \widehat{\mathcal{R}}$ *and* $\widehat{Y} = \text{diffspec } \widehat{\mathcal{S}}$. *If* $\phi : \widehat{\mathcal{R}} \to \widehat{\mathcal{S}}$ *is a* Δ-*homomorphism, then there is an induced morphism of LDR spaces*

$$(^a\phi, \phi^\#) : (\widehat{Y}, \mathcal{O}_{\widehat{Y}}) \to (\widehat{X}, \mathcal{O}_{\widehat{X}}).$$

Conversely if

$$(f, f^\#) : (\widehat{Y}, \mathcal{O}_{\widehat{Y}}) \to (\widehat{X}, \mathcal{O}_{\widehat{X}})$$

is a morphism of LDR spaces, then there is a Δ-*homomorphism* $\phi : \widehat{\mathcal{R}} \to \widehat{\mathcal{S}}$ *such that* $f = {}^a\phi$ *and* $f^\# = \phi^\#$.

Proof. In light of the previous theorem, the proof is the same as that of Hartshorne [11, Proposition 2.3 (c), p.73]. □

9 AAD modules

We observed above (following Definition 7.1) that annihilators are not necessarily Δ-ideals.

Definition 9.1 A Δ-$\mathcal{R}$-module $\mathcal{M}$ is said to be an AAD module (Annihilators Are Differential) if for every $m \in \mathcal{M}$, $\text{ann}(m)$ is a Δ-ideal of $\mathcal{R}$.

Any ring with trivial derivations ($\delta r = 0$ for all $r \in \mathcal{R}$ and $\delta \in \Delta$) is AAD. So the notion of AAD includes all the rings (and modules) of algebra.

Proposition 9.2 *If* $\mathcal{M}$ *is AAD, then, for every* $m \in \mathcal{M}$, $\{\text{ann}(m)\} = \sqrt{\text{ann}(m)}$. *If* $\mathcal{R}$ *is AAD, then the nil radical of* $\mathcal{R}$ *is a* Δ-*ideal.*

Proof. We must show that $\sqrt{\text{ann}(m)}$ is a Δ-ideal. Suppose that $a \in \sqrt{\text{ann}(m)}$, so $a^e \in \text{ann}(m)$ for some $e \in \mathbb{N}$. For $\delta \in \Delta$ we claim that $(\delta a)^e \in \text{ann}(m)$. Indeed, $a \in \text{ann}(a^{e-1}m)$ implies that $\delta a \in \text{ann}(a^{e-1}m)$ which implies that $a^{e-1} \in \text{ann}(\delta a\, m)$. Now use induction. For the last statement, observe that the nil radical of $\mathcal{R}$ is $\sqrt{(0)} = \sqrt{\text{ann}(1)}$. □

Proposition 9.3 *If* $\mathcal{M}$ *is AAD, then* $\mathfrak{Z}(\mathcal{M}) = 0$.

Proof. If $m \in \mathfrak{Z}(\mathcal{M})$, then $1 \in \{\text{ann}(m)\} = \sqrt{\text{ann}(m)}$, hence $1 \in \text{ann}(m)$. □

Proposition 9.4 *Suppose that* $\mathcal{M}$ *is AAD and* $\mathcal{N}$ *is a* Δ-*submodule of* $\mathcal{M}$, *then* $\mathcal{N}$ *is AAD. If* $S \subset \mathcal{R}$ *is a multiplicative set, then* $\mathcal{M}S^{-1}$ *is AAD.*

Proof. The first statement is easy. If $a/b \in \text{ann}(m/d)$, then there exists $s \in S$ with $a \in \text{ann}(sm)$. Hence $\delta a \in \text{ann}(sm)$ and $\delta(a/b) \in \text{ann}(m/d)$. $\qquad\square$

Proposition 9.5 *Let $\mathcal{N}$ be a Δ-submodule of $\mathcal{M}$. The following are equivalent.*

(a) *The quotient module $\mathcal{M}/\mathcal{N}$ is AAD.*

(b) *If $a \in \mathcal{R}$, $m \in \mathcal{M}$ are such that $am \in \mathcal{N}$, then $\delta a\, m \in \mathcal{N}$ for every $\delta \in \Delta$.*

Proof. Let $\pi : \mathcal{M} \to \mathcal{M}/\mathcal{N}$ be the quotient homomorphism. The proposition follows easily from the observation that $am \in \mathcal{N}$ if and only if $a\,\pi m = 0$, i.e. $a \in \text{ann}(\pi m)$. $\qquad\square$

Definition 9.6 Denote by $\mathfrak{A}(\mathcal{M})$ the intersection of all the Δ-submodules $\mathcal{N}$ having the property that $\mathcal{M}/\mathcal{N}$ is AAD.

Theorem 9.7 *For any Δ-$\mathcal{R}$-module $\mathcal{M}$, $\mathcal{M}/\mathfrak{A}(\mathcal{M})$ is AAD and $\mathfrak{A}(\mathcal{M})$ is the smallest Δ-submodule of $\mathcal{M}$ having that property.*

Proof. Condition (b) of the previous proposition shows that the set of submodules $\mathcal{N}$ such that $\mathcal{M}/\mathcal{N}$ is AAD is closed under arbitrary intersections. The result follows. $\qquad\square$

Corollary 9.8 *Let $\phi : \mathcal{M} \to \mathcal{N}$ be a Δ-homomorphism of Δ-$\mathcal{R}$-modules. Then $\phi(\mathfrak{A}(\mathcal{M})) \subset \mathfrak{A}(\mathcal{N})$.*

Proof. By Theorem 9.7, $\mathcal{N}/\mathfrak{A}(\mathcal{N})$ is AAD, so the kernel of $\mathcal{M} \to \mathcal{N} \to \mathcal{N}/\mathfrak{A}(\mathcal{N})$ must contain $\mathfrak{A}(\mathcal{M})$. $\qquad\square$

Proposition 9.9 *Every element of $\mathfrak{A}(\mathcal{R})$ is nilpotent. In particular a reduced Δ-ring is AAD.*

Proof. We prove the second statement first. Suppose that $\mathcal{R}$ is reduced. If $a \in \text{ann}(r)$, then $ar = 0$ and for every $\delta \in \Delta$, $0 = r\delta(ar) = \delta a\, r^2$, so $\delta a\, r = 0$.

For the first statement observe that $\mathcal{R}/\mathfrak{n}$ is reduced, where $\mathfrak{n}$ is the nil radical. Therefore, by the previous corollary, $\mathfrak{A}(\mathcal{R}) \subset \mathfrak{n}$. $\qquad\square$

Thus the category of AAD rings includes all the reduced rings – the rings used in "classical" differential algebraic geometry.

Example 9.10 Consider the ordinary Δ-ring $\mathcal{R} = \mathbb{Q}\{\eta\} = \mathbb{Q}\{y\}/\mathfrak{a}$, where y is a Δ-indeterminate over $\mathbb{Q}$, and $\mathfrak{a}$ is the Δ-ideal generated by $y^{(i)}y^{(j)}$ for every $i, j \in \mathbb{N}$. $\mathcal{R}$ is not reduced ($\eta^2 = 0$), however it is AAD. Any element of $\mathcal{R}$ can be written (uniquely) in the form

$$a = a_{-1} + \sum_{i \geq 0} a_i \eta^{(i)}.$$

Let $b \in \mathcal{R}$, $b \neq 0$. If $b_{-1} \neq 0$, then $a \in \text{ann}(b)$ if and only if

$$0 = ab = a_{-1}b_{-1} + a_{-1}\sum_i b_i\eta^{(i)} + b_{-1}\sum_i a_i\eta^{(i)},$$

which happens if and only if $a_{-1} = 0$ and $\sum a_i\eta^{(i)} = 0$, that is, if and only if $a = 0$. On the other hand if $b_{-1} = 0$, then $a \in \text{ann}(b)$ if and only if

$$0 = ab = a_{-1}\sum_i b_i\eta^{(i)},$$

that is, if and only if $a_{-1} = 0$, and again $\text{ann}(b)$ is a Δ-ideal.

10 Global sections of AAD rings

Proposition 10.1 *If $\mathcal{M}$ is AAD, then the canonical mapping $\mathcal{M} \to \widehat{\mathcal{M}}$ is injective. More generally, if $b \in \mathcal{R}$, then the canonical mapping $\mathcal{M}_b \to \widetilde{\mathcal{M}}(D(b))$ is injective.*

Proof. By Propositions 9.3 and 9.4, $3(\mathcal{M}_b) = (0)$. The result follows from Proposition 7.2. □

Proposition 10.2 *The mapping $\phi : \mathcal{R} \to \mathcal{R}/\mathfrak{A}(\mathcal{R})$ induces a homeomorphism from diffspec $(\mathcal{R}/\mathfrak{A}(\mathcal{R}))$ onto diffspec $\mathcal{R}$.*

Proof. By Corollary 3.10, $^a\phi$ is a homeomorphism of diffspec $\mathcal{R}/\mathfrak{A}(\mathcal{R})$ onto $V(\mathfrak{A}(\mathcal{R}))$. Since $\mathfrak{A}(\mathcal{R})$ is contained in the nilradical of $\mathcal{R}$ (Proposition 9.9), $V(\mathfrak{A}(\mathcal{R})) = X$. □

However, this mapping does not induce an isomorphism of schemes. But then neither does $\mathcal{R} \to \mathcal{R}/\mathfrak{n}$, where $\mathfrak{n}$ is the nil radical of $\mathcal{R}$.

We have not discussed surjectivity of the canonical mapping $\iota_{\mathcal{M}} : \mathcal{M} \to \widehat{\mathcal{M}}$ because it is not needed; the following theorem takes its place.

Theorem 10.3 *Suppose that $\mathcal{M}$ is AAD. Let $\iota_{\mathcal{R}} : \mathcal{R} \to \widehat{\mathcal{R}}$ and $\iota_{\mathcal{M}} : \mathcal{M} \to \widehat{\mathcal{M}}$ be the canonical homomorphisms and let $s \in \widehat{\mathcal{M}}$. Then, for some $n \in \mathbb{N}$, there exist $m_1, \ldots, m_n \in \mathcal{M}$ and $b_1, \ldots b_n \in \mathcal{R}$ such that $1 \in \{b_1, \ldots, b_n\}$ and $\iota_{\mathcal{R}}(b_i)s = \iota_{\mathcal{M}}(m_i)$ for each $i = 1, \ldots, n$.*

Proof. By Proposition 4.5, there exist $m_1, \ldots, m_n \in \mathcal{M}$, $b_1, \ldots, b_n \in \mathcal{R}$, such that $s(\mathfrak{p}) = m_i/b_i \in \mathcal{M}_\mathfrak{p}$ whenever $\mathfrak{p} \in D(b_i)$.

For any $\mathfrak{q} \in D(b_i) \cap D(b_j) = D(b_ib_j)$, we have $s(\mathfrak{q}) = m_i/b_i = m_j/b_j \in \mathcal{R}_\mathfrak{q}$. Therefore $b_im_j - b_jm_i$ is in the kernel of $\mathcal{M}_{b_ib_j} \to \widetilde{\mathcal{M}}(D(b_ib_j))$, which, by

Proposition 10.1, is (0). Therefore

$$(b_i b_j)^e (b_i m_j - b_j m_i) = 0 \in \mathcal{M}$$

for some $e \in \mathbb{N}$. We may assume that e is independent of i, j. Replace b_i by b_i^{e+1} and m_i by $b_i^e m_i$ to obtain

$$b_i m_j = b_j m_i, \ 1 \le i, j \le n.$$

For any $\mathfrak{p} \in X$, choose j so that $b_j \notin \mathfrak{p}$, then

$$\iota_{\mathcal{R}}(b_i) s(\mathfrak{p}) = \frac{b_i}{1} \frac{m_j}{b_j} = \frac{b_i m_j}{b_j} = \frac{b_j m_i}{b_j} = \frac{m_i}{1} = \iota_{\mathcal{M}}(m_i). \qquad \square$$

Beware, this result does *not* say that $s(\mathfrak{p}) = m_i / b_i \in \mathcal{M}_{\mathfrak{p}}$ for every $\mathfrak{p}$, for it may happen that $b_i \in \mathfrak{p}$. In the case of spec, as opposed to diffspec, it is well-known that we may always take $n = 1$. That is not the case here; see Kovacic [20] for an example.

The canonical mapping $\mathcal{R} \to \widehat{\mathcal{R}}$ induces a morphism $\widehat{X} = \mathrm{diffspec}\,\widehat{\mathcal{R}} \to X$ by Proposition 5.4. The next two propositions culminate in a theorem showing that it is an isomorphism if $\mathcal{R}$ is AAD.

Proposition 10.4 *Suppose that $\mathcal{R}$ is AAD. Define $\phi : X \to \widehat{X}$ by*

$$\phi(\mathfrak{p}) = \widehat{\mathfrak{p}} = \{ s \in \widehat{\mathcal{R}} \mid s(\mathfrak{p}) \in \mathfrak{p} \mathcal{R}_{\mathfrak{p}} \}.$$

Then ϕ is continuous.

Proof. Let $\mathfrak{a} \subset \widehat{\mathcal{R}}$ be a radical Δ-ideal. If $\mathfrak{p} \in \phi^{-1}(V(\mathfrak{a}))$ then $\phi\mathfrak{p} = \widehat{\mathfrak{p}} \supset \mathfrak{a}$ so, by Proposition 8.1,

$$\mathfrak{p} = \iota_{\mathcal{R}}^{-1} \widehat{\mathfrak{p}} \supset \iota_{\mathcal{R}}^{-1} \mathfrak{a}.$$

Therefore $\phi^{-1} V(\mathfrak{a}) \subset V(\iota_{\mathcal{R}}^{-1} \mathfrak{a})$.

Now let $\mathfrak{p} \in V(\iota_{\mathcal{R}}^{-1} \mathfrak{a})$; we need to show that $\mathfrak{a} \subset \widehat{\mathfrak{p}}$. Let $s \in \mathfrak{a}$. By Theorem 10.3, there exist $a, b \in \mathcal{R}$, $b \notin \mathfrak{p}$, such that $\iota_{\mathcal{R}} b s = \iota_{\mathcal{R}} a$. This implies that $\iota_{\mathcal{R}} a \in \mathfrak{a}$, so $a \in \iota_{\mathcal{R}}^{-1} \mathfrak{a} \subset \mathfrak{p}$, and therefore $s \in \widehat{\mathfrak{p}}$. $\qquad \square$

Proposition 10.5 *Suppose that $\mathcal{R}$ is AAD. Then $^a\iota_{\mathcal{R}} : \widehat{X} \to X$ is a homeomorphism.*

Proof. By Proposition 3.8, $^a\iota_{\mathcal{R}}$ is continuous. By the previous proposition $\phi : X \to \widehat{X}$ is continuous. We also know that $^a\iota_{\mathcal{R}} \circ \phi : X \to X$ is the identity. We claim that $\phi \circ {}^a\iota_{\mathcal{R}} : \widehat{X} \to \widehat{X}$ is also the identity.

Let $\mathfrak{q} \in \widehat{X}$ and set $\mathfrak{p} = {}^a\iota_{\mathcal{R}}(\mathfrak{q}) = \iota_{\mathcal{R}}^{-1}(\mathfrak{q})$. If $s \in \phi(^a\iota_{\mathcal{R}}(\mathfrak{q}))$, then $s(\mathfrak{p}) \in \mathfrak{p}\mathcal{R}_{\mathfrak{p}}$, so, by Proposition 10.3, we may choose $a, b \in \mathcal{R}$, $b \notin \mathfrak{p}$, with $\iota_{\mathcal{R}}(b) s = \iota_{\mathcal{R}}(a)$.

It follows that $\iota_{\mathcal{R}}(a) \in \mathfrak{p}\mathcal{R}_{\mathfrak{p}}$, $a \in \mathfrak{p}$ and $\iota_{\mathcal{R}}(a) \in \mathfrak{q}$. Since $b \notin \mathfrak{p}$, $\iota_{\mathcal{R}}(b) \notin \mathfrak{q}$, so $s \in \mathfrak{q}$. Hence $\phi \circ {}^a\iota_{\mathcal{R}}(\mathfrak{q}) \subset \mathfrak{q}$. For the converse, simply note that for $s \in \mathfrak{q}$, $\iota_{\mathcal{R}}(b)s = \iota_{\mathcal{R}}(a)$ implies that $\iota_{\mathcal{R}}(a) \in \mathfrak{q}$ and $a \in \mathfrak{p}$. $\qquad\square$

Theorem 10.6 *Suppose that $\mathcal{R}$ is AAD. Then the canonical mapping $\iota_{\mathcal{R}}$: $\mathcal{R} \to \widehat{\mathcal{R}}$ induces an isomorphism $\widehat{X} \to X$.*

Proof. By the previous proposition ${}^a\iota_{\mathcal{R}} : \widehat{X} \to X$ is a homeomorphism. We must show that the morphism on stalks, $\iota_{\mathcal{R},\mathfrak{q}}^{\#} : \mathcal{O}_{X,\mathfrak{p}} \to \mathcal{O}_{\widehat{X},\mathfrak{q}}$, where $\mathfrak{q} \in \widehat{X}$ and $\mathfrak{p} = {}^a\iota_{\mathcal{R}}(\mathfrak{q}) = \iota_{\mathcal{R}}^{-1}(\mathfrak{q})$, is an isomorphism. Recall that $\iota_{\mathcal{R},\mathfrak{q}}^{\#}$ is induced by the mapping $i : \mathcal{R}_{\mathfrak{p}} \to \widehat{\mathcal{R}}_{\mathfrak{q}}$, with $i(a/b) = \iota_{\mathcal{R}}(a)/\iota_{\mathcal{R}}(b)$.

If $i(a/b) = 0$, then there is $t \in \widehat{\mathcal{R}}$, $t \notin \mathfrak{q}$, with $t\,\iota_{\mathcal{R}}(a) = 0 \in \widehat{\mathcal{R}}$, in which case $t(\mathfrak{p})a = 0 \in \mathcal{R}_{\mathfrak{p}}$. Writing $t(\mathfrak{p}) = c/d$, we see that $c \notin \mathfrak{p}$ (since $t \notin \mathfrak{q}$), so $a = 0 \in \mathcal{R}_{\mathfrak{p}}$.

Now let $s/t \in \widehat{\mathcal{R}}_{\mathfrak{q}}$. Choose $a, b, c, d \in \mathcal{R}$, $b, d \notin \mathfrak{p}$, with $\iota_{\mathcal{R}}(b)s = \iota_{\mathcal{R}}(a)$ and $\iota_{\mathcal{R}}(d)t = \iota_{\mathcal{R}}(c)$. Observe that $c \notin \mathfrak{p}$ since $t \notin \mathfrak{q}$. Then

$$i\left(\frac{ac}{bd}\right) = \frac{\iota_{\mathcal{R}}(a)\,\iota_{\mathcal{R}}(c)}{\iota_{\mathcal{R}}(b)\,\iota_{\mathcal{R}}(d)} = \frac{\iota_{\mathcal{R}}(b)s\,\iota_{\mathcal{R}}(d)t}{\iota_{\mathcal{R}}(b)\,\iota_{\mathcal{R}}(d)} = st. \qquad\square$$

Proposition 5.4 states that a Δ-homomorphism $\mathcal{R} \to \mathcal{S}$ induces a morphism $Y = \operatorname{diffspec} \mathcal{S} \to X$. We remarked that the converse is false. Indeed, for an AAD Δ-ring, diffspec $\mathcal{R}$ is isomorphic to diffspec $\widehat{\mathcal{R}}$, however $\mathcal{R}$ and $\widehat{\mathcal{R}}$ are not necessarily isomorphic.

Theorem 10.7 *Let $\mathcal{R}$ and $\mathcal{S}$ be AAD Δ-rings. Let $X = \operatorname{diffspec} \mathcal{R}$ and $Y = \operatorname{diffspec} \mathcal{S}$. A Δ-homomorphism $\phi : \widehat{\mathcal{R}} \to \widehat{\mathcal{S}}$ induces a morphism*

$$({}^a\phi, \phi^{\#}) : (Y, \mathcal{O}_Y) \to (X, \mathcal{O}_X)$$

of LDR spaces. Conversely if $(f, f^{\#}) : (Y, \mathcal{O}_Y) \to (X, \mathcal{O}_X)$ is a morphism of LDR spaces, then there is a Δ-homomorphism $\phi : \widehat{\mathcal{R}} \to \widehat{\mathcal{S}}$ inducing $(f, f^{\#})$.

Proof. Immediate from Proposition 8.4 and the previous theorem. $\qquad\square$

Suppose that $(Y, \mathcal{O}_Y)$ is a Δ-scheme. Denote the set of morphisms of Δ-schemes $(X, \mathcal{O}_X) \to (Y, \mathcal{O}_Y)$ by $\operatorname{hom}_{\operatorname{sch}}(X, Y)$. A morphism $(f, f^{\#})$ induces a Δ-homomorphism $f^{\#}(Y) : \mathcal{O}_Y(Y) \to \mathcal{O}_X(X) = \widehat{\mathcal{R}}$, and therefore we have a mapping $\operatorname{hom}_{\operatorname{sch}}(X, Y) \to \operatorname{hom}(\Gamma(Y, \mathcal{O}_Y), \widehat{\mathcal{R}})$.

Theorem 10.8 *Suppose that $\mathcal{R}$ is AAD. Let Y be any Δ-scheme. Then the mapping*

$$\operatorname{hom}_{\operatorname{sch}}(X, Y) \to \operatorname{hom}(\Gamma(Y, \mathcal{O}_Y), \widehat{\mathcal{R}}),$$

defined above, is bijective.

Proof. Just as in Grothendieck and Dieudonné [10, Proposition (1.6.3), p.210], the mapping $\mathrm{hom}_{\mathrm{sch}}(\widehat{X}, Y) \to \mathrm{hom}(\Gamma(Y, \mathcal{O}_Y), \widehat{\mathcal{R}})$ is bijective. By the previous theorem, the canonical mapping $\widehat{X} \to X$ is an isomorphism. □

11 AAD schemes

Definition 11.1 A Δ-scheme $(Y, \mathcal{O}_Y)$ is said to be AAD if $\mathcal{O}_Y(U)$ is AAD for every open set $U \subset Y$. A sheaf $\mathcal{F}$ of Δ-modules on Y is said to be AAD if $\mathcal{F}(U)$ is AAD for every U.

Proposition 11.2 *Let $(Y, \mathcal{O}_Y)$ be a Δ-scheme, and let $\mathcal{F}$ be a sheaf of Δ-modules on Y. Then $\mathcal{F}$ is AAD if and only if every stalk $\mathcal{F}_y$, $y \in Y$, is AAD.*

Proof. Suppose that $\mathcal{F}$ is AAD. Let $y \in Y$, $m_y \in \mathcal{F}_y$ and $a_y \in \mathrm{ann}(m_y) \subset \mathcal{O}_{Y,y}$. Choose a neighborhood U of y and representatives $m \in \mathcal{F}(U)$, $a \in \mathrm{ann}(b) \subset \mathcal{O}_Y(U)$. Then, for $\delta \in \Delta$, $\delta a \in \mathrm{ann}(m)$, so $\delta a_y \in \mathrm{ann}(m_y)$.

For the converse, let U be any open set, $m \in \mathcal{F}(U)$ and $a \in \mathrm{ann}(m)\mathcal{O}_Y(U)$. Then, for $\delta \in \Delta$, and any $y \in U$, $(am)_y = 0 \in \mathcal{F}_y$ implies that $(\delta am)_y = 0$. Therefore $\delta a\,m = 0 \in \mathcal{F}(U)$. □

Proposition 11.3 *If $\mathcal{M}$ is AAD, then so is $\widetilde{\mathcal{M}}$.*

Proof. This follows immediately from Proposition 9.4. □

Proposition 11.4 *If $\mathcal{R}$ is AAD, then so is X. Conversely, if X is AAD, then there is an AAD Δ-ring S with diffspec S isomorphic to X.*

Proof. The first statement is contained in the previous proposition. Suppose that X is AAD, then $\widehat{\mathcal{R}} = \mathcal{O}_X(X)$ is an AAD Δ-ring. By Proposition 7.2, the kernel of the canonical mapping $\mathcal{R} \to \widehat{\mathcal{R}}$ is $3(\mathcal{R})$. By Corollary 9.3 and Proposition 9.7, $3(\mathcal{R}) = \mathfrak{A}(\mathcal{R})$. Then Proposition 7.5 states that X is isomorphic to diffspec $(\mathcal{R}/3(\mathcal{R}))$. □

12 AAD reduction

Let (X, O_X) be a Δ-scheme and $\mathcal{M}$ a sheaf of Δ-modules on X. For each open subset $U \subset X$, we may form the AAD module $\mathcal{M}(U)/\mathfrak{A}(\mathcal{M}(U))$. If $V \subset U$ and $\rho : \mathcal{M}(U) \to \mathcal{M}(V)$ is the restriction, we can define, using Corollary 9.8, an induced mapping $\mathcal{M}(U)/\mathfrak{A}(\mathcal{M}(U)) \to \mathcal{M}(V)/\mathfrak{A}(\mathcal{M}(V))$. This makes $U \mapsto \mathcal{M}(U)/\mathfrak{A}(\mathcal{M}(U))$ into a presheaf.

Definition 12.1 Let $(X, \mathcal{O}_X)$ be a Δ-scheme and $\mathcal{M}$ a sheaf of Δ-modules on X. Denote by $\mathcal{M}^{aad}$ the sheaf associated with the above presheaf. The sheaf associated to $U \mapsto \mathcal{O}_X(U)/\mathfrak{A}(\mathcal{O}_X(U))$ is denoted by $\mathcal{O}_X^{aad}$.

Proposition 12.2 Let $(X, \mathcal{O}_X)$ be a Δ-scheme. Then $(X, \mathcal{O}_X^{aad})$ is an AAD Δ-scheme. There is a morphism $(X, \mathcal{O}_X^{aad}) \to (X, \mathcal{O}_X)$ of Δ-schemes which is the identity on the underlying spaces.

Proof. Let U be an open affine subset of X, say $U = \operatorname{diffspec} \mathcal{R}$. The homomorphism $\mathcal{R} \to \mathcal{R}/\mathfrak{A}(\mathcal{R})$ induces, by Proposition 10.2, a homeomorphism on U, and therefore a mapping $(U, \mathcal{O}_X^{aad}|U) \to (U, \mathcal{O}_X|U)$ that is the identity on U. □

Proposition 12.3 Let $(X, \mathcal{O}_X)$ be an AAD Δ-scheme and $(Y, \mathcal{O}_Y)$ be any Δ-scheme. Suppose that $f : (X, \mathcal{O}_X) \to (Y, \mathcal{O}_Y)$ is a morphism of Δ-schemes. Then there is a unique morphism $g : (X, \mathcal{O}_X) \to (Y, \mathcal{O}_Y^{aad})$ of Δ-schemes such that

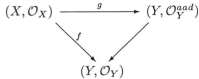

commutes, where the unlabelled arrow is as in the previous proposition.

Proof. Evidently the mapping on the underlying spaces, $g : X \to Y$, is f. For every open subset $U \subset Y$, $f^\#(U) : \mathcal{O}_Y(U) \to \mathcal{O}_X(f^{-1}(U))$. But the image is AAD so $f^\#(U)$ factors through $\mathcal{O}_Y(U)/\mathfrak{A}(\mathcal{O}_Y(U))$:

$$\mathcal{O}_Y(U) \longrightarrow \mathcal{O}_Y(U)/\mathfrak{A}(\mathcal{O}_Y(U))$$
$$\mathcal{O}_X(f^{-1}(U)).$$

□

13 Based schemes

Definition 13.1 Let B be a Δ-scheme. By a Δ-scheme *over* B we mean a Δ-scheme X together with a morphism of Δ-schemes $X \to B$. The morphism is called the *structure morphism* of X over B. B is called the *base Δ-scheme*.

Let $\mathcal{R}$ and $\mathcal{B}$ be Δ-rings and set $X = \operatorname{diffspec} \mathcal{R}$ and $B = \operatorname{diffspec} \mathcal{B}$. If $\mathcal{R}$ is a Δ-$\mathcal{B}$-algebra, then X is a Δ-scheme over B. The morphism $X \to B$ is induced by $\mathcal{B} \to \mathcal{R}$, $b \mapsto b \cdot 1$ (Proposition 3.8). The converse is false. In fact we know by Theorem 10.6 that X is isomorphic to $\widehat{X} = \text{diffspec}\,\widehat{\mathcal{R}}$, so each is a Δ-scheme over the other. $\widehat{\mathcal{R}}$ is indeed an $\mathcal{R}$-algebra but, in general, $\mathcal{R}$ is not a $\widehat{\mathcal{R}}$-algebra.

Proposition 13.2 *Let $\mathcal{R}$ and $\mathcal{B}$ be AAD Δ-rings. Let $X = \text{diffspec}\,\mathcal{R}$ and $B = \text{diffspec}\,\mathcal{B}$. Then X is a Δ-scheme over B if and only if $\widehat{\mathcal{R}}$ is a Δ-$\widehat{\mathcal{B}}$-algebra.*

Proof. This is Theorem 10.7. □

Definiton 13.3 Let X and Y be Δ-schemes over B. By a *morphism over B* we mean a morphism f of Δ-schemes such that the diagram

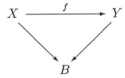

commutes, where the diagonal arrows are the structure morphisms.

14 Products

In the category of schemes (not Δ-schemes), products exist. In fact, for affine schemes $X = \text{spec}\,R$, $Y = \text{spec}\,S$ over a base $\text{spec}\,B$, the product is given by $\text{spec}\,(R \otimes_B S)$. The proof relies on the fact that morphisms between affine schemes are induced by ring homomorphisms. We have such a correspondence for AAD Δ-rings, and an analogue of the radical: $\mathfrak{A}(\mathcal{R})$ (Definition 9.6).

Definition 14.1 Let X and Y be AAD Δ-schemes over an AAD Δ-scheme B. By the *product $X \times_B Y$*, we mean an AAD Δ-scheme over B together with morphisms of Δ-schemes $\rho_1 : X \times_B Y \to X$ and $\rho_2 : X \times_B Y \to Y$ over B which satisfies the following universal property: Given any AAD Δ-scheme Z over B and morphisms of Δ-schemes $f : Z \to X$, $g : Z \to Y$ over B there is a unique morphism of Δ-schemes $h : Z \to X \times_B Y$ over B such that

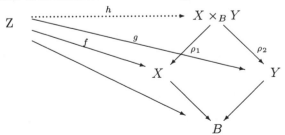

commutes, where the unlabelled arrows are the structure morphisms.

If the product exists, it is unique up to a unique isomorphism.

Theorem 14.2 *Let $\mathcal{R}$, $\mathcal{S}$ and $\mathcal{B}$ be AAD Δ-rings. Let $X = \text{diffspec}\,\mathcal{R}$, $Y = \text{diffspec}\,\mathcal{S}$ and $B = \text{diffspec}\,\mathcal{B}$. Suppose that X and Y are Δ-schemes over B. Then the product $X \times_B Y$ exists.*

Proof. Let $\mathcal{T} = \widehat{\mathcal{R}} \otimes_{\widehat{\mathcal{B}}} \widehat{\mathcal{S}}$ and $T = \text{diffspec}\,(\mathcal{T}/\mathfrak{A}(\mathcal{T}))$. Then T is an AAD Δ-scheme over B. We claim that $X \times_B Y = T$. Define

$$\rho_1 : \widehat{\mathcal{R}} \xrightarrow{\ i_1\ } \mathcal{T} \xrightarrow{\ \pi\ } \mathcal{T}/\mathfrak{A}(\mathcal{T})$$

and

$$\rho_2 : \widehat{\mathcal{S}} \xrightarrow{\ i_2\ } \mathcal{T} \xrightarrow{\ \pi\ } \mathcal{T}/\mathfrak{A}(\mathcal{T}),$$

where $i_1(r) = r \otimes 1$, $i_2(s) = 1 \otimes s$ and π is the quotient homomorphism. We have morphisms ${}^a\rho_1 : T \to X$ and ${}^a\rho_2 : T \to Y$, which are easily seen to be over B.

Let Z be any AAD Δ-scheme over B, with morphisms $Z \to X$ and $Z \to Y$ over B. By Proposition 10.8, these morphisms are induced by Δ-homomorphisms of $\widehat{\mathcal{B}}$-algebras

$$\phi : \widehat{\mathcal{R}} \to \Gamma(Z, \mathcal{O}_Z)$$

and

$$\psi : \widehat{\mathcal{S}} \to \Gamma(Z, \mathcal{O}_Z).$$

There is a unique Δ-homomorphism $\chi : \mathcal{T} \to \Gamma(Z, \mathcal{O}_Z)$ such that

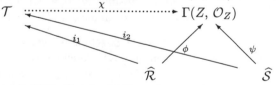

commutes. Because $\Gamma(Z, \mathcal{O}_Z)$ is AAD, χ factors, by Proposition 12.3,

$$\mathcal{T} \xrightarrow{\ \chi\ } \Gamma(Z, \mathcal{O}_Z)$$

$$\mathcal{T}/\mathfrak{A}(\mathcal{T}),$$

and we obtain the desired morphism $Z \to T$ by Proposition 10.8. $\qquad\square$

Theorem 14.3 *Let X and Y be AAD Δ-schemes over an AAD Δ-scheme B. Then the product $X \times_B Y$ exists.*

Proof. This is done by "patching", exactly as in Hartshorne [11, Theorem 3.3, p. 87]. $\qquad\square$

References

1. Bourbaki, N. *Elements of Mathematics, Commutative Algebra, Chapters 1-7,* Springer-Verlag, Berlin, 1989.
2. Buium, A. *Ritt schemes and torsion theory,* Pacific J. Math. **92** (1982), 281–293.
3. Buium, A. *Differential Function Fields and Moduli of Algebraic Varieties,* Lect. Notes in Math. **1226**, Springer-Verlag, Berlin-New York, 1986.
4. Buium, A., Cassidy, P. *Differential algebraic geometry and differential algebraic groups: from algebraic differential equations to diophantine geometry* in *Selected Works of Ellis Kolchin with Commentary,* H. Bass, A. Buium, P. Cassidy, eds., Amer. Math. Soc., Providence, RI, 1999, 567–636.
5. Carra' Ferro, G. *Sullo spettro differenziale di un anello differenziale,* Le Matematiche (Catania) **33** (1978), 1–17.
6. Carra' Ferro, G. *The ring of global sections of the structure sheaf on the differential spectrum,* Rev. Roumaine Math. Pures Appl. **30** (1985), 809–814.
7. Carra' Ferro, G. *Kolchin schemes,* J. Pure and Applied Algebra **63** (1990), 13–27.
8. Cassidy, P. *Differential algebraic groups,* Amer. J. Math. **94** (1972), 891–954.
9. H. Gorman, *Differential rings and modules,* Scripta Math. **29** (1973), 25–35.
10. Grothendieck, A., Dieudonné, J. *Eléments de Géométrie Algé-brique,* Springer-Verlag, Berlin, 1971.
11. Hartshorne, R. *Algebraic Geometry,* Springer-Verlag, New York, 1977.
12. Kaplansky, I. *An Introduction to Differential Algebra,* Hermann, Paris, 1957.
13. Keigher, W. *Adjunctions and comonads in differential algebra,* Pacific J. Math. **59** (1975), 99–112.
14. Keigher, W. *Prime differential ideals in differential rings, Contributions to Algebra: A Collection of Papers dedicated to Ellis Kolchin,* Bass, Cassidy, Kovacic, eds., Academic Press, New York, 1977, 239–249.
15. Keigher, W. *On the structure presheaf of a differential ring,* J. Pure and Applied Algebra **27** (1983), 163–172.
16. Keigher, W. *On the quasi-affine scheme of a differential ring,* Adv. Math. **42** (2) (1981), 143–153.
17. Keigher, W. *Differential schemes and premodels of differential fields,* J. of Algebra **79** (1982), 37–50.

94

18. Kolchin, E. *Differential Algebra and Algebraic Groups*, Academic Press, New York, 1973.
19. Kolchin, E. *Differential Algebraic Groups*, Academic Press, New York, 1985.
20. Kovacic, J. J. *Global sections of diffspec*, to appear in J. Pure and Applied Algebra.
21. Okugawa, K. *Differential Algebra of Nonzero Characteristic*, Lectures in Mathematics **16**, Kinokuniya Company Ltd., Tokyo, 1987.

Differential Algebra and Related Topics, pp. 95–123
Proceedings of the International Workshop
Eds. L. Guo, P. J. Cassidy, W. F. Keigher & W. Y. Sit
© 2002 World Scientific Publishing Company

DIFFERENTIAL ALGEBRA
A SCHEME THEORY APPROACH

HENRI GILLET

*Department of Mathematics, Statistics, and Computer Science (m/c 249),
University of Illinois at Chicago, 851 S. Morgan Street,
Chicago, IL 60607, USA
E-mail: henri@math.uic.edu*

Two results in Differential Algebra, Kolchin's Irreducibility Theorem, and a result
on descent of projective varieties (due to Buium) are proved using methods of
"modern" or "Grothendieck style" algebraic geometry.

Introduction

The goal of this paper is to approach some results in differential algebra from
the perspective, and using the results of, modern algebraic geometry and
commutative algebra. In particular we shall see new proofs of two results:
Kolchin's Irreducibility Theorem, and Buium's result describing the minimal
field of definition of a projective variety over an algebraically closed field of
characteristic zero.

The first section of the paper describes the construction of prolongations
(which associate to an algebraic variety X over a differential field, the ring
of differential polynomial functions on X) from the point of view of adjoint
functors. This allows us to give simple proofs of several properties of the
"prolongation" operation, especially how the functor behaves with respect
to formally smooth and formally étale morphisms. In particular we observe
that the prolongation functor is a quasi-coherent sheaf of rings in the étale
topology.

In Section 2, we discuss the proof of Kolchin's theorem:

Theorem (Kolchin's irreducibility theorem [14, Chapter IV, Proposition 10])
*Let A be an integral domain, of finite type over a differential field k; then the
associated differential variety is irreducible.*

Note that Proposition 10 of Kolchin also contains a computation of the
differential dimension polynomial of the associated differential variety. Also
here we only consider rings with a single derivation. The key point in the proof
given here is that discrete valuation rings containing a field of characteristic
zero are formally smooth over that field, and hence have rings of differential
polynomial functions which are integral domains. It is interesting to notice

that for a more general valuation ring it is still true that the associated ring of differential polynomial functions is an integral domain; however the only proof that I know of this result uses Zariski's uniformization theorem.

In Section 3, we turn to the result of Buium:

Theorem[4,3] *Let X be a variety, proper over an algebraically closed field K. Then X is defined over the fixed field of the set of all derivations of K which lift to derivations of the structure sheaf of X.*

The main tools in the proof given here are Grothendieck's theorem on algebraizability of morphisms between projective varieties over formal schemes, and Artin's approximation theorem.

Finally in Section 4, two questions which arise from the techniques used in the paper are posed.

The initial genesis for this paper was a seminar at UIC organized by David Marker, Lawrence Ein and myself, in which we and some graduate students read the book [2] of Buium. In addition to my talk at the Rutgers Newark workshop, I gave talks on this material at the conference on model theory, algebraic and arithmetic at MSRI in 1998, and Columbia University and CCNY in 1999.

I would like to thank David Marker, Lawrence Ein, and Phyllis Cassidy for discussions about various aspects of this work, and I would like to thank the referees for a very careful job.

1 Differential Rings

1.1 Some Commutative Algebra

In this section we shall review some basic facts about commutative rings and derivations.

All rings are commutative with unit. If k is a ring, then a k-algebra A is simply a ring A together with a homomorphism $k \to A$.

Étale homomorphisms. See [19] for details on this section.

Definition 1.1 Recall that a ring homomorphism $R \to S$ is *formally étale* if, given a ring C and a square zero ideal $I \lhd C$, together with a commutative diagram of ring homomorphisms:

$$
\begin{array}{ccc}
S & \longrightarrow & C/I \\
\uparrow & & \uparrow \\
R & \longrightarrow & C
\end{array}
$$

there is a *unique* homomorphism $S \to C$ making the diagram commute. If instead there exists *at least one* such homomorphism, we say that $R \to S$ is *formally smooth*, while if there exists *at most one* such homomorphism, we say that $R \to S$ is *formally unramified*. If in addition to satisfying one of the above conditions, S is an R-algebra of finite type, we remove the adjective "formally" and say that the morphism is *étale, smooth,* or *unramified* as appropriate.

The following exercises are straightforward.

Exercise 1.2 The composition of two formally étale (resp. unramified, resp. smooth) maps is again formally étale, (resp. unramified, resp. smooth).

Exercise 1.3 If $S \subset R$ is a multiplicative subset of R, and $S^{-1}R$ is the localization of R with respect to S, then the localization map $R \to S^{-1}R$ is formally étale, and is étale if S is finitely generated. This follows immediately from the fact that an element $x \in C$ is a unit if and only if it is a unit modulo I, since $I \lhd C$ is a nilpotent ideal.

Exercise 1.4 If $S = R[t]/(f(t))$ with f monic, and (the image of) $f'(t)$ is a unit in S, then S/R is étale.

Exercise 1.5 If $R \to S$ is formally étale, then so is $A \otimes_R R \to A \otimes_R S$ for all R-algebras A.

Exercise 1.6 If $(R_n, \theta_n : R_n \to R_{n+1})$ for $n \geq 1$ is a direct system of rings, with R_{n+1} formally smooth over R_n for all n, then the direct limit $\varinjlim_n R_n$ is formally smooth over R.

Derivations. If R is a commutative ring, and M an R-module, then recall that a derivation $\delta : R \to M$ is an additive map $\delta : R \to M$, such that $\delta(ab) = a\delta(b) + b\delta(a)$.

To give a derivation $\delta : R \to M$ is equivalent to giving a homomorphism of rings,

$$\delta_* : R \to R \oplus M\varepsilon,$$
$$r \longmapsto r + \delta(r)\varepsilon$$

where $R \oplus M\varepsilon$, with $\varepsilon^2 = 0$, is the ring of dual numbers over R with coefficients in M, *i.e.* as an abelian group $R \oplus M\varepsilon$ is just the direct sum $R \oplus M$, and the multiplication law is defined by $(r + m\varepsilon).(r' + m'\varepsilon) = (rr' + (rm' + r'm)\varepsilon)$. Note that we require that the composition of the augmentation

$$R \oplus M\varepsilon \to R,$$
$$r + m\varepsilon \longmapsto r$$

with δ_* is the identity. The set $\mathcal{D}er(R, M)$, of all derivations $\delta : R \to M$ is an R-module, indeed a sub-module of the module of all functions from R (viewed as a set) to M (viewed as an R-module), with addition and scalar multiplication defined pointwise. If R is a k-algebra, i.e., there is a homomorphism $\phi : k \to R$, we write $\mathcal{D}er_k(R, M)$ for the submodule of $\mathcal{D}er(R, M)$ consisting of all δ for which $\delta(\phi(x)) = 0$ for all $x \in k$; note that $\mathcal{D}er(R, M) = \mathcal{D}er_{\mathbb{Z}}(R, M)$. Given a derivation, the set $\{r|\delta(r) = 0\}$ is clearly a subring of R (resp. a subfield if R is a field) which is called the *ring (resp. field) of constants* of δ. Observe that if $\delta \in \mathcal{D}er_k(R, M)$, then k is contained in the constants of δ, so that $\mathbb{Z}$ is always in the constants, and also that if R is a $\mathbb{Q}$-algebra, then the ring of constants is a $\mathbb{Q}$-algebra.

The covariant functor $M \mapsto \mathcal{D}er_k(R, M)$ from R-modules to R-modules is represented by the R-module $\Omega_{R/k}$ of Kähler differentials of R over k, [17, §25], i.e. $\mathcal{D}er_k(R, M) \simeq \mathrm{Hom}_R(\Omega_{R/k}, M)$. This isomorphism is induced by the universal derivation:

$$R \to \Omega_{R/k}$$
$$r \mapsto dr.$$

If R is a k-algebra, $f : R \to S$ is a homomorphism of k-algebras, and M is an S-module, there is an exact sequence of S-modules:

$$0 \to \mathcal{D}er_R(S, M) \to \mathcal{D}er_k(S, M) \to \mathcal{D}er_k(R, M) \qquad (1.7)$$

which is obtained by applying the functor $\mathrm{Hom}(-, M)$ to the natural exact sequence:

$$S \otimes_R \Omega_{R/k} \xrightarrow{v_{S/R/k}} \Omega_{S/k} \to \Omega_{S/R} \to 0.$$

Proposition 1.8 *If S is formally smooth over R, this sequence becomes split exact (i.e. the map $v_{S/R/k}$ has a left inverse), as does the sequence (1.7) for any S module M.*

Proof. This is a standard result, see [10, EGA 0, Théorème 20.5.7] for example. However, for completeness, we shall sketch the proof here. Suppose that S is formally smooth over R. If M is an S-module, a derivation $d : R \to M$ (where we view M as an R-module by restriction of scalars) is equivalent to a ring homomorphism $R \to R \oplus M\varepsilon$ (compatible with the augmentation from $R \oplus M\varepsilon \to R$). This induces a ring homomorphism from $R \to S \oplus M\varepsilon$. Thus we have a commutative square:

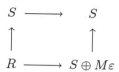

$$S \longrightarrow S$$
$$\uparrow \qquad \uparrow$$
$$R \longrightarrow S \oplus M\varepsilon$$

and hence, since $R \to S$ is formally smooth, there is a lifting of the identity on S to a map $S \to S \oplus M\varepsilon$ compatible with the map $R \to S \oplus M\varepsilon$. I.e., the derivation $d : R \to M$ has an extension to a derivation $S \to M$. Note that if S is formally étale over R, this extension is unique. Thus $\mathcal{D}er_k(S, M) \to \mathcal{D}er_k(R, M)$ is surjective.

Now consider the universal derivation $R \to \Omega_{R/k}$. This induces a derivation from R to the S-module $S \otimes \Omega_{R/k}$, which by the previous discussion extends to a derivation $S \to S \otimes \Omega_{R/k}$, and which, by the universal property of the Kähler differentials, is therefore induced by a homomorphism of S-modules $\Omega_{S/k} \to S \otimes \Omega_{R/k}$. It is easily checked that since this extends the universal derivation of R, that it is a left inverse of the map $v_{S/R/k}$, and that, by taking $\mathrm{Hom}(-, M)$ it induces the splitting of (1.7). $\qquad\square$

Proposition 1.9 *If $R \to S$ is a formally unramified homomorphism of k-algebras, and M is an S-module, then $\mathcal{D}er_R(S, M) \simeq 0$.*

Proof. Again consider the square

$$S \longrightarrow S$$
$$\uparrow \qquad \uparrow$$
$$R \longrightarrow S \oplus M\varepsilon$$

where now the map $R \to S \oplus M$ is $r \mapsto r + 0\varepsilon$. The zero derivation $S \to M$ provides one lifting in the square, and since R is unramified over S, this is the only lifting; *i.e.*, there are no non-trivial derivations $S \to M$ which induce the zero derivation on R. $\qquad\square$

Combining these two results we immediately get:

Proposition 1.10 *If $R \to S$ is a formally étale homomorphism of k-algebras, and M is an S-module, then*

$$\mathcal{D}er_k(S, M) \simeq \mathcal{D}er_k(R, M) .$$

The reader may observe that there is a resemblance between the definitions, via lifting properties, of a formally smooth morphism and of a projective module. This is not a total coincidence:

Proposition 1.11 *Let A be a ring, V a projective A-module. Then the symmetric algebra $B = \mathbb{S}_A(V)$ is formally smooth over A.*

Proof. This follows immediately from the fact that

$$\mathrm{Hom}_{A-\mathrm{algebras}}(\mathbb{S}_A(V), C) \simeq \mathrm{Hom}_{A-\mathrm{modules}}(V, C) \,,$$

so that a lifting exists, even when the ideal $I \subset C$ is not nilpotent. $\square$

Proposition 1.12 *Let A be a ring, B an A-algebra. Then if B is formally smooth over A, $\Omega_{B/A}$ is a projective (but not necessarily finitely generated) B-module.*

Proof. This follows from [11, EGA IV, Proposition 16.10.2], where this is deduced from [10, EGA 0, Corollary 19.5.4]. $\square$

Definition and Elementary Properties.

Definition 1.13 A *differential ring* consists of a pair (R, δ) in which R is a commutative ring with unit, and $\delta : R \to R$ is a derivation.

If (R, δ) and (R', δ') are differential rings, then a ring homomorphism $f : R \to R'$ is said to be a *differential* homomorphism if it is compatible with the two derivations, *i.e.* if $\delta' \cdot f = f \cdot \delta$. The kernel of a differential homomorphism $(R, \delta) \to (R', \delta')$ is a *differential ideal*, *i.e.* an ideal $\mathfrak{p} \lhd R$ which is closed under δ. Note that the intersection of a family of differential ideals in a differential ring (R, δ) is again a differential ideal, and hence any subset $X \subset R$ is contained in a smallest differential ideal, denoted by $[X]$.

The following lemma, which is easily proved by induction on k, will be useful:

Lemma 1.14 *If $\delta : R \to R$ is a derivation, then for all $a, b \in R$, and $k \geq 1$,*

$$\delta^k(ab) = \sum_{j=0}^{k} \binom{k}{j} \delta^j(a) \delta^{k-j}(b) \,.$$

From this we deduce:

Proposition 1.15 *If R is a $\mathbb{Q}$-algebra, then a derivation $\delta : R \to R$ induces a ring homomorphism:*

$$\exp(\delta) : R \to R[[t]] \,,$$

$$r \mapsto \sum_{k=0}^{\infty} \frac{\delta^k(r)}{k!} t^k \,.$$

Proof. Since δ, and therefore δ^k for any k, is additive, we need only check that $\exp(\delta)$ preserves products, which follows immediately from the previous lemma. $\qquad\square$

For any R, not just a $\mathbb{Q}$-algebra, recall that giving a derivation $\delta : R \to R$ is equivalent to giving a homomorphism of rings (augmented towards R):

$$\delta_* = \exp_{\leq 1}(\delta) : R \to R[\varepsilon] \, ,$$

$$r \longmapsto r + \delta(r)\varepsilon$$

where $R[\varepsilon]$, with $\varepsilon^2 = 0$, is the ring of dual numbers over R. More generally, if $n!$ is a unit in R, then we get a truncated exponential $\exp_{\leq n}(\delta) : R \to R[t]/t^{n+1}$ by dropping all the terms of degree greater than n from the exponential map.

This interpretation leads to the following lemma:

Lemma 1.16 *Let $f : R \to S$ be a formally étale ring homomorphism. Then if $\delta_R : R \to R$ is a derivation, there is a unique derivation $\delta_S : S \to S$ extending δ. Furthermore if $\sigma : S \to \Lambda$ is a ring homomorphism, with $\Lambda = (\Lambda, \delta_\Lambda)$ a differential ring, such that $\rho = \sigma \cdot f : R \to \Lambda$ is a differential homomorphism, then σ is a differential homomorphism with respect to δ_S and δ_Λ.*

Proof. The first assertion follows from Proposition 1.10. The second assertion follows by considering the diagram

$$
\begin{array}{ccc}
S & \xrightarrow{\ \sigma\ } & \Lambda \\
f\big\uparrow & & \big\uparrow \\
R & \xrightarrow{\ \tilde\rho \cdot \exp_{\leq 1}(\delta_R)\ } & \Lambda[\epsilon]
\end{array}
$$

Here we use $\tilde\rho$ to denote the map $R[\varepsilon] \to \Lambda[\varepsilon]$ induced by ρ. There are two possible ways of filling this diagram in with a homomorphism from S to $\Lambda[\varepsilon]$:

(a) $\exp_{\leq 1}(\delta_\Lambda) \cdot \sigma$;

(b) $\tilde\sigma \cdot \exp_{\leq 1}(\delta_S)$ (here $\tilde\sigma$ denotes the map $S[\varepsilon] \to \Lambda[\varepsilon]$ induced by σ).

Since f is formally étale, and in particular formally unramified, these maps must be equal, and we are done. $\qquad\square$

Since localization is étale, we automatically get:

Corollary 1.17 *If $R = (R, \delta_R)$ is a differential ring, and $S \subset R$ is a multiplicative set, the derivation δ_R has a unique extension to the localization $S^{-1}R$.*

Of course this result can also be proved explicitly, by using the quotient rule to define the derivation.

Lemma 1.18 *If R is a differential $\mathbb{Q}$-algebra, then its nilradical is a differential ideal.*

Proof. This is a standard result, which may be found in [2] for example. Here is a short proof using the exponential map. We must show that if $r \in R$ is nilpotent, then so is $\delta(r)$. Since $\exp(\delta)$ is a ring homomorphism, $\exp(\delta)(r)$ is nilpotent in $R[[t]]$. Therefore $\exp(\delta)(r) - r$ is also nilpotent in $R[[t]]$. But $\exp(\delta)(r) - r = \delta(r)t + O(t^2)$ and therefore if $(\exp(\delta)(r) - r)^N = 0$, we have:

$$(\exp(\delta)(r) - r)^N = (\delta(r)t + O(t^2))^N = \delta(r)^N t^N + O(t^{N+1}) = 0$$

and hence $\delta(r)^N = 0$. □

Lemma 1.19 *If R is a differential $\mathbb{Q}$-algebra, and $\mathfrak{p} \lhd R$ is a minimal prime ideal in R, then $\mathfrak{p}$ is a differential ideal.*

Proof. $\mathfrak{p}$ is the inverse image of $\mathfrak{p}R_{\mathfrak{p}}$ under the natural map $R \to R_{\mathfrak{p}}$. The derivation in R extends to $R_{\mathfrak{p}}$, and $\mathfrak{p}R_{\mathfrak{p}}$ is the unique minimal prime ideal in $R \to R_{\mathfrak{p}}$, hence is equal to the nilradical of $R_{\mathfrak{p}}$, and is therefore a differential ideal in $R_{\mathfrak{p}}$. Since the inverse image of a differential ideal is a differential ideal, it follows that $\mathfrak{p}$ is a differential ideal. □

1.2 Prolongation

Existence. If (k, δ) is a differential ring we may consider the category $\mathbf{Diff}_{(k,\delta)}$ of differential (k, δ)-algebras. There is clearly a forgetful functor $U : \mathbf{Diff}_{(k,\delta)} \to \mathbf{Alg}_k$, which associates to the differential (k, δ)-algebra $(R, \tilde{\delta})$ the k-algebra R.

Theorem 1.20 *The forgetful functor $U : \mathbf{Diff}_{(k,\delta)} \to \mathbf{Alg}_k$ has a left adjoint.*

Proof. It is a general fact in universal algebra that the forgetful functor between two categories of algebras induced by forgetting one or more operations or equations has a left adjoint. More generally, it is a result of Lawvere that "algebraic functors" have left adjoints. It is difficult to give a nice reference for the proof, in part because of variations in the way that a *variety of algebras* can be defined. However one can find, in various references, proofs of the existence of free algebras, i.e. of left adjoints for the forgetful functors from a category of algebras, such as differential k-algebras or k-algebras, to the category of sets.

For example, see [15, Section V.6], and in particular the discussion following Theorem 3. The more general case of the theorem then follows immediately from the existence of free algebras via [13, Theorem 28.12].

Let us give a more detailed version of this proof, which will also be useful later. First, we need two lemmas:

Lemma 1.21 *Let k be a ring, X a set, $k[X]$ the associated polynomial ring, and M a $k[X]$-module. Suppose given a derivation $\delta : k \to M$. Let $Der_{(k,\delta)}(k[X], M)$ be the set of derivations $k[X] \to M$ which extend δ. Then the restriction map*

$$Der_{(k,\delta)}(k[X], M) \to Hom_{Sets}(X, M)$$

which associates to a derivation $\tilde{\delta}$ extending δ, the restriction of $\tilde{\delta}$ to X, is a bijection.

Proof. Giving a derivation $\tilde{\delta} : k[X] \to M$ extending δ is equivalent to giving a homomorphism of k-algebras $k[X] \to k[X] \oplus M\varepsilon$ (of which the first component is the identity), where $k[X] \oplus M\varepsilon$ is a k-algebra via the homomorphism $k \to k[X] \oplus M\varepsilon$ determined by δ. Since $k[X]$ is the free k-algebra on the set X, the restriction map from the set of such homomorphisms to the set of functions $X \to k[X] \oplus M\varepsilon$ which have as their first component the inclusion $X \to k[X]$, i.e., the set of functions $X \to M$, is bijective. $\square$

Lemma 1.22 *If X is a set, the functor from (k, δ)-algebras to sets:*

$$(R, \delta_R) \mapsto R^X$$

is representable - i.e. **Diff**$_{(k,\delta)}$ *has free objects, and hence the forgetful functor from (k, δ)-algebras to sets has a left adjoint.*

Proof. Define a differential ring $(k\{X\}, \tilde{\delta})$ as follows. Let $\mathbb{Z}_{\geq 0}$ be the set of non-negative integers, and set $k\{X\}$ equal to the polynomial ring on $X \times \mathbb{Z}_{\geq 0}$. By the lemma, the set of derivations $k\{X\} \to k\{X\}$ extending $\delta : k \to k \subset k\{X\}$ is bijective with the set of functions $X \times \mathbb{Z}_{\geq 0} \to k\{X\}$. We fix $\tilde{\delta}$ to be the derivation induced by the function

$$X \times \mathbb{Z}_{\geq 0} \to X \times \mathbb{Z}_{\geq 0},$$
$$(x, n) \mapsto (x, n + 1).$$

Now we identify X with a subset of $k\{X\}$ via the function

$$\eta : X \to k\{X\},$$
$$x \mapsto (x, 0).$$

We must show that $\eta : X \to k\{X\}$ is universal for maps from X to (k, δ)-algebras. Given a (k, δ)-algebra (R, δ_R), and a function $f : X \to R$, let $\tilde{f}$ be the function $\tilde{f} : X \times \mathbb{Z}_{\geq 0} \to R$ by $f : (x, n) \mapsto \delta_R^n(f(x))$; note that via the identification of X with the subset $X \times \{0\} \subset X \times \mathbb{Z}_{\geq 0}$ $\tilde{f}$ extends f. The function $\tilde{f}$ determines a unique ring homomorphism $\rho_f : k\{X\} \to R$. To check that ρ_f is a homomorphism of differential rings, we must check that the two derivations

$$\rho_f \cdot \tilde{\delta} : k\{X\} \to R$$

and

$$\delta_R \cdot \rho_f : k\{X\} \to R$$

coincide. By Lemma 1.21 it is enough to observe that they both induce the same function $X \times \mathbb{Z}_{\geq 0} \to R$:

$$\rho_f \cdot \tilde{\delta}((x, n)) = \rho_f((x, n+1)) = \delta_R^{n+1}(f(x))$$

and

$$\delta_R \cdot \rho_f((x, n)) = \delta_R(\delta_R^n(f(x))) = \delta_R^{n+1}(f(x)) .$$

It remains to show that ρ_f is the unique extension of f to a differential homomorphism. If σ is another extension, since polynomial rings are free, $\rho_f = \sigma$ if they have the same restriction to $X \times \mathbb{Z}_{\geq 0}$, but then the requirement that ρ_f and σ be differential homomorphisms forces $\rho_f((x, n)) = \rho_f(\tilde{\delta}^n(x)) = \delta_R^n(x)$ which equals $\sigma((x, n))$ by symmetry. $\qquad\square$

Note that one usually writes $x^{(n)}$ in place of (x, n), and we shall do so in the rest of the paper.

Let us now return to the proof of the theorem:

Given any k-algebra R, we form the ring of differential polynomials $k\{R\}$, *i.e.* the free (k, δ)-algebra on the *set* R. Let $\mathfrak{I}_R$ be the ideal which is the kernel of the canonical surjection $k[R] \to R$. Notice that the triple

$$\mathfrak{I}_R \subset k[R] \subset k\{R\}$$

is functorial in R and hence we get a functor $R \mapsto k\{R\}/[\mathfrak{I}_R]$ from k-algebras to (k, δ)-algebras. We claim that this is the left adjoint that we seek. First note that the inclusion $k[R] \subset k\{R\}$ induces a homomorphism $\theta_R : R \simeq k[R]/\mathfrak{I}_R \to k\{R\}/[\mathfrak{I}_R]$ which is functorial in R, *i.e.* it is a natural transformation of functors from k-algebras to k-algebras. We must show that, given a (k, δ)-algebra (A, δ_A) and a homomorphism of k-algebras $f : R \to A$, it has a unique factorization $f = \sigma_f \cdot \theta_R$ through a homomorphism of (k, δ)-algebras σ_f.

Viewing f simply as a function between sets, it has a unique factorization via a differential homomorphism $\rho_f : k\{R\} \to A$ and through the function $R \to k\{R\}$, by the earlier discussion of free algebras. Since $f : R \to A$ is a homomorphism of *rings*, ρ_f must map the image of $\mathfrak{I}_R$ in $k\{R\}$ to zero in A, and hence ρ_f factors through a homomorphism $\sigma_f : k\{R\}/[\mathfrak{I}_R] \to A$. Since ρ_f is uniquely determined by f, so is σ_f, and we are done. $\qquad\square$

Let

$$(\ )^\infty : R \mapsto R^\infty$$

denote the left adjoint of the theorem. Thus if (k, δ) is a differential ring, $(\ )^\infty$ associates to each k-algebra R a differential (k, δ)-algebra R^∞, and a k-algebra homomorphism $\eta : R \to R^\infty$ with the following universal property: given a homomorphism $\phi : (k, \delta) \to (S, \delta)$ of differential rings, and a k-algebra homomorphism $f : R \to S$, there is a unique homomorphism $f^\infty : R^\infty \to S$ of (k, δ)-algebras which makes the following diagram commute:

$$
\begin{array}{ccc}
R^\infty & \xrightarrow{\ f^\infty\ } & S \\
\eta \uparrow & & \| \\
R & \xrightarrow{\ f\ } & S
\end{array}
$$

Lemma 1.23 *The functor* $(\ )^\infty : \mathbf{Alg}_k \to \mathbf{Diff}_{(k,\delta)}$ *commutes with direct (or inductive) limits.*

Proof. This is a standard property of left adjoint functors. See [15, V. 6], for example. $\qquad\square$

Lemma 1.24 *Let k be a differential field and let X be a set. Then*

$$(k[X])^\infty \simeq k\{X\}.$$

Proof. The composition of the forgetful functor from (k, δ)-algebras to k-algebras with the forgetful functor from k-algebras to sets is the forgetful functor from (k, δ)-algebras to sets, Hence the left adjoint of the composition is the composition of the two left adjoints. $\qquad\square$

Definition 1.25 $k\{X\}$ is called the *ring of differential polynomials* on the set X over (k, δ).

An Alternative Construction. We can also give another construction of R^∞ using the universal properties of the module of differentials. An advantage

106

of this approach will be that when R is a formally smooth k-algebra and an integral domain, then R^∞ is also an integral domain.

We start with:

Lemma 1.26 *Suppose that A is a commutative ring, that B is an A-algebra, and that we are given a derivation $\delta : A \to B$. The functor from the category of B-algebras to the category of sets, which assigns to the B-algebra R the set of all derivations $\tilde\delta : B \to R$ which extend the derivation δ, is representable. If B is formally smooth over A, the object representing this functor is isomorphic to the symmetric algebra on a projective B-module.*

Proof. If we omit the requirement that $\tilde\delta : B \to R$ extend δ, then we are simply representing the set of all derivations from B to R. The set of such derivations is represented by the set of B-module homomorphisms $\Omega_B \to R$. Since R is a B-algebra, this is equivalent to the set of all B-algebra homomorphisms $\mathbb{S}_B(\Omega_B) \to R$. The condition that $\tilde\delta : B \to R$ extend the derivation δ is that the homomorphism $\mathbb{S}_B(\Omega_B) \to R$ map all elements of the form da for $a \in A$ to $\delta(a)$. Thus we divide $\mathbb{S}_B(\Omega_B)$ by the ideal I generated by all elements of the form $da - \delta(a)$ for $a \in A$. If we now assume that B is formally smooth over A, then we know that $\Omega_{B/A}$ is a projective B-module, and that Ω_B is non-canonically isomorphic to the direct sum of $\Omega_{B/A}$ and Ω_A. Therefore the quotient $\mathbb{S}_B(\Omega_B)/I$ is isomorphic to $\mathbb{S}_B(\Omega_{B/A})$, and is formally smooth over B. $\qquad\square$

Suppose that (k, δ) is a differential ring, and that R is a k-algebra. Consider the category $\mathbf{T}_{(k,\delta,R)}$ with objects sequences $(R^{\{n\}}, \rho_n, \delta^{\{n\}})$ for $n \geq -1$, in which:

(a) $R^{\{-1\}} = k$.

(b) $R^{\{0\}} = R$.

(c) For all $n \geq -1$, ρ_n is a ring homomorphism $R^{\{n\}} \to R^{\{n+1\}}$, and $\delta^{\{n\}}$ is a derivation : $R^{\{n\}} \to R^{\{n+1\}}$ (with respect to the $R^{\{n\}}$ module structure on $R^{\{n+1\}}$ via ρ_n). ρ_{-1} is the structure map for the k-algebra R, and $\delta^{\{-1\}}$) is the composition $\rho_{-1} \cdot \delta$.

(d) For each $n \geq 0$, we have a commutative diagram:

$$\begin{array}{ccc} R^{\{n\}} & \xrightarrow{\delta^{\{n\}}} & R^{\{n+1\}} \\ \uparrow & & \uparrow \\ R^{\{n-1\}} & \xrightarrow{\delta^{\{n-1\}}} & R^{\{n\}} \end{array}$$

A morphism between two sequences $(R^{\{n\}}, \rho_n, \delta^{\{n\}})$ and $(R'^{\{n\}}, \rho'_n, \delta'^{\{n\}})$ is a sequence of ring homomorphisms $f_n : R^{\{n\}} \to R'^{\{n\}}$ for $n \geq 1$, compatible with the homomorphisms and derivations in each sequence.

Note that if $g : R \to A$ is a homomorphism of k-algebras with (A, δ_A) a (k, δ)-algebra, then we get an object in the category by setting $R^{\{n\}} := A$ for $n \geq 1$, $\delta^{\{n\}} := \delta_A$ for $n \geq 1$, and $\delta^{\{0\}} := \delta_A \cdot g$ when $n = 0$. Finally observe that the $\delta^{\{n\}}$ induce a derivation on $\varinjlim_n R^{\{n\}}$ which makes it a (k, δ)-algebra.

Proposition 1.27 *The category* $T_{(k,\delta,R)}$ *has an initial object of the form* $(R^{\{n\}}, \rho_n, \delta^{\{n\}})$, *and there is an isomorphism* $\varinjlim_n R^{\{n\}} \simeq R^\infty$. *If R is formally smooth over k, then $R^{\{n\}}$ is formally smooth over $R^{\{n-1\}}$ for all $n \geq 1$, and hence R^∞ is formally smooth over R by Exercise 1.5 in Section 1.1.*

Note that $R^{\{n\}}$ is isomorphic to the symmetric algebra on a projective $R^{\{n-1\}}$ module.

Proof. We proceed by induction on $n \geq 0$. The case $n = 0$ is already given to us. For the inductive step, given a sequence of algebras and derivations, $(R^{\{k\}}, \rho_k, \delta^{\{k\}})$, for $k < n$ (which we could refer to as a truncated sequence of length n), there is, by Lemma 1.26, a $R^{\{n-1\}}$-algebra $R^{\{n\}}$ together with a derivation $R^{\{n-1\}} \to R^{\{n\}}$ which is universal and therefore defines a universal truncated sequence of length $n + 1$. If R is formally smooth over k, then again by Lemma 1.26, and induction on n, $R^{\{n\}}$ is formally smooth over $R^{\{n-1\}}$.

Since a homomorphism $g : R \to A$ of k-algebras with (A, δ_A) a (k, δ)-algebra defines an object in the category $T_{(k,\delta,R)}$, there is unique map of sequences from the universal sequence $(R^{\{n\}}, \rho_n, \delta^{\{n\}})$ to the constant sequence defined by (A, δ_A), and hence a map of R-algebras $\varinjlim_n R^{\{n\}} \to A$, which is also a differential homomorphism of (k, δ) algebras. $\square$

Corollary 1.28 *If (k, δ) is a differential ring, and R is an integral domain which is formally smooth over k, then $R^{\{n\}}$ is an integral domain for all $n \geq 0$, and hence R^∞ is an integral domain. Furthermore $R \to R^\infty$ is injective.*

Proof. Since the direct limit of a sequence $(R^{\{n\}}, \rho_n : R^{\{n\}} \to R^{\{n+1\}})_{n \geq 1}$ of ring homomorphisms between integral domains is an integral domain, it suffices to show that $R^{\{n\}}$ is an integral domain for all n. Since we assume that R is an integral domain, and all the ρ_n are formally smooth, it is enough to know, following Lemma 1.26, that if A is a commutative ring, and P is a projective A-module, then the symmetric algebra on P over A is an integral domain. But we know that a projective module is a direct summand of a free module. Hence the symmetric algebra on P is isomorphic to a subring of a polynomial ring over A, and is therefore an integral domain. Furthermore, $\rho_n : R^{\{n\}} \to R^{\{n+1\}}$ is then clearly injective for all n. □

Etale Base Change.

Lemma 1.29 *Let* (k, δ) *be a differential ring. Suppose that* $f : R \to S$ *is a formally étale homomorphism of* k*-algebras. Then*

$$S^\infty \simeq S \otimes_R R^\infty .$$

Proof. Suppose that we are given a (k, δ) algebra (Λ, δ) and ring homomorphism $\phi : S \to \Lambda$. We wish to show that this has a unique factorization through a differential homomorphism $S \otimes_R R^\infty \to \Lambda$. Writing $i : R \to S$ for the inclusion, the induced homomorphism $\phi \cdot i : R \to \Lambda$ has a canonical factorization $\phi \cdot i = (\phi \cdot i)^\infty \cdot \eta_R$, where $(\phi \cdot i)^\infty : R^\infty \to \Lambda$ is a differential homomorphism. Hence we have a commutative diagram:

$$
\begin{array}{ccc}
S & \xrightarrow{\phi} & \Lambda \\
{\scriptstyle i}\uparrow & & {\scriptstyle (\phi \cdot i)^\infty}\uparrow \\
R & \xrightarrow{\eta_R} & R^\infty
\end{array}
$$

and an induced homomorphism of commutative rings

$$S \otimes_R R^\infty \to \Lambda . \qquad (1.30)$$

Since $R^\infty \to S \otimes_R R^\infty$ is formally étale and $R^\infty \to \Lambda$ is a differential homomorphism, it follows from Lemma 1.16 that the induced map (1.30) is a differential homomorphism. The rest of the details are left to the reader. □

The conclusion of the lemma could be phrased as "the functor $R \mapsto R^\infty$ is a quasi-coherent sheaf in the étale topology". Thus we can give the following definition.

Definition 1.31 *For any scheme* X, *define the sheaf (in the étale topology)* $\mathcal{O}_X^\infty$ *of differential polynomial functions on* X *(if* $U = \mathrm{Spec}(R)$ *is affine and étale over* X*) by setting:*

$$\Gamma(U, \mathcal{O}_X^\infty) := R^\infty.$$

Definition 1.32 *Let* $k = (k, \delta)$ *be a differential ring. Suppose that* X *is a scheme over* k. *Then we define* $X^\infty := \mathbf{Spec}(\mathcal{O}_X^\infty)$. *(See Ch. II, Exercise 5.17 in [12] for the construction of* **Spec**.)

X^∞ is a "differential scheme" in the sense that $\mathcal{O}_X^\infty$ is equipped with a derivation. The passage from X to X^∞ is known as *prolongation*; when the derivation δ on k is zero, and X is smooth over k, X^∞ is the (infinite) jet bundle over X. However, X^∞ is not a differential scheme in the sense of other authors – *i.e.*, it does *not* use the "differential spectrum" as a local model.

Prolongation of fields and valuation rings.

Proposition 1.33 *If* $k = (k, \delta)$ *is a differential field, and* $k \subset K$ *is a finitely generated separable field extension of* k, *then* K^∞ *is a polynomial ring over* K. *If we assume in addition that* $k = (k, \delta)$ *has characteristic zero, then we can remove the assumption of finite generation.*

Proof. By [6, Chapter V, §16.7 Corollary to Theorem 5], we know that K has a separating transcendence basis $\{x_1, \ldots, x_n\}$ over k. Thus K is a finite separable algebraic (and hence étale) extension of $k(x_1, \ldots, x_n)$. Since localizations (including passage to the field of fractions) are formally étale, it follows that K is formally étale over the polynomial ring $k[x_1, \ldots, x_n]$. Hence, by Lemmas 1.22 and 1.29,

$$K^\infty \simeq K \otimes_{k[x_1, \ldots, x_n]} k\{x_1, \ldots, x_n\} \simeq K[\{x_1, \ldots, x_n\} \times \mathbb{N}] .$$

If k has characteristic zero, then by Steinitz Theorem [6, Chapter V, Theorem 1 of §14.2], K has a transcendence basis $X \subset K$ over k, and the extension $k(X) \subset K$ is algebraic, hence (since we assume characteristic zero) a composite of finite separable extensions, and hence we again have that K is formally étale over the polynomial ring $k[X]$ (see also *op. cit.* V.16.3.). $\square$

If k has characteristic p, and the extension is not finitely generated, then the extension need not have a separating transcendence basis; see [6, Chapter V §16, Exercises 1 and 2]. We will ignore the question of whether the lemma might be still true in the absence of a separating basis.

This proposition, combined with Lemma 1.29, gives another way of proving Corollary 1.28, though under the stronger hypothesis that R is a smooth (not just formally smooth) k-algebra:

Corollary 1.34 *If* R *is a smooth* k-algebra, *with* (k, δ) *a differential ring, and if* R *is an integral domain, then* R^∞ *is an integral domain.*

Proof. Since R is smooth over k, $X = \operatorname{Spec}(R)$ is locally isomorphic in the étale topology to affine space $\mathbb{A}_k^n$ (where n is the dimension of R over k). Hence by Lemma 1.22, R^∞ is locally isomorphic, again in the étale topology, to the ring of differential polynomials $k\{x_1 \ldots x_n\}$. Since flatness is a local condition, and any polynomial ring is flat over its ring of coefficients, R^∞ is flat over R. Since R is a domain, it injects into its fraction field F. Since R^∞ is flat over R, we have that R^∞ injects into $R^\infty \otimes_R F$. But we know that $R \mapsto R^\infty$ commutes with localization, hence $R^\infty \otimes_R F \simeq F^\infty$, which by Proposition 1.33 is a polynomial ring over F, and is therefore a domain. Thus R^∞ injects into a domain, and is itself a domain. $\qquad\square$

We can globalize this as follows – the proof is left to the reader:

Corollary 1.35 *If X is a variety smooth and integral over a differential field of characteristic zero, then the differential algebraic variety X^∞ associated by prolongation to X is integral, and in particular irreducible.*
The following will be useful in the next section:

Corollary 1.36 *Let k be a differential field of characteristic zero, and let $E \subset F$ be extensions of k. Then the natural map $E^\infty \to F^\infty$ is injective.*

Proof. Since any transcendence basis of E can be extended to a transcendence basis of F, the corollary follows immediately from Proposition 1.33. $\qquad\square$

We can also give an alternate proof of the first assertion of Proposition 1.33 using Proposition 1.27.

Proposition 1.37 *Suppose that (k, δ) is a differential field of characteristic zero, and that $k \subset K$ is an extension of fields. Then K^∞ (formed relative to (k, δ)) is an integral domain.*

Proof. By a theorem of Cohen, we know that K is formally smooth over k (see [10, Ch. 0, Théorème 19.6.1]) and so we can apply Proposition 1.27. $\qquad\square$

We turn now from fields to valuation rings, and in particular to discrete valuation rings. First a technical lemma:

Lemma 1.38 *Let k be a field of characteristic zero, and R a k-algebra which is a discrete valuation ring. Then R is a formally smooth k algebra.*

Proof. Let K be the residue field of R. The induced map from k to K is an inclusion, which makes K a separable extension of k, and hence using Cohen's theorem again, (see EGA Ch. 0, Corollary §19.6.1, in [10]) K is formally smooth over k. If M is the maximal ideal of R, then $M/M^2 \simeq K$ and is therefore a projective K-module. Therefore by Corollary §19.5.4, in *op. cit.*, R is formally smooth over k. $\qquad\square$

From the lemma, together with Proposition 1.27 we immediately get:

Corollary 1.39 *Let k be a differential field of characteristic zero, and R a k-algebra which is a discrete valuation ring. Then R^∞ is an integral domain.*

More generally, by using uniformization, we have:

Lemma 1.40 *Let k be a differential field of characteristic zero, and R a k algebra which is a valuation ring. Then R^∞ is an integral domain.*

Proof. By the *Uniformization Theorem* of Zariski, [20], see also Popescu [18] for a more general result, we know that any valuation ring R containing a field k of characteristic zero is the direct limit of smooth k-algebras (which we may assume to be integral domains):

$$R = \varinjlim A_\alpha \ .$$

Since $R \mapsto R^\infty$ commutes with direct limits, we have that

$$R^\infty = \varinjlim (A^\alpha)^\infty$$

is a direct limit of integral domains, and is therefore an integral domain. $\quad\square$

2 Kolchin's Irreducibility Theorem

In the last section we saw that if X is a variety smooth and integral over a differential field, then the differential algebraic variety X^∞ associated by prolongation to X is integral, and in particular irreducible. Kolchin's Irreducibility Theorem tells us that in characteristic zero, the irreducibility conclusion remains true even if X is not smooth. Note however that X^∞ will not in general be reduced, *i.e.*, its coordinate ring may contain nilpotent elements.

Theorem 2.1 *Let $K = (K, \delta)$ be a differential field of characteristic zero, and suppose that R is a K-algebra which is an integral domain. Then R^∞ has a unique minimal prime ideal.*

Remark In Kolchin [14, Proposition 10 of Chapter IV], this is phrased as: "Let $\mathfrak{p}_0$ be a prime ideal of $\mathcal{F}[y_1, \ldots, y_n] \ldots$. Then $\{\mathfrak{p}_0\}$ is a prime differential ideal $\ldots$." Here $\{\mathfrak{p}_0\}$ denotes the smallest radical differential ideal containing $\mathfrak{p}_0$. If we set $R := \mathcal{F}[y_1, \ldots, y_n]/\mathfrak{p}_0$, then the image of $\{\mathfrak{p}_0\}$ in $R^\infty \simeq \mathcal{F}\{y_1, \ldots, y_n\}/[\mathfrak{p}_0]$ is the nilradical of R^∞. Thus $\mathfrak{p}_0$ is prime if and only if R^∞ has a unique minimal prime. In general, the nilradical of a ring is prime if and only if the spectrum of the ring is irreducible; thus the theorem tells us that if X is a reduced irreducible affine scheme, then X^∞ is also irreducible.

The following definition will be useful in the proof of the theorem:

Definition 2.2 Let R be a K-algebra, with K a differential field of characteristic zero. Suppose that $\mathfrak{p}$ is a prime ideal in R. Denote by $[[\mathfrak{p}]]$ the prime differential ideal in R^∞ which is the kernel of the homomorphism $R^\infty \to k(\mathfrak{p})^\infty$ where $k(\mathfrak{p})$ denotes the residue field of $\mathfrak{p}$.

Note that $[[\mathfrak{p}]]$ is a prime ideal because $k(\mathfrak{p})^\infty$ is an integral domain by Proposition 1.33. With this definition, we can rephrase Kolchin's theorem as asserting that any prime ideal $\mathfrak{p} \lhd R^\infty$ contains $[[0]]$.

2.1 Proof of the theorem

We start by assuming that R is noetherian. The proof will then consist of the following five steps:

(a) Show that it is enough to prove that any prime *differential* ideal $\mathfrak{p} \lhd R^\infty$ contains $[[0]]$.

(b) Show that if $\eta : R \to R^\infty$ is the natural map, then for any prime differential ideal $\mathfrak{p} \lhd R^\infty$, $[[\eta^{-1}(\mathfrak{p})]] \subset \mathfrak{p}$. Therefore it will be enough to prove that for any prime ideal $\mathfrak{q} \lhd R$, $[[0]] \subset [[\mathfrak{q}]] \lhd R^\infty$.

(c) Show that if R is a one dimensional local domain, with maximal ideal $\mathfrak{p}$, then $[[0]] \subset [[\mathfrak{p}]] \lhd R^\infty$.

(d) Show that if $\mathfrak{p} \subset \mathfrak{q}$ are prime ideals in R, with $\mathfrak{q}$ of height one above $\mathfrak{p}$, then $[[\mathfrak{p}]] \subset [[\mathfrak{q}]] \lhd R^\infty$.

(e) Show that if $\mathfrak{p} \subset \mathfrak{q}$ are arbitrary prime ideals in R, then $[[\mathfrak{p}]] \subset [[\mathfrak{q}]] \lhd R^\infty$. In particular for any prime ideal $\mathfrak{q} \lhd R$, $[[0]] \subset [[\mathfrak{q}]] \lhd R^\infty$.

Proof of Step 1. By Zorn's lemma any prime ideal in R^∞ contains a minimal prime ideal, which by Lemma 1.19 is a differential ideal. Thus it suffices to show that any *prime* differential ideal contains $[[0]]$.

Proof of Step 2.

Lemma 2.3 *Let K and R be as above. Suppose that $\mathfrak{p}$ is a prime differential ideal in R^∞. If we write $\mathfrak{p}_0$ for the prime ideal $\eta^{-1}\mathfrak{p}$, where η is the natural map from R to R^∞, then $[[\mathfrak{p}_0]] \subset \mathfrak{p}$.*

Proof. Since the map $R \to R^\infty/\mathfrak{p}$ factors through $R/\mathfrak{p}_0$, there is an induced map $\theta : (R/\mathfrak{p}_0)^\infty \to R^\infty/\mathfrak{p}$, and the quotient map $R^\infty \to R^\infty/\mathfrak{p}$ factors through θ. Since the prolongation functor $(\ )^\infty$ commutes with localization, if k is the fraction field of $R/\mathfrak{p}_0$, the natural map $(R/\mathfrak{p}_0)^\infty \otimes_{R/\mathfrak{p}_0} k \to k^\infty$ is an isomorphism. Therefore there is a commutative diagram:

$$
\begin{array}{ccccc}
R & \xrightarrow{\ \eta\ } & R^\infty & = & R^\infty \\
\downarrow & & \downarrow & & \downarrow \\
R/\mathfrak{p}_0 & \longrightarrow & (R/\mathfrak{p}_0)^\infty & \xrightarrow{\ \theta\ } & R^\infty/\mathfrak{p} \\
\downarrow & & \downarrow & & \downarrow \\
k & \longrightarrow & k^\infty \simeq (R/\mathfrak{p}_0)^\infty \otimes_{R/\mathfrak{p}_0} k & \longrightarrow & R^\infty/\mathfrak{p} \otimes_{R/\mathfrak{p}_0} k
\end{array}
$$

Since the bottom left hand vertical map in this diagram is a localization, the other two bottom vertical maps are also localizations. Therefore, since $R^\infty/\mathfrak{p}$ is an integral domain, the right bottom vertical map is injective. Hence:

$$
\mathfrak{p} = \mathrm{Ker}(R^\infty \to R^\infty/\mathfrak{p}) = \mathrm{Ker}(R^\infty \to R^\infty/\mathfrak{p} \otimes_{R/\mathfrak{p}_0} k)
$$

and therefore:

$$
[[\mathfrak{p}_0]] = \mathrm{Ker}(R^\infty \to k^\infty) \subset \mathfrak{p} = \mathrm{Ker}(R^\infty \to R^\infty/\mathfrak{p} \otimes_{R/\mathfrak{p}_0} k).
$$

$\square$

Proof of Step 3.

Lemma 2.4 *Let K and R be as above, and assume in addition that R is a one dimensional noetherian local domain. If $\mathfrak{p}$ is the maximal ideal in R, then with the notation above,*

$$
[[0]] \subset [[\mathfrak{p}]].
$$

Proof. Let A be the normalization of R, *i.e.* the integral closure of R in its fraction field F. By [5, V§2.1, Proposition 1], for any maximal ideal $\mathfrak{m} \lhd A$, $\mathfrak{m} \cap R = \mathfrak{p}$. Then by the Krull-Akizuki Theorem (Proposition 5 of Chapter VII, §2.5 of [5]), A is a Dedekind domain. Hence for any maximal ideal $\mathfrak{m} \lhd A$, $A_\mathfrak{m}$ is a discrete valuation ring. Therefore, we have a commutative diagram:

$$
\begin{array}{ccc}
R & \longrightarrow & R/\mathfrak{p} = k_0 \\
\downarrow & & \downarrow \\
A & \longrightarrow & A/\mathfrak{m} = k \\
\downarrow & & \\
F & &
\end{array}
$$

114

in which all the vertical maps are injective, with $k_0 \subset k$ a separable field extension, and the inclusion $A \subset F$ is a localization. Since A is a discrete valuation ring, A^∞ is an integral domain, by Lemma 1.38, and $A^\infty \to F^\infty = A^\infty \otimes_A F$ is injective. Therefore $[[0]] = \mathrm{Ker}(R^\infty \to F^\infty) = \mathrm{Ker}(R^\infty \to A^\infty)$. Since $k_0 \to k$ is injective and k has characteristic zero, the field extension k/k_0 is formally smooth, has finite degree by the Krull-Akizuki Theorem, and is therefore étale. Hence, by étale base change, $k^\infty \simeq (k_0)^\infty \otimes_{k_0} k$ and so the map $(k_0)^\infty \to k^\infty$ is injective. Therefore $[[\mathfrak{p}]] = \mathrm{Ker}(R^\infty \to k^\infty)$. Since the map $R^\infty \to k^\infty$ factors through A^∞, we get that $[[0]] \subset [[\mathfrak{p}]]$. $\square$

Proof of Step 4.

Lemma 2.5 *Suppose that* $\mathfrak{p} \subset \mathfrak{b} \lhd R$ *when* $\mathfrak{b}$ *has height 1 above* $\mathfrak{p}$*. Then* $[[\mathfrak{p}]] \subset [[\mathfrak{b}]] \lhd R^\infty$*.*

Proof. We know by the previous lemma that, since $R_\mathfrak{b}/(\mathfrak{p}R_\mathfrak{b}) \simeq (R/\mathfrak{p})_\mathfrak{b}$ is a one dimensional local ring, there is an inclusions of kernels:

$$\mathrm{Ker}((R_\mathfrak{b}/(\mathfrak{p}R_\mathfrak{b}))^\infty \to k(\mathfrak{p})^\infty) \subset \mathrm{Ker}((R_\mathfrak{b}/(\mathfrak{p}R_\mathfrak{b}))^\infty \to k(\mathfrak{b})^\infty) \,.$$

Hence there is also an inclusion,

$$\mathrm{Ker}(R^\infty \to k(\mathfrak{p})^\infty) \subset \mathrm{Ker}(R^\infty \to k(\mathfrak{b})^\infty)$$

as desired. $\square$

Proof of Step 5. Since we are assuming that R is noetherian, by Krull's principal ideal theorem, (Theorem §13.5 of [17] or Section 12.E of [16]) the prime ideals in R satisfy the descending chain condition. In particular, if $\mathfrak{p} \lhd R$ is a prime ideal, there are only finitely many prime ideals between $0 \lhd R$ and $\mathfrak{p}$, and so Step 5 follows by induction from Step 4.

The non-noetherian case. We shall give two proofs of the non-noetherian case. The first replaces Steps 3 and 5 with the following argument:

Lemma 2.6 *Let* Λ *be a differential ring, and* R *be a* Λ-*algebra which is a local domain. If* $\mathfrak{p}$ *is the maximal ideal in* R*, then with the notation above,*

$$[[0]] \subset [[\mathfrak{p}]] \,.$$

Proof. Now, rather than using integral closure and the Krull-Akizuki Theorem, we shall use Lemma 1.40 (which depends on uniformization), together with the existence of valuations with a given center.

Let F be the fraction field of R. By [5, Chapter VI, the Corollary in §1.2], there is a valuation ring $A \subset F$ which dominates R, *i.e.*, $R \subset A$, and if $\mathfrak{m} \lhd A$

is the maximal ideal of A, $\mathfrak{m} \cap R = \mathfrak{p}$. Then, we have a commutative diagram:

$$
\begin{array}{ccc}
R & \longrightarrow & R/\mathfrak{p} = k_0 \\
\downarrow & & \downarrow \\
A & \longrightarrow & A/\mathfrak{m} = k \\
\downarrow & & \\
F & &
\end{array}
$$

in which all the vertical maps are injective, and the inclusion $A \subset F$ is a localization. Since A is a valuation ring, A^∞ is an integral domain, by Lemma 1.40, and $A^\infty \to F^\infty = A^\infty \otimes_A F$ is injective. Therefore

$$[[0]] = \mathrm{Ker}(R^\infty \to F^\infty) = \mathrm{Ker}(R^\infty \to A^\infty).$$

Since $k_0 \to k$ is injective and k is characteristic zero, the map $(k_0)^\infty \to k^\infty$ is injective by Lemma 1.29. Therefore $[[\mathfrak{p}]] = \mathrm{Ker}(R^\infty \to k^\infty)$. Since the map $R^\infty \to k^\infty$ factors through A^∞, we get that $[[0]] \subset [[\mathfrak{p}]]$. □

The second proof does not use general valuation rings, but rather deduces the non-noetherian case from the finitely generated case. Suppose therefore that R is an arbitrary integral domain in the category of k-algebras, i.e., R is not necessarily finitely generated over k. Since the functor $(\ )^\infty$ commutes with direct limits, R^∞ is the direct limit of A^∞, as A runs through the partially ordered set of finitely generated subalgebras $A \subset R$. Since each such subalgebra A is an integral domain which is finitely generated over k, we know that each A^∞ has a single minimal prime. Note that the assertion that a ring has a single minimal prime is equivalent to saying that its nilradical is prime. Suppose now that a pair of elements $a, b \in R^\infty$ have nilpotent product: $(ab)^n = 0$ for some $n \in \mathbb{N}$. Then there exists a finitely generated $A \subset R$ such that $a, b \in A^\infty$, and $(ab)^n = 0$ in A^∞. By the finite type case of the theorem, it follows that one or other of a or b is nilpotent in A^∞, and hence in R^∞, and we are done.

This completes the proof of Kolchin's theorem.

3 Descent for Projective Varieties

Theorem 3.1 *Let X be a proper variety over an algebraically closed field K of characteristic zero. Let Δ be the K-vector space $H^0(X, \mathcal{D}er(\mathcal{O}_X))$ of global sections of the sheaf of derivations of the structure sheaf of X. Let K^Δ be the (algebraically closed) subfield of K consisting of elements fixed*

116

under the action of Δ. Then there exists a variety Y, proper over K^Δ, and an isomorphism

$$X \simeq Y \otimes_{K^\Delta} K$$

i.e., K^Δ is a field of definition for X. Furthermore, K^Δ is the minimal field of definition for X, in the sense that any other algebraically closed subfield $L \subset K$, which is a field of definition for X, contains K^Δ.

3.1 Proof of the theorem

Without loss of generality, we assume that X is connected. The strategy of the proof is as follows:

- Reduce to the case where K has finite transcendence degree over $\mathbb{Q}$.

- Reduce to the case where X is the generic fibre of a smooth proper family $\mathcal{X}$ over the spectrum of a Henselian discrete valuation ring $R^h \subset K$. The special fiber X_0 of $\mathcal{X}$ will then be defined over a field with smaller transcendence degree than that of K, and therefore it will be enough to show that if there is a horizontal vector field on $\mathcal{X}$, then

$$\mathcal{X} \simeq X_0 \times \operatorname{Spec}(R^h).$$

- Use the horizontal vector field to construct an isomorphism of formal schemes over the completion of R^h.

- Algebraize this formal isomorphism, *i.e.*, show that it is induced by an isomorphism of schemes (not just formal schemes) over the completion of R^h.

- Use Artin Approximation to deduce the existence of an isomorphism $\mathcal{X} \simeq X_0 \times \operatorname{Spec}(R^h)$ over R^h.

Step 1.

Lemma 3.2 K^Δ *is contained in all other fields of definition for X.*

(Hence it is enough to show that X is defined over K^Δ.)

Proof. Suppose that $F \subset K$ is a field of definition for X. Thus there is a variety Y, proper over F, and an isomorphism

$$X \simeq Y \otimes_F K .$$

On the category of F-algebras, $A \to \Omega_{A/F}$ commutes with direct limits (see, for example, [7, Theorems 16.5 and 16.8]).

Thus we have an isomorphism of sheaves on $X \simeq Y \otimes_F K$:

$$\Omega_{(\mathcal{O}_Y \otimes_F K)/F} \simeq (\mathcal{O}_Y \otimes_F \Omega_{K/F}) \oplus (K \otimes_F \Omega_{\mathcal{O}_Y/F})$$

and therefore there is an isomorphism

$$H^0(X, \mathcal{D}er_F(\mathcal{O}_X)) \simeq \mathcal{D}er_F(K) \oplus (K \otimes_F H^0(Y, \mathcal{D}er_F(\mathcal{O}_Y))) \, .$$

It follows that $\mathcal{D}er_F(K) \subset \Delta = H^0(X, \mathcal{D}er(\mathcal{O}_X))$, and it therefore acts naturally on $K \simeq H^0(X, \mathcal{O}_X)$ with fixed field F, and hence $K^\Delta \subset F = K^{\mathcal{D}er_F(K)}$. $\quad\square$

Lemma 3.3 *There exists a field of definition F for X which is of finite transcendence degree over the prime field $\mathbb{Q}$.*

Proof. This is a standard argument: as a variety of finite type over K, X may be defined by a finite set of equations. We may then take F to be the algebraic closure of the subfield of K generated by these equations. $\quad\square$

Step 2. From the two lemmas of the previous section, we see that we need only consider fields of definition K which have finite transcendence degree over $\mathbb{Q}$, and if such a field is not equal to K^Δ, then it must contain K^Δ as a subfield with $\mathrm{trdeg}(K/K^\Delta) > 0$.

Lemma 3.4 *In order to show that X may be defined over K^Δ it suffices to show that if there is a nonzero element $\xi \in H^0(X, \mathcal{D}er(\mathcal{O}_X))$, which acts non-trivially on $K \simeq H^0(X, \mathcal{O}_X)$ with X defined over a field K which is finitely generated over the prime field, then X is defined over a field F which has strictly smaller transcendence degree over the prime field than K does.*

Proof. Since, by the previous lemma, there exist fields of definition for X which have finite transcendence degree over $\overline{\mathbb{Q}}$, we know that we can choose a field of definition F for X which has minimal transcendence degree over K^Δ. Since F is a field of definition for X, we may write $X \simeq Y \otimes_F K$, with Y proper over F, If $F \neq K^\Delta$, then by definition of K^Δ there exists a nonzero element $\delta \in H^0(Y, \mathcal{D}er(\mathcal{O}_Y))$ which has a non-trivial action on F. Hence if we can show that this implies that Y is defined over a smaller field than F, this will contradict the minimality of F. $\quad\square$

Step 3. Next we "spread X out" over a discrete valuation ring $R \subset K$.

Proposition 3.5 *Suppose that X is proper and connected over an algebraically closed field K of finite transcendence degree over $\overline{\mathbb{Q}}$, and that there is a $\delta \in H^0(X, \mathcal{D}er(\mathcal{O}_X))$, such that $\mathrm{trdeg}(K/K^\delta) > 0$. (Note that δ acts on $K = H^0(X, \mathcal{O}_X)$). Then there exists a subfield $F \subset K$, with $\mathrm{trdeg}(K/F) > 0$ such that X is defined over F.*

In order to prove the proposition, we first need a lemma:

Lemma 3.6 *Let X and K be as above. Then there exists a discrete valuation ring $R \subset K$, such that:*

(a) *R is a localization of an algebra which is smooth and of finite type over $\overline{\mathbb{Q}}$.*

(b) *there is a scheme $\mathcal{X}$ which is proper, with geometrically connected fibres, over $S = \mathrm{Spec}(R)$ such that $\mathcal{X} \otimes_R K = X$.*

(c) *there is a $\xi \in H^0(\mathcal{X}, \mathcal{D}er(\mathcal{O}_{\mathcal{X}}))$ which restricts to $\delta \in H^0(X, \mathcal{D}er(\mathcal{O}_X))$.*

(d) *If we write $\overline{\xi}$ for the derivation of R induced by ξ, and f for the generator of the maximal ideal of R, then $\overline{\xi}(f)$ is a unit in R.*

(e) *R contains a subfield E, such that the residue field k of R is a finite algebraic extension of E under the natural inclusion of $E \subset k$.*

Proof. Arguing as in the proof of Lemma 3.3, we first assume that there is a subring $\Lambda \subset K$ which is of finite type over $\overline{\mathbb{Q}}$, and a scheme $\mathcal{X}$ which is proper over $S = \mathrm{Spec}(\Lambda)$ such that $\mathcal{X} \otimes_\Lambda K = X$ and that δ extends to a derivation ξ of $\mathcal{O}$.

By generic smoothness of varieties over algebraically closed fields, we may assume after localization that $S = \mathrm{Spec}(\Lambda)$ is smooth over $\overline{\mathbb{Q}}$. Since S is smooth, and the geometric generic fibre X of $\mathcal{X}$ over S is connected, we know that $H^0(\mathcal{X}, \mathcal{O}_{\mathcal{X}}) = \Lambda$, and hence ξ induces a derivation $\overline{\xi}$ of Λ (which agrees with δ via the inclusion $\Lambda \subset K$). If $\overline{\xi}$ is trivial, then $\Lambda \subset K^\delta$ and we can take $F = K^\delta$.

If $\overline{\xi}$ is non-zero, then there is a closed point $x \in S$ at which $\overline{\xi}$ (which may now be viewed as a section of the tangent bundle of S over $\overline{\mathbb{Q}}$) does not vanish. Since S is smooth at x, there is a regular system of parameters $\{z_1, \ldots, z_d\}$ of $\mathcal{O}_{S,x}$, and since $\overline{\xi}$ does not vanish at x, there is at least one i for which $\overline{\xi}(z_i)$ does not vanish at x. Without loss of generality, we may take $i = 1$. After further localization, we may assume that all the z_i are regular on S, that the induced map $\pi = (z_1, \ldots, z_d) : S \to \mathbb{A}^d_{\overline{\mathbb{Q}}}$ is étale, and that $\overline{\xi}(z_1)$ vanishes nowhere on S.

Let R be the discrete valuation ring which is the localization of Λ at the prime ideal generated by z_1. We may then take $f = z_1$. Then the map $\pi : S \to \mathbb{A}^d_{\overline{\mathbb{Q}}}$ realizes R as an étale extension of the discrete valuation ring $\overline{\mathbb{Q}}[z_1, z_2, \ldots, z_d]_{(z_1)}$, and the residue field k of R is a finite separable extension of $\overline{\mathbb{Q}}(z_2, \ldots, z_d)$ since π is étale. Since R contains $\overline{\mathbb{Q}}[z_2, \ldots, z_d]$ as a subring which injects into the residue field of R, it follows that there is an inclusion $\overline{\mathbb{Q}}(z_2, \ldots, z_d) \subset R$. Finally we set E equal to the algebraic closure of $\overline{\mathbb{Q}}(z_2, \ldots, z_d)$ in R. $\qquad \square$

Step 4. Let R^h be a strict Henselization of R. Then the algebraic closure of E in R^h maps isomorphically onto the residue field of R^h. Since the fraction field of R^h is an algebraic extension of the fraction field of R, and K is algebraically closed, there is an embedding $R^h \subset K$ extending $R \subset K$.

The isomorphism $\mathcal{X} \otimes_R K \simeq X$ extends uniquely to an isomorphism $\mathcal{X} \otimes_R R^h \otimes_{R^h} K \simeq X$. Since $R \subset R^h$ is formally étale, ξ lifts uniquely to a derivation of the structure sheaf of $\mathcal{X} \otimes_R R^h$, and this lift is compatible with base change from R^h to K. Note that $f \in R \subset R^h$ is also the generator of the maximal ideal of R^h.

Lemma 3.7 *Let (A, δ) be a differential ring, and $I = fA$ a principal ideal in A. Let $B := A/I$, and for $n \geq 0$, write A_n for A/I^n. Suppose that $\delta(f)$ is a unit mod I; then the composition of the ring homomorphisms:*

$$\phi : A \xrightarrow{\exp(\delta)} A[[t]] \to B[[t]] \ .$$

induces a compatible system of ring isomorphisms:

$$A_n \to B[t]/(t^n).$$

Proof. First observe that the composition of ϕ with the map $B[[t]] \to B$ sending t to 0 maps I to 0, and hence $\phi(I) \subset tB[[t]]$. Let us write $J = tB[[t]]$. To prove the lemma it suffices to show that for all $k \geq 0$ the induced map $I^k/I^{k+1} \to J^k/J^{k+1}$ is an isomorphism for all $k \geq 0$. We start with the case $k = 0$, where it is clear that $A/I \simeq B \to B[[t]]/tB[[t]] \simeq B$ is an isomorphism. For $k \geq 1$, I^k/I^{k+1} is a free rank one B-module generated by $f^k + I^{k+1}$ while J^k/J^{k+1} is a free rank one B-module generated by $t^k + J^{k+1}$. Since $\phi(f) = \delta(f)t + O(t^2)$, with $\delta(f)$ a unit, $\phi(f^k) = (\delta(f))^k t^k + O(t^{k+1})$ is a generator of J^k and it follows that ϕ induces an isomorphism $I^k/I^{k+1} \to J^k/J^{k+1}$ as desired. $\square$

Algebraization. We now show the existence of a formal isomorphism between $\mathcal{X}$ and a variety defined over a smaller field.

We may apply the lemma to $\mathcal{X}$ and the sheaf of ideals on $\mathcal{X}$ generated by f. The vector field ξ on $\mathcal{X}$ induces maps

$$\phi_n : \mathcal{X}_n \to Y \otimes_{F=R/(f)} A[t]/(t^n) \ ,$$

where Y is the fibre of $\mathcal{X}$ over the closed point of $\mathrm{Spec}(R)$, which by the lemma are isomorphisms, *i.e.*, ξ induces an isomorphism of formal schemes

$$\widehat{\phi} = \exp(\xi) : \widehat{\mathcal{X}} \to Y \widehat{\otimes}_{F=R/(f)} \widehat{\mathbb{A}}^1$$

where $\widehat{\mathcal{X}}$ is the formal scheme which is the formal completion of $\mathcal{X}$ with respect to the sheaf of ideals generated by f. This isomorphism is an isomorphism of formal schemes over the formal scheme $\mathrm{Specf}(\widehat{R} \simeq F[[t]])$, and hence by [9, Ch. III, Th. 5.4.1], since $\mathcal{X}$ and Y are both proper over $\widehat{R} \simeq F[[t]]$, $\exp(\xi)$ is algebraizable, *i.e.*, it is induced by an isomorphism of schemes (not just formal schemes!)

$$\phi : \mathcal{X} \otimes_R \widehat{R} \to Y \otimes_{F=R/(f)} (\widehat{R} \simeq F[[t]]).$$

Approximation. Now Artin Approximation [1] implies, since R is Henselian, that there exists an isomorphism

$$\phi' : \mathcal{X} \to Y \otimes_{F=R/(f)} R.$$

(Note that ϕ' may be chosen to agree with ϕ (and hence $\widehat{\phi}$) to any given finite order, but does not necessarily induce the map $\widehat{\phi}$.) Finally we observe that by taking $\otimes_R K$, we obtain an isomorphism

$$X = \mathcal{X} \otimes_R K \to Y \otimes_F K$$

and we are done.

3.2 Remark

The only place that we used that X is proper was in the algebraization of isomorphisms in the category of formal schemes over $\mathrm{Specf}(F[[t]])$. Thus Buium's result will hold for any subcategory of the category of schemes with this property.

4   Complements and Questions

4.1 Hasse-Schmidt Differentiation

It is natural to ask to what extent the methods used above to prove Buium's theorems can be used in characteristic p.

Note that the proof of Lemma 3.7 does not use that δ is a derivation, but only that ϕ is a ring homomorphism. In general a homomorphism from a ring R to the ring $R[[t]]$ of formal power series over R which, when composed with the augmentation $R[[t]] \to R$ which sends t to 0, is the identity, is known as a *Hasse-Schmidt Differentiation*. More generally we, define:

Definition 4.1 A *Hasse-Schmidt Differentiation* or *flow* ϕ on a scheme X is a map of formal schemes $\phi : X \widehat{\otimes}_{\mathbb{Z}} \mathbb{Z}[[t]] \to X$, which when composed with the map $X \to X \widehat{\otimes}_{\mathbb{Z}} \mathbb{Z}[[t]]$ given by setting $t = 0$ is the identity.

In order for the differentiation to be the exponential of a derivation, you need additional information:

Definition 4.2 Let X be a scheme over a base S, and let $\mathcal{G}$ be a one parameter formal group over S. A flow ϕ on a scheme X over S is said to be $\mathcal{G}$-*iterative*, with respect to S, if ϕ is an action of $\mathcal{G}$ on X.

Note that in characteristic zero, all one parameter formal groups are isomorphic to the formal additive group $\mathcal{G}_a$, in which case a flow is iterative if and only if it is the integral of a vector field. See [17].

In Lemma 3.7, we are given a scheme $p : \mathcal{X} \to S$, proper, and geometrically connected over the spectrum $S = \mathrm{Spec}(\Lambda)$ of a discrete valuation ring, and a flow ϕ on $\mathcal{X}$ with the property that if ξ is the associated vector field, and π is the generator of the maximal ideal in Λ then $\xi(\pi)$ is a unit modulo π. Of course once one no longer assumes that the flow is the exponential of the associated vector field, the flow is not determined by a finite amount of data; thus one cannot argue as in Section 3.1, Step 3, that one can reduce to a field of finite transcendence degree over the prime field.

Notice that given a flow $\phi : A \to A[[t]]$, if we write the flow as:

$$\phi : a \mapsto \sum_i D_i(a)t^i$$

then each D_i is a differential operator of order i. (See [11, §16] for the definition of a differential operator.) If ϕ is the exponential of a vector field, then $D_i = (D_1)^i/i!$.

Thus one can ask:

Question 4.3 *Suppose that X is a variety, projective and geometrically connected over a field k, possibly of characteristic greater than zero. Consider the algebra $\mathcal{D} = H^0(X, \mathcal{D}iff_k(\mathcal{O}_X, \mathcal{O}_X))$ of global sections of the sheaf $\mathcal{D}iff_k(\mathcal{O}_X, \mathcal{O}_X)$ of differential operators on $\mathcal{O}_X$. Then $\mathcal{D}$ acts on $H^0(X, \mathcal{O}_X) \simeq k$, and we ask whether X is defined over the field of constants $k^{\mathcal{D}}$, or at least over its algebraic closure.*

4.2 Derivations and Valuation Rings

Recall that a valuation ring is a local domain R such that if $x \in F$, F being the fraction field of R, then $x \notin R$ if and only if $1/x \in \mathfrak{m}$, $\mathfrak{m}$ being the maximal ideal of R. The quotient F^*/R^* is an abelian group which is totally ordered by $v(x) \geq 0$ if and only if $x \in R$, where $v : F^* \to F^*/R^*$ is the quotient map. Given a totally ordered abelian group Γ, a *valuation* v on F with values in Γ is a homomorphism $v : F^* \to \Gamma$, such that $v(x + y) \geq \mathrm{Min}(v(x), v(y))$. Given such a valuation, the set $\{x \in F | v(x) \geq 0\}$ is a subring R_v, which is easily seen to be a valuation ring with maximal ideal $\mathfrak{m}_v = \{x \in F | v(x) > 0\}$, and the valuation induces an (order preserving) injection $F^*/R_v^* \subset \Gamma$. Generally when we speak of a valuation on a field we assume that this inclusion is the identity. See [5, Chapter VI], [21], and [8] for more details on valuation rings.

Valuation rings are in general *not* noetherian, however they do still have some nice properties:

Lemma 4.4 *Let R be a valuation ring. Then*

(a) *Any finitely generated torsion free R-module is free.*

(b) *Any torsion free R-module is flat.*

Proof. Exercise; see [5, Lemma 1 of VI.3.6]. $\qquad\qquad\qquad\square$

Question 4.5 *Let R be a valuation ring containing a differential field of characteristic zero. Is there a "simple" proof (or at least a proof that is not equivalent to proving uniformization) that R^∞ is an integral domain?*

For example it would be enough to know that $\Omega_{R/k}$ is torsion free. For by Lemma 4.4 this would imply that $\Omega_{R/k} = \bigcup_\alpha F_\alpha$ is a union of free submodules, and hence that $R^{\{1\}} = \varinjlim \mathbb{S}_R^*(F_\alpha)$ is a direct limit of polynomial rings over R, where $R^{\{1\}}$ is the first stage in the sequence of rings of Section 1.2. $R^{\{1\}}$ is therefore an integral domain which is formally smooth over R. It is then easy to show that $R^{\{n\}}$ is formally smooth over R for all n. However I am told by at least one of the experts that this statement is quite probably equivalent to uniformization.

Another remark is that since R is a domain, we know by Kolchin's theorem that R^∞ has a unique minimal ideal. Hence it would also suffice to show that R^∞ is reduced.

References

1. Artin, M. *Algebraic approximation of structures over complete local rings*, Inst. Hautes Études Sci. Publ. Math. **36** (1969) 23–58.
2. Buium, A. *Differential algebra and Diophantine geometry*, Actualités Mathématiques, Hermann, Paris, 1994.
3. Buium, A. *Differential function fields and moduli of algebraic varieties*, Lect. Notes in Math. **1226**. Springer-Verlag, Berlin, 1986.
4. Buium, A. *Fields of definition of algebraic varieties in characteristic zero*, Compositio Math. **61**(3) (1987), 339–352.
5. Bourbaki, N. *Commutative Algebra. Chapters 1–7*, Translated from the French, Springer-Verlag, Berlin, 1998.

6. Bourbaki, N. *Algebra. Volume I. Chapters 1–3, Volume II, Chapters 4–7*, Translated from the French, Springer-Verlag, Berlin, 1989.

7. Eisenbud, D. *Commutative Algebra with a View toward Algebraic Geometry*, Graduate Texts in Mathematics **150**, Springer-Verlag, New York, 1995.

8. Endler, O. *Valuation Theory*, Universitext. Springer-Verlag, New York-Heidelberg, 1972.

9. Grothendieck, A. *Éléments de géométrie algébrique. III. Étude cohomologique des faisceaux cohérents, (Première Partie)*, Inst. Hautes Études Sci. Publ. Math. No. **11**, 1961.

10. Grothendieck, A. *Éléments de géométrie algébrique. IV. Étude locale des schémas et des morphismes de schémas, (Première Partie)*, Inst. Hautes Études Sci. Publ. Math. No. **20**, 1964.

11. Grothendieck, A. *Éléments de géométrie algébrique. IV. Étude locale des schémas et des morphismes de schémas, (Quatrième Partie)*, Inst. Hautes Études Sci. Publ. Math. No. **32**, 1967.

12. Hartshorne, R. *Algebraic Geometry*, Graduate Texts in Mathematics **52**, Springer-Verlag, New York-Heidelberg, 1977.

13. Herrlich, H., Strecker, G. E. *Category Theory: an Introduction*, Allyn and Bacon Series in Advanced Mathematics. Allyn and Bacon Inc., Boston, Mass., 1973.

14. Kolchin, E. R. *Differential Algebra and Algebraic Groups*, Pure and Applied Mathematics, Vol. **54**, Academic Press, New York-London, 1973.

15. MacLane, S. *Categories for the Working Mathematician*, Graduate Texts in Mathematics, Vol. **5**, Springer-Verlag, New York-Berlin, 1971.

16. Matsumura, H. *Commutative Algebra*, W. A. Benjamin, Inc., New York 1970.

17. Matsumura, H. *Commutative Ring Theory*, Cambridge University Press, Cambridge, 1986.

18. Popescu, D. *On Zariski's uniformization theorem, Algebraic Geometry, Bucharest 1982 (Bucharest, 1982)*, Lect. Notes in Math. **1056**, Springer-Verlag, Berlin, 1984, 264–296.

19. Raynaud, M. *Anneaux locaux henséliens*, Lect. Notes in Math. **169**, Springer-Verlag, Berlin-New York, 1970.

20. Zariski, O. *Local uniformization on algebraic varieties*, Ann. of Math. (2) **41** (1940), 852–896.

21. Zariski, O., Samuel, P. *Commutative Algebra. Vol. II*, The University Series in Higher Mathematics. D. Van Nostrand Co., Inc., Princeton, N. J.-Toronto-London-New York, 1960.

Differential Algebra and Related Topics, pp. 125–150
Proceedings of the International Workshop
Eds. L. Guo, P. J. Cassidy, W. F. Keigher & W. Y. Sit
© 2002 World Scientific Publishing Company

MODEL THEORY AND DIFFERENTIAL ALGEBRA

THOMAS SCANLON

Department of Mathematics, Evans Hall
University of California, Berkeley
Berkeley, CA 94720-3480, USA
E-mail: scanlon@math.berkeley.edu

We survey some of the main points of contact between model theory and differential algebra. We will discuss the development of differentially closed fields, which stands at the center of the relationship between the two branches of mathematics. We will also explore recent developments in model theory of real valued function, difference algebra, and other related structures.

1 Introduction

The origins of model theory and differential algebra (respectively, those of the foundations of mathematics and real analysis) may be starkly different in character, but in recent decades large parts of these subjects have developed symbiotically. Abraham Robinson recognized that the broad view of model theory could supply differential algebra with universal domains and differentially closed fields [33]. Not long after Robinson's insight, Blum observed that differentially closed fields instantiated Morley's abstruse totally transcendental theories [3]. Since then, differentially closed fields have served as proving grounds for pure model theory. In some cases, significant theorems of pure model theory were proven in the service of a deeper understanding of differential equations.

In this survey we discuss some of the main points of contact between model theory and differential algebra. As mentioned in the previous paragraph, the development of differentially closed fields stands at the center of this relationship. However, there have been other significant developments. Notably, differential algebra has been instrumental in the model theory of real valued functions. In related developments, model theorists have investigated difference algebra and more complicated structures in which derivations and valuations are connected.

2 Notation and conventions in differential algebra

We refer the reader to the introductory articles in this volume for more details on differential rings. In general, we use standard notations and conventions.

Definition 2.1 By a *differential ring* we mean a commutative, unital ring R given together with a distinguished nonempty finite set Δ of commuting derivations. If $|\Delta| = 1$, then we say that R is an *ordinary* differential ring. Otherwise, R is a *partial* differential ring. A *differential field* is a differential ring which is also a field.

Definition 2.2 Let R be a differential ring. The differential ring of differential polynomials over R in the variable X, denoted by $R\{X\}$, is the free object on the one generator X in the category of differential rings over R. More concretely, if $M(\Delta)$ is the free commutative monoid generated by Δ, then as an R-algebra, $R\{X\}$ is the polynomial ring $R[\{\mu(X)\}_{\mu \in M(\Delta)}]$. If $\partial \in \Delta$, then the action of ∂ on $R\{X\}$ is determined by $\partial \upharpoonright R = \partial$, $\partial(\mu(X)) = (\partial \cdot \mu)(X)$, and the sum and Leibniz rules for derivations. Similarly we can construct the differential polynomial ring $R\{X_1, \ldots, X_n\}$ in n variables.

If L/K is an extension of differential fields and $a \in L$, then $K\langle a \rangle$ is the differential subfield of L generated by a over K. If (R, Δ) is a differential ring, then we write $\mathcal{C}_R = \{r \in R \mid (\forall \partial \in \Delta)\ \partial(r) = 0\}$ for the ring of constants. In the case that R is a field, the ring of constants is also a field. If (R, Δ) is a differential ring and n is a positive integer, then a *Kolchin closed* subset of R^n is a set of the form $X(R) = \{a \in R^n \mid f(a) = 0 \text{ for all } f(x_1, \ldots, x_n) \in \Sigma\}$ for some set of differential polynomials in n variables $\Sigma \subset R\{X_1, \ldots, X_n\}$. In the case that (K, Δ) is a differential field, the Kolchin closed subsets of K^n comprise the closed sets of a topology, the *Kolchin topology*, on K^n. A finite Boolean combination of Kolchin closed sets is said to be *Kolchin constructible*.

3   What is model theory?

If you know the answer to this question, then you may want to skip or skim this section referring back only to look up conventions. If you are unfamiliar with model theory, then you want to consult a textbook on logic, such as [5,8,9], for more details.

Model theory is the systematic study of *models*. Of course, this answer invites the question: What is a model? While in common parlance, a model is a mathematical abstraction of some real system, problem or event; to a logician a model is the real object itself and *models* in the sense that it is a concrete realization of some abstract theory. More formally, a model $\mathfrak{M}$ is a nonempty set M given together with some distinguished elements, functions defined on certain powers of M, and relations on certain powers of M.

Example 3.1 A unital ring R is a model when we regard $0, 1 \in R$ as distinguished elements, and $+ : R \times R \to R$ and $\cdot : R \times R \to R$ as distinguished functions, and we have no distinguished relations.

Example 3.2 A non-empty partially ordered set $(X, <)$ is a model with no distinguished elements or functions but one distinguished binary relation, namely $<$.

The definition of *model* given above could stand refinement. The important feature of a model is not merely that it has some distinguished structure but that its extra structure is tied to a formal language.

Given a *signature*, σ, that is a choice of names for distinguished functions, relations and constants, we build the corresponding language by the following procedure. First, we construct all the σ-*terms*, the meaningful compositions of the distinguished function symbols applied to variables and constant symbols. For example, in Example 3.1, the expressions $+(x, 0)$ and $\cdot(+(x, y), 1)$ are σ-terms. Usually, for the sake of readability, we write these as $x + 0$ and $(x + y) \cdot 1$. Secondly, we form all the *atomic formulas* as the set of expressions of the form $t = s$ or $R(t_1, \ldots, t_n)$ where $t, s, t_1, \ldots, t_n$ are all terms and R is a distinguished (n-place) relation symbol of σ. In Example 3.1 the atomic formulas would be essentially equations between (not necessarily associative) polynomials while in Example 3.2 the atomic formulas would take the form $x = y$ and $x < y$ for variables x and y. By closing under finite Boolean operations (& (and), $\vee$ (or), $\to$ (implies), $\leftrightarrow$ (if and only if), and $\neg$ (not)), we form the set of *quantifier-free formulas*. By closing under existential and universal quantification over elements, we obtain the *language* $\mathcal{L}(\sigma)$. We say that the formula ψ is *universal* if it takes the form $(\forall x_0) \cdots (\forall x_m) \phi$ for some quantifier-free formula ϕ. In Example 3.1, the expression $(\forall x)(\exists y)[(x \cdot y = 1 + z)\ \&\ (x \cdot 0 = y)]$ is a formula. If M is a set on which each of the distinguished symbols of σ has been interpreted by actual functions and relations (M with such interpretations is called an $\mathcal{L}(\sigma)$-*structure*, which we denote by the corresponding symbol $\mathfrak{M}$), then every formula in the language $\mathcal{L}(\sigma)$ has a natural interpretation on $\mathfrak{M}$.

Formulas which take a truth value (under a specific interpretation of the distinguished function, relation, and constant symbols) are called *sentences*. For example, $(\forall x)[x + 1 = 0]$ is a sentence while $x + 1 = 0$ is not. If ϕ is an $(\mathcal{L}(\sigma))$-sentence and $\mathfrak{M}$ is an $\mathcal{L}(\sigma)$-structure in which ϕ is interpreted as true, then we write $\mathfrak{M} \models \phi$ and say that $\mathfrak{M}$ *models* ϕ. If Σ is a set of $\mathcal{L}(\sigma)$-sentences and $\mathfrak{M}$ is an $\mathcal{L}(\sigma)$-structure, then we say that $\mathfrak{M}$ is a model of Σ, written $\mathfrak{M} \models \Sigma$, if $\mathfrak{M} \models \phi$ for each $\phi \in \Sigma$. A set Σ of $\mathcal{L}(\sigma)$-sentences is called a (*consistent*) *theory* if there is some model of Σ. We say that the theory Σ is *complete* if for each $\mathcal{L}(\sigma)$-sentence φ, either in every model of Σ the sentence φ is true or in every model of Σ the sentence φ is false. The theory of the $\mathcal{L}(\sigma)$-structure $\mathfrak{M}$ is the set of all $\mathcal{L}(\sigma)$-sentences true in $\mathfrak{M}$. Note that the theory of a structure is necessarily complete. We say that $\Sigma \subseteq T$ is a set of *axioms* for T if $\mathfrak{M} \models \Sigma \Rightarrow \mathfrak{M} \models T$.

Let $\mathfrak{N}$ and $\mathfrak{M}$ be $\mathcal{L}(\sigma)$-structures. If $A \subseteq M$ is a subset, then there is a natural expansion of $\mathfrak{M}$ to a language, written $\mathcal{L}(\sigma)_A$, in which every element of A is treated as a distinguished constant.

Let ϕ be an $\mathcal{L}(\sigma)_M$-formula having (free) variables among $x_1, \ldots, x_n$ and let $\phi(a_1, \ldots, a_n)$ be the result of substituting a_i for each (free) occurrence of the variable x_i. A *definable set* in an $\mathcal{L}(\sigma)$-structure $\mathfrak{M}$ is a set of the form $\phi(M) := \{(a_1, \ldots, a_n) \in M^n \mid \mathfrak{M} \models \phi(a_1, \ldots, a_n)\}$. For example, if R is a commutative ring and $f(x, y) \in R[x, y]$ is a polynomial in two variables, then the set $\{(a, b) \in R^2 \mid f(a, b) = 0\}$ is definable. If $A \subseteq M$ is a subset and $X \subseteq M^n$ is a definable set, we say that X is A-*definable* if there is some formula $\phi \in \mathcal{L}(\sigma)_A$ with $X = \{a \in M^n \mid \mathfrak{M} \models \phi(a)\}$.

If $\mathfrak{M}$ is a $\mathcal{L}(\sigma)$-sub-structure of $\mathfrak{N}$, we say the inclusion $\mathfrak{M} \subseteq \mathfrak{N}$ is *elementary*, written $\mathfrak{M} \preceq \mathfrak{N}$, if $\mathfrak{M} \models \phi \Leftrightarrow \mathfrak{N} \models \phi$ for each $\mathcal{L}(\sigma)_M$-sentence ϕ. More generally, if $A \subseteq N$ is a subset of N and $\iota : A \to M$ is a function, we say that ι is *elementary* if $\mathfrak{N} \models \phi(a_1, \ldots, a_n) \Leftrightarrow \mathfrak{M} \models \phi(\iota(a_1), \ldots, \iota(a_n))$ for each formula $\phi \in \mathcal{L}(\sigma)$ and each n-tuples $(a_1, \ldots, a_n) \in A^n$.

In any $\mathcal{L}(\sigma)$-structure $\mathfrak{N}$, the definable sets form a sub-basis of clopen (closed and open) sets for a topology. We say that $\mathfrak{N}$ is *saturated* if, for every subset $A \subset M$ of strictly smaller cardinality, the class of A-definable sets has the finite intersection property. Under some mild set theoretic hypotheses it can shown that for any structure $\mathfrak{M}$ there is a saturated structure $\mathfrak{N}$ with $\mathfrak{M} \preceq \mathfrak{N}$.

4 Differentially closed fields

4.1 Universal domains and quantifier elimination

In Weil's approach to the foundations of algebraic geometry [42], a central role is played by the notion of a universal domain: an algebraically closed field into which every "small" field of its characteristic admits an embedding and for which every isomorphism between "small" subfields extends to an automorphism. One might ask whether a given system of polynomial equations has a solution in some extension field. This question is equivalent to the syntactically simpler question of whether the same system has a solution in the universal domain. While the foundations of algebraic geometry have shifted, these properties of algebraically closed fields remain at the heart of the subject. Anyone attempting to duplicate the success of algebraic geometry for differential algebraic geometry runs into the question of whether there are analogous universal domains for differential algebra.

It is not hard to state a version of the conditions on Weil's universal domains for general first-order theories.

Definition 4.1 Let $\mathcal{L}$ be a first-order language and T be a consistent $\mathcal{L}$-theory. We say that the model $\mathfrak{U} \models T$ is a *universal domain* for T if the following conditions hold:

- $|U| > |\mathcal{L}|$,

- If $\mathfrak{M} \models T$ and $|M| < |U|$, then there is an $\mathcal{L}$-embedding $f : \mathfrak{M} \to \mathfrak{U}$.

- If $\mathfrak{M} \subseteq \mathfrak{U}$, $|M| < |U|$, and $g : \mathfrak{M} \to \mathfrak{U}$ is an $\mathcal{L}$-embedding, then there is an $\mathcal{L}$-automorphism $\tilde{g} : \mathfrak{U} \to \mathfrak{U}$ with $\tilde{g} \restriction \mathfrak{M} = g$.

For example, if T is the theory of fields of characteristic zero expressed in $\mathcal{L}(0, 1, +, \cdot)$, then $\mathbb{C}$ is a universal domain. However, for many natural theories there are no universal domains. For example, the theory of groups has no universal domain and even the theory of formally real fields (fields in which -1 is *not* a sum of squares) admits no universal domain.

However, some of these theories which lack universal domains in the sense of Definition 4.1 admit a weaker completion. In our generalization of the notion of universal domain we have taken a category-theoretic approach; universality is defined by the existence of certain morphisms. We noted above that algebraically closed fields have the property that if some variety could have a point rational over some extension field, then it already has a point. By extending this principle of *everything that could happen, does* to general first order theories, Abraham Robinson arrived at the notion of model completeness (and the related notions of model completion and model companion, which we now define).

Definition 4.2 The theory T' is a *model companion* of the theory T if

- T and T' are *co-theories*: every model of T may be embedded in a model of T' and *vice versa* and

- every extension of models of T' is elementary: if $\mathfrak{M}, \mathfrak{N} \models T'$ and $\mathfrak{M} \subseteq \mathfrak{N}$, then $\mathfrak{M} \preceq \mathfrak{N}$.

There is a more refined notion of a model completion.

Definition 4.3 For a structure $\mathfrak{M}$, the *diagram* $\mathrm{diag}(\mathfrak{M})$ of $\mathfrak{M}$ is the set of quantifier-free sentences of $\mathcal{L}(\sigma)_M$ that are true in $\mathfrak{M}$. We say that T' is a *model completion* of T if T' is a model companion of T and for every model $\mathfrak{M} \models T$ of T, the theory $T' \cup \mathrm{diag}(\mathfrak{M})$ is complete and consistent.

If T has a model companion, then it has only one. In particular, if T has a model completion, it has only one.

Example 4.4 The theory of algebraically closed fields is the model completion of the theory of fields.

Example 4.5 The theory of real closed fields is the model companion of the theory of formally real fields. Considered with the signature of ordered rings: $(\{0, 1\}, \{+, \cdot\}, \{<\})$ it is the model completion of the theory of ordered fields.

Notably, the theory of differential fields of characteristic zero has a model completion.

Theorem 4.6 *The model completion of the theory of differential domains of characteristic zero is the theory of differentially closed fields of characteristic zero,* DCF_0.

We will define differentially closed fields later. For now, Theorem 4.6 takes a geometric form.

Proposition 4.7 *Let K be a differentially closed field of characteristic zero, let $X \subseteq K^n$ be Kolchin-constructible, let $m \leq n$, and let $\pi : K^n \to K^m$ be a projection onto m-coordinates. Then $\pi(X) \subseteq K^m$ is also Kolchin-constructible.*

The implication from Theorem 4.6 to Proposition 4.7 follows from a general result in logic. We say that a theory T is *universal* if it has a set of universal sentences as axioms. We say that the $\mathcal{L}$-theory T *eliminates quantifiers* if for any model $\mathfrak{M} \models T$ and any $\mathcal{L}_M$-formula ϕ there is some quantifier-free $\mathcal{L}_M$-formula ψ such that $\mathfrak{M} \models (\phi \leftrightarrow \psi)$. As a general result, if T' is a model completion of a universal theory T, then T' eliminates quantifiers.

The standard algebraic axioms for differential domains are universal so that this general result applies to DCF_0. A Kolchin constructible set in a differentially closed field is nothing more nor less than a definable set defined by a quantifier free formula. The projection of such a set is naturally defined by a formula with a string of $(m - n)$ existential quantifiers. As DCF_0 eliminates quantifiers, this set is also defined by a quantifier free formula and is therefore also Kolchin constructible.

For the remainder of this paper, we restrict our attention to *ordinary* differentially closed fields of characteristic zero. McGrail and Pierce have developed, independently, considerably more complicated axioms for partial differentially closed fields with n commuting derivations [26,29]. Many (but not all) of the theorems discussed here are valid for partial differentially closed fields as well.

There are a few reasonable ways to axiomatize DCF_0. There is a general procedure for finding axioms for the model companion of a given theory (if the model companion exists), but in practice, this procedure does not give a useful system of axioms. In general, it may force one to consider formulas with existential quantifiers ranging over arbitrarily many variables [3]. The system of axioms presented in Definition 4.9 is concise and requires only one existentially quantified variable. We begin by recalling notion of one differential polynomial being simpler than another.

Definition 4.8 If (K, ∂) is a differential ring and $f \in K\{X\}$ is a differential polynomial over K, then $f = F(X, \partial X, \ldots, \partial^n X)$ for some polynomial $F(X_0, \ldots, X_n)$ in $(n+1)$-variables. The least n for which f has this form is called the *order* of f. The *degree* of f is the degree of F as a polynomial in X_n. We say that the differential polynomial g is *simpler* than f if the order of g is less than the order of f or their orders are equal but g has smaller degree than does f.

Definition 4.9 A differential field of characteristic zero K is differentially closed if for each pair $f, g \in K\{x\}$ of differential polynomials with f irreducible and g simpler than f, there is some $a \in K$ with $f(a) = 0$ and $g(a) \neq 0$.

There are also systems of axioms for differentially closed fields based on geometric conditions. Before we can state the geometric axioms, developed by Hrushovski, Pierce and Pillay [30], we need to discuss jet spaces.

We use jet spaces to reduce problems in differential algebraic geometry to algebraic geometry. Informally, if X is a Kolchin closed set, then the nth jet space $\nabla_n X$ of X is the algebraic locus of the set of sequences of points in X together with all of their derivatives of order less than or equal to n. Let us give a more formal, though still naïve, definition. We need a bit of notation.

Definition 4.10 If Δ is a finite set of derivations, then by $M_n(\Delta)$ we mean the subset of derivative operators of order at most n in the free commutative monoid $M(\Delta)$ generated by Δ. Let $\|M_n(\Delta)\|$ be its cardinality.

Definition 4.11 If $X \subseteq K^m$ is a Kolchin closed subset of some Cartesian power of a differentially closed field (K, Δ), then the nth *jet space* $\nabla_n X$ of

X is the Zariski closure in $K^{m\|M_n(\Delta)\|}$ of $\{(\mu(a))_{\mu \in M_n(\Delta)} \mid a \in X\}$. The inclusions $M_n(\Delta) \subseteq M_k(\Delta)$ for $n \leq k$ induces corresponding projections $\pi_{k,n} : \nabla_k X \to \nabla_n X$. We identify X with $\nabla_0 X$ and we write π_k for $\pi_{k,0}$.

Our naïve definition of the jet spaces suffices for our present purposes. However, a more functorial version has proven its worth in many applications.

Proposition 4.12 *A differential field of characteristic zero K is differentially closed if and only if for any irreducible affine variety X over K and Zariski constructible set $W \subseteq \nabla_1 X$ with $\pi_1 \upharpoonright_W : W \to X$ dominant, there is some point $a \in X(K)$ with $(a, \partial a) \in W(K)$.*

4.2 Totally transcendental theories, Zariski geometries, and ranks

Model theory's contribution to differential algebra is not merely foundational. While model theory encompasses all first-order theories, strong theorems require strong hypotheses. Some of the deepest results are known for *totally transcendental* theories (one of which is the theory of differentially closed fields of characteristic zero).

Definition 4.13 Let T be a theory in the language $\mathcal{L}$. Let $\mathfrak{M} \models T$ be a universal domain for T. Let $\psi(x_1, \ldots, x_n)$ be a consistent $\mathcal{L}_M$ formula in the free variables $x_1, \ldots, x_n$. Recall that $\psi(M) = \{a \in M^n \mid \mathfrak{M} \models \psi(a)\}$. The *Morley rank* $\mathrm{RM}(\psi)$ of ψ is defined by the following recursion.

- $\mathrm{RM}(\psi) = -1$ if $\psi(M) = \varnothing$
- $\mathrm{RM}(\psi) \geq 0$ if $\psi(M) \neq \varnothing$
- $\mathrm{RM}(\psi) \geq \alpha + 1$ if there is some way to split $\psi(M)$ into infinitely many disjoint sets each of rank at least α. More precisely, the Morley rank of ψ is at least $\alpha + 1$ if there is some sequence $\{\phi_i\}_{i=1}^{\infty}$ of $\mathcal{L}_M$-formulas such that $\phi_i(M) \subseteq \psi(M)$ for each i, $\phi_i(M) \cap \phi_j(M) = \varnothing$ for $i \neq j$, and $\mathrm{RM}(\phi_i) \geq \alpha$ for all i.
- $\mathrm{RM}(\psi) \geq \lambda$ for a limit ordinal λ if $\mathrm{RM}(\psi) \geq \alpha$ for all $\alpha < \lambda$.
- $\mathrm{RM}(\psi) := \min\{\alpha : \mathrm{RM}(\psi) \geq \alpha \text{ but } \mathrm{RM}(\psi) \not\geq \alpha + 1\} \cup \{+\infty\}$.

We say T is *totally transcendental* if for every $\mathfrak{M} \models T$, and every consistent $\mathcal{L}_M$ formula $\psi = \psi(x)$, the definable set $\psi(M)$ has an ordinal valued Morley rank (that is, $\mathrm{RM}(\psi) \neq +\infty$).

Totally transcendental theories carry many other ordinal-valued ranks (Lascar, Shelah, local, *et cetera*). Applications of these ranks and their interconnections in differentially closed fields were studied in depth by Pong [32].

Hrushovski and Scanlon showed that these ranks are all distinct in differentially closed fields [19].

While many deep theorems have been proven about general totally transcendental theories, for all practical purposes, the theory of differentially closed fields is the only known mathematically significant theory to which the deeper parts of the general theory apply. For example, a theorem of Shelah on the uniqueness of prime models for totally transcendental theories implies the uniqueness of differential closures.

Definition 4.14 Let T be a theory, let $\mathfrak{M}$ be a model of T and let A be a subset of M. A *prime model of T over A* is a model $\mathfrak{P}$ of T with $A \subseteq \mathfrak{P} \subseteq \mathfrak{M}$ having the property that if $\iota : A \hookrightarrow N$ is an elementary map from A into any other model $\mathfrak{N} \models T$, then ι extends to an elementary embedding of $\mathfrak{P}$ into $\mathfrak{N}$.

Theorem 4.15 *If T is a totally transcendental theory, then for any model $\mathfrak{M} \models T$ and subset $A \subseteq M$ there is prime model over A. Moreover, the prime model is unique up to isomorphism over A.*

As a corollary, we have the existence and uniqueness of differential closures.

Corollary 4.16 *If K is a differential field of characteristic zero, then there is a differentially closed differential field extension K^{dif}/K, called the differential closure of K, which embeds over K into any differentially closed extension of K and which is unique up to K-isomorphism.*

The theory of algebraically closed fields is also totally transcendental and the prime model over a field K is its algebraic closure K^{alg}. The algebraic closure is also *minimal*. That is, if $K \subseteq L \subseteq K^{alg}$ with L algebraically closed, then $L = K^{alg}$. Kolchin, Rosenlicht, and Shelah independently showed that the differential closure does not share this property [23,34,38]

Theorem 4.17 *If K is a differential closure of $\mathbb{Q}$, then there are $\aleph_0$ differentially closed subfields of K.*

The non-minimality of differential closures results from the existence of *trivial* differential equations. In this context, *trivial* does not mean *easy* or *unimportant*. Rather, it means that an associated combinatorial geometry is degenerate.

Definition 4.18 Let $\mathcal{P}(S)$ denote the power set of a set S. A (*combinatorial*) *pregeometry* is a set S given together with a closure operator cl $: \mathcal{P}(S) \to \mathcal{P}(S)$ satisfying universally

- $X \subseteq \mathrm{cl}(X)$
- $X \subseteq Y \Rightarrow \mathrm{cl}(X) \subseteq \mathrm{cl}(Y)$
- $\mathrm{cl}(\mathrm{cl}(X)) = \mathrm{cl}(X)$
- If $a \in \mathrm{cl}(X \cup \{b\}) \setminus \mathrm{cl}(X)$, then $b \in \mathrm{cl}(X \cup \{a\})$.
- If $a \in \mathrm{cl}(X)$, then there is some finite $X_0 \subseteq X$ such that $a \in \mathrm{cl}(X_0)$.

If (S, cl) satisfies $\mathrm{cl}(\varnothing) = \varnothing$ and $\mathrm{cl}(\{x\}) = \{x\}$, then we say that (S, cl) is a (*combinatorial*) *geometry*.

Example 4.19 If S is any set and $\mathrm{cl}(X) := X$, then (S, cl) is a combinatorial geometry.

Example 4.20 If S is a vector space over a field K and $\mathrm{cl}(X) :=$ the K-span of X, then (S, cl) is a combinatorial pregeometry.

Example 4.21 If S is an algebraically closed field and $\mathrm{cl}(A)$ is the algebraic closure of the field generated by A, then (S, cl) is a combinatorial pregeometry.

Definition 4.22 The pregeometry (S, cl) is *trivial* if for any $X \in \mathcal{P}(S)$ one has $\mathrm{cl}(X) = \bigcup_{x \in X} \mathrm{cl}(\{x\})$.

Definition 4.23 If (S, cl) is a pregeometry, then a set $X \subseteq S$ is *independent* if for any $x \in X$ one has $x \notin \mathrm{cl}(X \setminus \{x\})$.

In a vector space, any two maximal linearly independent sets have the same size. Likewise, any two transcendence bases in an algebraically closed field have the same cardinality. These results are instances of a general principle for combinatorial pregeometries.

Proposition 4.24 *If (S, cl) is a pregeometry, $A \subseteq S$, and $X, Y \subseteq A$ are two maximal independent subsets of A, then $\|X\| = \|Y\|$. We define* $\dim(A) := \|X\|$.

Combinatorial pregeometries in which the dimension function is additive, called *locally modular*, are especially well-behaved.

Definition 4.25 A combinatorial pregeometry (S, cl) is *locally modular* if whenever $X, Y \subseteq S$ and $\dim(\mathrm{cl}(X) \cap \mathrm{cl}(Y)) > 0$ we have $\dim(\mathrm{cl}(X) \cap \mathrm{cl}(Y)) + \dim(\mathrm{cl}(X \cup Y)) = \dim(\mathrm{cl}(X)) + \dim(\mathrm{cl}(Y))$.

Example 4.26 If S is a vector space and cl : $\mathcal{P}(S) \to \mathcal{P}(S)$ is defined by $\text{cl}(X) :=$ the linear span of X, then the rank-nullity theorem of linear algebra shows that S is locally modular.

Example 4.27 If S is $\mathbb{C}$ and cl : $\mathcal{P}(S) \to \mathcal{P}(S)$ is defined by $\text{cl}(X) := \mathbb{Q}(X)^{alg}$, then S is *not* locally modular.

Totally transcendental theories supply some examples of combinatorial pregeometries as strongly minimal sets or as definable sets of Morley rank one having no infinite/co-infinite definable subsets. (We give a more down-to-earth definition below.) These strongly minimal sets are the backbone of (the finite rank part of) these theories.

Definition 4.28 Let $\mathfrak{M}$ be an $\mathcal{L}$-structure for some language $\mathcal{L}$. Let $\psi(x_1, \ldots, x_n)$ be some $\mathcal{L}$-formula with free variables among $x_1, \ldots, x_n$. We say that the set $D := \psi(M)$ is *strongly minimal* if $\psi(M)$ is infinite and for any $\mathfrak{M} \preceq \mathfrak{N}$ and any formula $\phi(x_1, \ldots, x_n) \in \mathcal{L}_N$ either $\psi(N) \cap \phi(N)$ is finite or $\psi(N) \cap (\neg\phi)(N)$ is finite.

Strongly minimal sets are the underlying sets of combinatorial pregeometries. The closure operator is given by model-theoretic algebraic closure.

Definition 4.29 Let $\mathfrak{M}$ be an $\mathcal{L}$-structure for some language $\mathcal{L}$. Let $A \subseteq M$ and let n be positive integer. We say that $a \in M^n$ is (*model theoretically*) *algebraic* over A if there is a formula $\psi(x) \in \mathcal{L}_A$ such that $\mathfrak{M} \models \psi(a)$ but $\psi(M)$ is finite. We denote by $\text{acl}(A)$ the set of all elements of M which are algebraic over A. If $X \subseteq M^n$, then $\text{acl}(X)$ is the algebraic closure of the set of coordinates of elements of X.

Example 4.30 If K is a differentially closed field and A is a subset of K, then $\text{acl}(A) = \mathbb{Q}\langle A \rangle^{alg}$.

Proposition 4.31 *Let D be a strongly minimal set. Let* cl : $\mathcal{P}(D) \to \mathcal{P}(D)$ *be defined by $X \mapsto \text{acl}(X) \cap D$. Then (D, cl) is a combinatorial pregeometry.*

Model theorists have been accused of obsession with algebraically closed fields. Zilber conjectured that for strongly minimal sets, the only interesting examples are algebraically closed fields.

Conjecture 4.32 (Zilber) *If D is a strongly minimal set whose associated pregeometry is not locally modular, then D interprets an algebraically closed field.*

Nevertheless, Hrushovski presented a procedure for producing families of counterexamples to Zilber's conjecture [15].

Theorem 4.33 *Zilber's conjecture is false in general.*

While the strong form of Zilber's conjecture about general strongly minimal sets is false, Hrushovski and Zilber salvaged it by imposing topological conditions [21].

Theorem 4.34 *Zilber's conjecture holds for Zariski geometries (strongly minimal sets satisfying certain topological and smoothness properties).*

Theorem 4.34 is especially relevant to differential algebra as Hrushovski and Sokolović showed that every strongly minimal set in a differentially closed field of characteristic zero is (essentially) a Zariski geometry [20].

Theorem 4.35 *Every strongly minimal set in a differentially closed field is a Zariski geometry after finitely many points are removed. Hence, Zilber's conjecture is true for strongly minimal sets in differentially closed fields. In fact, if D is a non-locally modular strongly minimal set defined in some differentially closed field K, then there is a differential rational function f for which $f(D) \cap \mathcal{C}_K$ is infinite.*

Theorem 4.35 is instrumental in the analysis of the structure of differential algebraic groups.

On general grounds, as shown by Hrushovski and Pillay, groups connected with locally modular strongly minimal sets have very little structure [17].

Theorem 4.36 *Suppose that $D_1, \ldots, D_n$ are locally modular strongly minimal sets, G is a definable group, and $G \subseteq \operatorname{acl}(D_1 \cup \cdots \cup D_n)$. Then every definable subset of any power of G is a finite Boolean combination of cosets of definable subgroups.*

We call a group satisfying the conclusion of Theorem 4.36 *modular*. While no infinite algebraic group is modular, there are modular differential algebraic groups. One can find these exotic groups as subgroups of abelian varieties.

Definition 4.37 An *abelian variety* is a projective connected algebraic group. A *semi-abelian variety* is a connected algebraic group S having a subalgebraic group T which (over an algebraically closed field) is isomorphic to a product of multiplicative groups with S/T being an abelian variety.

In his proof of the function field version of the Mordell conjecture, Manin introduced a differential algebraic group homomorphism on the points of an abelian variety rational over a finitely generated field [25]. Buium saw that Manin's homomorphisms are best understood in terms of differential algebraic geometry [4].

Theorem 4.38 *If A is an abelian variety of dimension g defined over a differentially closed field of characteristic zero K, then there is a differential rational homomorphism $\mu : A(K) \to K^g$ having a kernel with finite Morley rank.*

The kernel of μ is denoted by $A^\sharp$ and is called the *Manin kernel* of A. Generically, Manin kernels are modular [13].

Theorem 4.39 *If A is an abelian variety defined over an ordinary differentially closed field K and A admits no non-zero algebraic homomorphisms to abelian varieties defined over $\mathcal{C}_K$, then $A^\sharp(K)$ is modular.*

The modularity of Manin kernels has a diophantine geometric interpretation. We need some notation to state the theorem properly.

Definition 4.40 If G is a commutative group, then the torsion subgroup of G is $G_{tor} := \{g \in G \mid ng = 0 \text{ for some positive integer } n\}$.

Theorem 4.41 (Function field Manin-Mumford conjecture) *If A is an abelian variety defined over a field K of characteristic zero, A does not admit any nontrivial algebraic homomorphisms to abelian varieties defined over $\mathbb{Q}^{alg}$, and $X \subseteq A$ is an irreducible variety for which $X(K) \cap A(K)_{tor}$ is Zariski dense in X, then X is a translate of an algebraic subgroup of A.*

Proof: We can find a finite set Δ of derivations on K for which there are no non-trivial homomorphisms of algebraic groups from A to an abelian variety defined over K^Δ. Replacing K by a differential closure, we can show that the genericity condition on A continues to hold.

Since the additive group is torsion free, the Manin kernel $A^\sharp(K)$ must contain the torsion group of $A(K)$. Thus, if $X(K) \cap A(K)_{tor}$ is Zariski dense in X, then $X(K) \cap A^\sharp(K)$ is dense in X. However, as $A^\sharp(K)$ is modular, we know that $X(K) \cap A^\sharp(K)$ must be a finite Boolean combination of cosets of groups. Using the fact that X is closed and irreducible, we observe that $X(K) \cap A^\sharp(K)$ must be a translate of a group. By considering the stabilizer of X, we see that this implies that X itself is a translate of an algebraic subgroup of A. $\qquad\square$

The function field Mordell-Lang conjecture follows from Theorem 4.39 and a general result of Hrushovski on the structure of finite rank groups.

Definition 4.42 If G is a commutative group of finite Morley rank, then the *socle* $G^{\flat}$ of G is the

maximal connected definable subgroup of G for which $G^{\flat} \subseteq \mathrm{acl}(D_1 \cup \ldots \cup D_n)$ for some strongly minimal sets $D_1, \ldots, D_n$. In the above definition one assumes implicitly that G is saturated.

Example 4.43 If $G = A^{\sharp}$ is a Manin kernel, then $G^{\flat} = G$.

Definition 4.44 Let G be a group defined over some set A. We say that G is *rigid* if every subgroup of G is definable over $\mathrm{acl}(A)$.

Example 4.45 If G is an abelian variety, then $G^{\sharp}$ is rigid.

Assuming the rigidity of the socle, one can analyze the structure of a group G of finite Morley rank in terms of the structure on $G^{\flat}$ and on $G/G^{\flat}$ [13].

Proposition 4.46 *Let G be a group of finite Morley rank. Suppose that $G^{\flat}$ is rigid. If $X \subseteq G$ is a definable set with trivial (generic) stabilizer, then X is contained (up to a set of lower rank) in a coset of $G^{\flat}$.*

With Proposition 4.46 in place, we have all the main ingredients for a differential algebraic proof of the function field Mordell-Lang conjecture. For the sake of readability, we state a weaker version of the theorem than the one in [4,13].

Theorem 4.47 *Let A be an abelian variety defined over some field K of characteristic zero. Suppose that A is generic in the sense that there are no nontrivial homomorphisms of algebraic groups from A to any abelian variety defined over the algebraic numbers. If $\Gamma \leq A(K)$ is a finite dimensional subgroup $(\dim_{\mathbb{Q}}(\Gamma \otimes \mathbb{Q}) < \infty)$ and $X \subseteq A$ is an irreducible subvariety with $X(K) \cap \Gamma$ Zariski dense in X, then X is a translate of an algebraic subgroup of A.*

Proof: We observe that by passing to the quotient of A by the stabilizer of X we may assume that X has a trivial stabilizer. We are then charged with showing that X is a singleton.

As with the proof of Theorem 4.41, we replace K with a differentially closed field in such a way that A remains generic. That is, there are no nontrivial homomorphisms of algebraic groups from A to any abelian variety defined over $\mathcal{C}_K$.

Consider the Manin map $\mu : A(K) \to K^g$ (where $g = \dim A$). The finite dimensionality hypothesis on Γ implies that $\mu(\Gamma)$ is contained in a finite dimensional vector space over $\mathbb{Q}$. A *fortiori*, $\mu(\Gamma)$ is contained in a finite dimensional vector space, $\bar{\Gamma}$, over $\mathcal{C}_K$. All such vector spaces are definable by linear differential equations and have finite Morley rank. Let $\tilde{\bar{\Gamma}} := \mu^{-1}(\bar{\Gamma})$. Then $\tilde{\Gamma}$ is a group of finite Morley rank containing Γ and its socle is $A^\sharp(K)$.

If $X(K) \cap \Gamma$ is Zariski dense in X, then so is $X(K) \cap \tilde{\Gamma}$. By Proposition 4.46, $X(K) \cap \tilde{\Gamma}$ must be contained in a single coset of $A^\sharp(K)$ up to a set of lower rank. As in the proof of Theorem 4.41, the modularity of $A^\sharp(K)$ together with the irreducibility of X implies that X is a single point. $\square$

Of course, in stronger forms of Theorem 4.47, one can conlude that X is a translate of an algebraic subgroup of G [28]. However, the proof of Theorem 4.47 exhibits some uniformities not known to hold in the absolute case.

As a consequence of the geometric axioms for differentially closed fields, Proposition 4.46, and intersection theory, one can derive explicit bounds on the number of generic points on subvarieties of semiabelian varieties [18].

Theorem 4.48 *Let K be a finitely generated field extension of $\mathbb{Q}^{alg}$. Let G be a semiabelian variety defined over $\mathbb{Q}^{alg}$. Suppose that $X \subseteq G$ is an irreducible subvariety defined over $\mathbb{Q}^{alg}$ which cannot be expressed as $X_1 + X_2$ for some positive dimensional subvarieties X_1 and X_2 of G. If $\Gamma < G(K)$ is a finitely generated group, then the number of points in $\Gamma \cap (X(K) \setminus X(\mathbb{Q}^{alg}))$ is finite and may be bounded by an explicit function of geometric data.*

4.3 Generalized differential Galois theory

There is a general theory of definable automorphism groups in stable theories. Pillay observed that when specialized to the case of differentially closed fields, this theory gives a differential Galois theory which properly extends the Picard-Vessiot and Kolchin strongly normal Galois theories [31].

Definition 4.49 Let K be a differential field and X be a Kolchin constructible set defined over K. Let $\mathfrak{U} \supseteq K$ be a universal domain for differentially closed fields extending K. A differential field extension $K \subseteq L \subseteq \mathfrak{U}$ is called *X-strongly normal* if

- L is finitely generated over K as a differential field,

- $X(K) = X(L^{dif})$, and

- If $\sigma \in \text{Aut}(\mathfrak{U}/K)$ is a differential field automorphism of $\mathfrak{U}$ fixing K, then $\sigma(L) \subseteq L\langle X(\mathfrak{U})\rangle$.

The extension is called *generalized strongly normal* if it is X-strongly normal for some X.

Kolchin's strongly normal extensions are exactly the $C_\mathfrak{U}$-strongly normal extensions.

Theorem 4.50 *If L/K is an X-strongly normal extension, then there is a differential algebraic group $G_{L/K}$ defined over K and a group isomorphism $\mu : \mathrm{Aut}(L\langle X(\mathfrak{U})\rangle/K\langle X(\mathfrak{U})\rangle) \to G_{L/K}(\mathfrak{U})$. Moreover, there is a natural embedding $\mathrm{Aut}(L/K) \hookrightarrow \mathrm{Aut}(L\langle X(\mathfrak{U})\rangle/K\langle X(\mathfrak{U})\rangle)$ and with respect to this embedding we have $\mu(\mathrm{Aut}(L/K)) = G_{L/K}(K^{dif}) = G_{L/K}(K)$.*

As with Kolchin's differential Galois theory, we have a Galois correspondence between intermediate differential fields of $K \subseteq L$ and differential algebraic subgroups of $G_{L/K}$ defined over K.

Moreover, every differential algebraic group with finite rank may be realized as the differential Galois group of some generalized strongly normal differential field extension. Thus, as every differential Galois group of a Kolchin strongly normal extension is a group of constant points of an algebraic group over the constants and there are other differential algebraic groups of finite rank (Manin kernels, for example), differential Galois theory of generalized strongly normal extensions properly extends Kolchin's theory.

However, there are many finitely generated differential field extensions which are not generalized strongly normal. Trivial equations produce this phenomenon as well.

4.4 Classification of trivial differential equations

Theorem 4.34 implies that strongly minimal sets in differentially closed fields are either (essentially) algebraic curves over the constants or locally modular. On general grounds, locally modular strongly minimal sets are either (essentially) groups or trivial (in the sense of pregeometries). The theories of the field of constants and, as we have seen, locally modular groups are well-understood. We are left with the task of understanding trivial strongly minimal sets.

We begin by introducing the notion of *orthogonality* in order to give a precise sense to the parenthetical qualifier "essentially."

Definition 4.51 Let X and Y be strongly minimal sets. Let the projections from $X \times Y$ to X and Y be $\pi : X \times Y \to X$ and $\nu : X \times Y \to Y$. We say that X and Y are *non-orthogonal*, written $X \not\perp Y$, if there is an infinite definable set $\Gamma \subseteq X \times Y$ such that $\pi \upharpoonright_\Gamma$ and $\nu \upharpoonright_\Gamma$ are finite-to-one functions.

Theorem 4.35 may be restated as: *If X is a non-locally modular strongly minimal set in a universal domain $\mathfrak{U}$ for DCF_0, then $X \not\perp C_\mathfrak{U}$.*

Theorem 4.36 together with a general group existence theorem of Hrushovski implies that if X is a nontrivial, locally modular, strongly minimal set in a differentially closed field, then X is non-orthogonal to the Manin kernel of some simple abelian variety. Moreover, $A^\sharp \not\perp B^\sharp$ if and only if A and B are isogenous abelian varieties.

Question 4.52 How can one classify trivial strongly minimal sets in differentially closed fields up to nonorthogonality?

Question 4.53 Is there a structure theory for trivial strongly minimal sets in differentially closed fields analogous to the structure theory for locally modular groups?

It is possible for a general trivial strongly minimal set to have no structure whatsoever, but it is also possible for it to carry some structure. As an example, the set of natural numbers $\mathbb{N}$ given together with the successor function $S : \mathbb{N} \to \mathbb{N}$ defined by $x \mapsto x + 1$ is a trivial strongly minimal set.

The answers to these questions are unknown in general. In particular, it is not known whether there is some trivial strongly minimal set X definable in a differentially closed field having a definable function $f : X \to X$ with infinite orbits.

However, for *order one* trivial strongly minimal sets defined over the constants of an ordinary differentially closed field, there are satisfactory answers to these questions [16].

Definition 4.54 Let $K \subseteq \mathfrak{U}$ be a countable differential subfield of the universal domain. Let $X \subseteq \mathfrak{U}^n$ be a constructible set defined over K. We define the *order* of X to be the maximum of $\mathrm{tr.deg}_K K\langle x \rangle$ as x ranges over X.

Definition 4.55 Let X be a strongly minimal set defined over the set A. We say that X is *totally degenerate* if every permutation of X is induced by an element of $\mathrm{Aut}(\mathfrak{U}/A)$.

Generalizing a finiteness result of Jouanolou on hypersurface solutions to Pfaffian equations on certain compact complex manifolds [22], Hrushovski showed that order one sets are either essentially curves over the constants or essentially totally degenerate [11]. More precisely, we have the following theorem.

Theorem 4.56 *If X is an order one set defined over an ordinary differentially closed field K, then either $X \not\perp \mathcal{C}_K$ or there is some totally degenerate X' with $X \not\perp X'$.*

142

As a corollary of Theorem 4.56 we have a finiteness result on the number of solutions to order one equations.

Corollary 4.57 *Let $\mathfrak{U}$ be an ordinary differentially closed field. Let $f(x,y)$ be a nonzero polynomial in $\mathfrak{U}[x,y]$ with constant coefficients. Suppose we have $\{a \in \mathcal{U} : f(a,a') = 0\} \perp \mathcal{C}_{\mathfrak{U}}$. Then the number of solutions to $f(a,a') = 0$ in a differential field K is bounded by a function of $\mathrm{tr.deg}(K)$.*

Theorem 4.56 begs the question of whether there are any trivial sets. By directly analyzing differential equations, McGrail produced a family of trivial sets [27]. By producing a dictionary between properties of one forms on curves and properties of certain order one sets in ordinary differentially closed fields, Hrushovski and Itai produced families of examples of trivial order one sets [16].

4.5 Differential fields of positive characteristic

There has been significant development of the model theory of differential fields of positive characteristic. The theory of differential fields of characteristic p admits a model companion DCF_p, the theory of differentially closed fields of characteristic p [44]. This theory is not totally transcendental, but it shares some properties with totally transcendental theories. For example, Wood showed that positive characteristic differential closures exist and are unique [45].

However, differential fields satisfying fewer equations have proved to be more useful. The theory of separably closed fields of finite imperfection degree, which may be understood fruitfully in terms of differential algebra, underlies the proof of the positive characteristic Mordell-Lang conjecture.

5   O-minimal theories

Differential algebra has played a crucial role in the model theoretic analysis of well-behaved real-valued functions. The best behaved ordered structures are *o-minimal*: the definable subsets of the line are just finite Boolean combinations of points and intervals.

Definition 5.1 An *o-minimal* expansion of $\mathbb{R}$ is a σ-structure on $\mathbb{R}$ for some signature σ having a binary relation symbol $<$ interpreted in the usual manner such that for any $\mathcal{L}_{\mathbb{R}}(\sigma)$-formula $\psi(x)$ with one free variable x the set $\psi(\mathbb{R})$ is a finite union of intervals and points.

Example 5.2 $\mathbb{R}$ considered just as an ordered set is o-minimal. [Cantor]

Example 5.3 $\mathbb{R}$ considered as an ordered field is o-minimal. [Tarski [41]]

Remark 5.4 Tarski did not state his theorem on the real field in terms of o-minimality. Rather, he proved quantifier elimination for the real field in the language of ordered rings. O-minimality follows as an immediate corollary.

Theorem 5.5 (Wilkie [43]) *The expansion of $\mathbb{R}$ by the field operations and the exponential function is o-minimal.*

Behind the proof of Theorem 5.5 is a more basic theorem on expansions of $\mathbb{R}$ by restricted Pfaffian functions.

Definition 5.6 Let $f_1, \ldots, f_n$ be a sequence of differentiable real valued functions on $[0,1]^m$. We say that this sequence is a *Pfaffian chain* if $\frac{\partial f_i}{\partial x_j} \in \mathbb{R}[x_1, \ldots, x_m, f_1, \ldots, f_i]$ for each $i \leq n$ and $j \leq m$. We say that f is a *Pfaffian function* if f belongs to some Pfaffian chain.

Example 5.7 e^x restricted to the interval $[0,1]$ is Pfaffian.

Theorem 5.8 (Wilkie) *If $f_1, \ldots, f_n$ is a Pfaffian chain, then the structure $(\mathbb{R}, +, \cdots, <, f_1, \ldots, f_n)$ is o-minimal.*

Patrick Speisseger has generalized Wilkie's result to the case where the base structure is an arbitrary o-minimal expansion of $\mathbb{R}$ rather than simply the real field [40].

While the work on o-minimal expansions of $\mathbb{R}$ concerns, on the face of it, real valued functions of a real variable, it is often convenient to work with the ordered differential field of germs of functions at infinity.

Definition 5.9 A *Hardy field* is a subdifferential field H of the germs at $+\infty$ of smooth real-valued functions on the real line which is totally ordered by the relation $f < g \Leftrightarrow (\exists R \in \mathbb{R})(\forall x > R) f(x) < g(x)$.

If $\mathcal{R}$ is an o-minimal expansion of $\mathbb{R}$, then the set of germs at $+\infty$ of $\mathcal{R}$-definable functions forms a Hardy field $\mathcal{H}(\mathcal{R})$.

Hardy fields carry a natural differential valuation with the valuation ring being the set of germs with a finite limit and the maximal ideal being the set of germs which tend to zero.

Definition 5.10 Let (K, ∂) be a differential field. A *differential valuation* on K (in the sense of Rosenlicht) is a valuation v on K for which

- $v(x) = 0$ for any nonzero constant $x \in (K^\partial)^\times$,

- for any y with $v(y) \geq 0$ there is some ϵ with $\partial(\epsilon) = 0$ and $v(y - \epsilon) > 0$, and

- $v(x), v(y) > 0 \Rightarrow v(\frac{y\partial(x)}{x}) > 0$.

The logarithmic-exponential series $\mathbb{R}((t))^{LE}$ is obtained by closing $\mathbb{R}((t))$ under logarithms, exponentials, and generalized summation. $\mathbb{R}((t))^{LE}$ carries a natural derivation and differential valuation [7].

For many examples of o-minimal expansions $\mathcal{R}$ of $\mathbb{R}$, there is a natural embedding $\mathcal{H}(\mathcal{R}) \hookrightarrow \mathbb{R}((t))^{LE}$. These embeddings, which may be regarded as divergent series expansions, can be used to show that certain functions cannot be approximated by other more basic function. Answering a question of Hardy, one has the following theorem.

Theorem 5.11 *The compositional inverse to* $(\log x)(\log \log x)$ *is not asymptotic to any function obtained by repeated composition of semi-algebraic functions,* e^x, *and* $\log x$.

The empirical fact that many interesting Hardy fields embed into $\mathbb{R}((t))^{LE}$ suggests the conjecture that the theory of $\mathbb{R}((t))^{LE}$ is the model companion of the universal theory of Hardy fields.

Van der Hoeven has announced a sign change rule for differential polynomials over (his version of) $\mathbb{R}((t))^{LE}$. This result would go a long way towards proving the model completeness of $\mathbb{R}((t))^{LE}$ [10].

Aschenbrenner and van den Dries have isolated a class of ordered differential fields with differential valuations, H-fields, to which every Hardy field belongs. They show, among other things, that the class of H-fields is closed under Liouville extensions [1].

6 Valued differential fields

The model theory of valued differential fields, which serves as a framework for studying perturbed differential equations, has also been developed.

Definition 6.1 A D-*ring* is a commutative ring R together with an element $e \in R$ and an additive function $D : R \to R$ satisfying $D(1) = 0$ and $D(x \cdot y) = x \cdot D(y) + y \cdot D(x) + eD(x)D(y)$.

If (R, D, e) is a D-ring, then the function $\sigma : R \to R$ defined by $x \mapsto eD(x) + x$ is a ring endomorphism. If $e = 0$, then a D-ring is just a differential ring. If $e \in R^\times$ is a unit, then $Dx = \frac{\sigma(x) - x}{e}$ so that a D-ring is just a difference ring in disguise.

Definition 6.2 A valued D-field is a valued field (K, v) which is also a D-ring (K, D, e) and satisfies $v(e) \geq 0$ and $v(Dx) \geq v(x)$ for all $x \in K$.

Example 6.3 If (k, D, e) is a D-field and $K = k((\epsilon))$ is the field of Laurent series over k with D extended by $D(\epsilon) = 0$ and continuity, then K is a valued D-field.

Example 6.4 If (k, ∂) is a differential field of characteristic, we extend ∂ to $k((\epsilon))$ continuously with $\partial(\epsilon) = 0$. Let $\sigma : k((\epsilon)) \to k((\epsilon))$ be the map $x \mapsto \sum_{i=0}^{\infty} \frac{1}{n!} \partial^n(x) \epsilon^n$, and D be defined by $x \mapsto \frac{\sigma(x) - x}{\epsilon}$. Then $(k((\epsilon)), D, \epsilon)$ is a valued D-field.

Example 6.5 If k is a field of positive characteristic p and $\overline{\sigma} : k \to k$ is any automorphism of k, then there is a unique lifting of $\overline{\sigma}$ to an automorphism $\sigma : W(k) \to W(k)$ of the field of quotients of the Witt vectors of k. Define $D(x) := \sigma(x) - x$, then $(W(k), D, 1)$ is a valued D-field.

Definition 6.6 A valued D-field (K, v, D, e) is *D-henselian* if

- K has enough constants: $(\forall x \in K)(\exists \epsilon \in K) \ v(x) = v(\epsilon)$ and $D\epsilon = 0$ and
- K satisfies D-hensel's lemma: if $P(X_0, \ldots, X_n) \in \mathcal{O}_K[X_0, \ldots, X_n]$ is a polynomial with v-integral coefficients and for some $a \in \mathcal{O}_K$ and integer i we have $v(P(a, \ldots, D^n a)) > 0 = v(\frac{\partial P}{\partial X_i}(a, \ldots, D^n a))$, then there is some $b \in \mathcal{O}_K$ with $P(b, \ldots, D^n b) = 0$ and $v(a - b) > 0$.

D-henselian fields can serve as universal domains for valued D-fields [35,36].

Theorem 6.7 *The theory of D-henselian fields with $v(e) > 0$, densely ordered value group, and differentially closed residue field of characteristic zero is the model completion of the theory of equicharacteristic zero valued D-fields with $v(e) > 0$.*

There are refinements (with more complicated statements) of Theorem 6.7 with $v(e) \geq 0$ and restrictions on the valued group and residue field.

The relative theorem in the case of a lifting of a Frobenius on the Witt vectors, proved by Bélair, Macintyre, and Scanlon, may be the most important case [2,37].

Theorem 6.8 *In a natural expansion of the language of valued difference fields, the theory of the maximal unramified extension of $\mathbb{Q}_p$ together with an automorphism lifting the p-power Frobenius map eliminates quantifiers (in* an expansion of the language of valued D-fields having angular component functions and divisibility predicates on the value group) and is axiomatized by

- the axioms for D-henselian fields of characteristic zero,

- the assertion that the residue field is algebraically closed of characteristic p and that the distinguished automorphism is the map $x \mapsto x^p$, and

- the assertion that the valued group satisfies the theory of $(\mathbb{Z}, +, 0, <)$ with $v(p)$ being the least positive element.

There are a number of corollaries of Theorem 6.8, one of which is that the theory of the Witt vectors with the relative Frobenius is decidable.

7 Model theory of difference fields

Model theorists have also analyzed *difference algebra* in some depth. Since the main topic of this volume is differential algebra, we will contain ourself to a few highlights of the model theoretic work on difference algebra.

Definition 7.1 A *difference ring* is a ring R given together with a distinguished ring endomorphism $\sigma : R \to R$.

Difference algebra admits universal domains in a weaker sense than does differential algebra.

Proposition 7.2 *The theory of difference fields admits a model companion, ACFA. A difference field $(K, +, \cdot, \sigma, 0, 1)$ satisfies ACFA if and only if $K = K^{alg}$, $\sigma : K \to K$ is an automorphism, and for any irreducible variety X defined over K and irreducible Zariski constructible set $W \subseteq X \times \sigma(X)$ projecting dominantly onto X and onto $\sigma(X)$, there is some $a \in X(K)$ with $(a, \sigma(a)) \in W(K)$.*

Unlike DCF_0, the theory ACFA is *not* totally transcendental. However, it falls into the weaker class of *supersimple* theories for which many of the techniques and results of totally transcendental theories carry over. The analysis of ACFA preceded and stimulated the development of the general work on simple theories.

An analogue of Theorem 4.35 holds for ACFA [6]. As a consequence of this theorem, one can derive an effective version of the Manin-Mumford conjecture [14].

While it is essentially impossible to actually construct differentially closed fields, Hrushovski and Macintrye have shown that limits of Frobenius automorphisms provide models of ACFA [12,24].

Theorem 7.3 *Let $R := \prod_{n \in \omega,\, p \text{ prime}} \mathbb{F}_{p^n}^{alg}$. Let $\sigma : R \to R$ be defined by $(a_{p^n}) \mapsto (a_{p^n}^{p^n})$. If $\mathfrak{m} \subseteq R$ is a maximal ideal for which $R/\mathfrak{m}$ is not locally finite, then $(R/\mathfrak{m}, \overline{\sigma}) \models \text{ACFA}$.*

A slight strengthening of Theorem 7.3 based on the Chebotarev Density Theorem may be expressed more meaningfully, if less algebraically, as *The theory of the generic automorphism is the limit of the theories of the Frobenius.* This means that if ϕ is a sentence in the language of difference rings, then ϕ is true in some model of ACFA if and only if there are infinitely many prime powers q such that $(\mathbb{F}_q^{alg}, (x \mapsto x^q)) \models \phi$.

Acknowledgments

This work was partially supported by an NSF grant, DMS-0078190. This paper stems from lecture notes of the author for a talk given at Rutgers University in Newark on 3 November 2000 as part of the Workshop on Differential Algebra and Related Topics. The author thanks Phyllis Cassidy, Li Guo, William Keigher and William Sit for inviting him to speak and for organizing such a successful meeting of the disparate strands of the differential algebra community. He thanks also the referee for carefully reading an earlier version of this note and for suggesting many improvements.

References

1. Aschenbrenner, M., van den Dries, L. *H-fields and their Liouvillian extensions*, Mathematische Zeitschrift (to appear).
2. Bélair, L., Macintyre, A., Scanlon, T. *Model theory of Frobenius on Witt vectors*, preprint, 2001.
3. Blum, L. *Generalized algebraic structures: a model theoretical approach*, Ph.D. Thesis, MIT, 1968.
4. Buium, A. *Intersections in jet spaces and a conjecture of S. Lang*, Ann. of Math. (2) **136**(3) (1992), 557–567.
5. Chang, C. C., Keisler, J. *Model Theory*, 3rd ed., Stud. Logic Found. Math. **73**, North-Holland Publishing Co., Amsterdam, 1990.
6. Chatzidakis, Z., Hrushovski, E., Peterzil, Y. *Model theory of difference fields, II*, J. London Math. Soc, (to appear[1]).

---

[1] http://www.logique.jussieu.fr/www.zoe/papiers/ACFAp.dvi

7. van den Dries, L., Macintyre, A., Marker, D. *The elementary theory of restricted analytic fields with exponentiation*, Ann. of Math. (2) **140**(1) (1994), 183–205.

8. Enderton, H. *A Mathematical Introduction to Logic*, Academic Press, New York-London, 1972.

9. Hodges, W. *Model Theory*, Encycl. Math. Appl. **42**, Cambridge University Press, Cambridge, 1993.

10. van der Hoeven, J. *Asymptotique automatique*, Thèse, Université Paris VII, Paris, 1997.

11. Hrushovski, E. *ODEs of order one and a generalization of a theorem of Jouanolou*, manuscript, 1996.

12. Hrushovski, E. *The first-order theory of the Frobenius*, preprint, 1996.

13. Hrushovski, E. *The Mordell-Lang conjecture for function fields*, J. Amer. Math. Soc. **9**(3) (1996), 667–690.

14. Hrushovski, E. *The Manin-Mumford conjecture and the model theory of difference fields*, Annals of Pure and Applied Logic **112**(1) (2001), 43–115.

15. Hrushovski, E. *A new strongly minimal set*, Stability in Model Theory, III, Trento, 1991; Ann. Pure Appl. Logic **62** (2) (1993), 147–166.

16. Hrushovski, E. Itai, M. *On model complete differential fields*, preprint, 1998.

17. Hrushovski, E., Pillay, A. *Weakly normal groups*, Logic Colloquium '85 (Orsay, 1985), 233–244, Stud. Logic Found. Math., **122**, North-Holland, Amsterdam, 1987.

18. Hrushovski, E., Pillay, A. *Effective bounds for the number of transcendental points on subvarieties of semi-abelian varieties*, Amer. J. Math. **122** (3) (2000), 439–450.

19. Hrushovski, E., Scanlon, T. *Lascar and Morley ranks differ in differentially closed fields*. J. Symbolic Logic **64**(3) (1999), 1280–1284.

20. Hrushovski, E., Sokolović, Ž. *Strongly minimal sets in differentially closed fields*, Trans. AMS, (to appear).

21. Hrushovski, E., Zilber, B. *Zariski geometries*, J. Amer. Math. Soc. **9**(1) (1996), 1–56.

22. Jouanolou, J. P. *Hypersurfaces solutions d'une équation de Pfaff analytique*, Math. Ann. **232** (3) (1978), 239–245.

23. Kolchin, E. R. *Constrained extensions of differential fields*, Advances in Mathematics **12**(2) (1974), 141 – 170.

24. Macintyre, A. *Nonstandard Frobenius automorphisms*, manuscript, 1996.

25. Manin, Y. *Proof of an analogue of Mordell's conjecture for algebraic curves over function fields*, Dokl. Akad. Nauk SSSR **152** (1963), 1061–1063.

26. McGrail, T. *The model theory of differential fields with finitely many commuting derivations*, J. Symbolic Logic **65**(2) (2000), 885–913.
27. McGrail, T. *The search for trivial types*, Illinois J. Math. **44**(2) (2000), 263–271.
28. McQuillan, M. *Division points on semi-abelian varieties*, Invent. Math. **120**(1) (1995), 143–159.
29. Pierce, D. *Differential forms in the model theory of differential fields*, preprint[2], 2001.
30. Pierce, D., Pillay, A. *A note on the axioms for differentially closed fields of characteristic zero*, J. Algebra **204**(1) (1998), 108–115.
31. Pillay, A. *Differential Galois theory I*, Illinois J. Math. **42**(4) (1998), 678–699.
32. Pong, W. Y. *Some applications of ordinal dimensions to the theory of differentially closed fields*, J. Symbolic Logic **65**(1) (2000), 347–356.
33. Robinson, A. *On the concept of a differentially closed field*, Bull. Res. Council Israel Sect. F 8F (1959), 113–128.
34. Rosenlicht, M. *The nonminimality of the differential closure*, Pacific J. Math. **52** (1974), 529–537.
35. Scanlon, T. *Model theory of valued D-fields*, Ph.D. thesis, Harvard University, 1997.
36. Scanlon, T. *A model complete theory of valued D-fields*, J. Symbolic Logic **65**(4) (2000), 1758–1784.
37. Scanlon, T. *Quantifier elimination for the relative Frobenius*, Proceedings of the First International Conference on Valuation Theory, Saskatoon, 1999 (to appear).
38. Shelah, S. *Differentially closed fields*, Israel J. Math. **16** (1973), 314–328.
39. Shelah, S. *Classification Theory and the Number of Nonisomorphic Models*, 2nd ed., Stud. Logic Found. Math. **92**, North-Holland Publishing Co., Amsterdam, 1990.
40. Speissegger, P. *The Pfaffian closure of an o-minimal structure*, J. Reine Angew. Math. **508** (1999), 189–211.
41. Tarski, A. *Sur les clases d'ensembles définissables de nombres reéls*, Fundamenta Mathematicae **17** (1931), 210–239.
42. Weil, A. *Foundations of Algebraic Geometry*, A.M.S. Colloq. Publ. **29**. Amer. Math. Soc., New York, 1946.
43. Wilkie, A. *Model completeness results for expansions of the ordered field of real numbers by restricted Pfaffian functions and the exponential function*, J. Amer. Math. Soc. **9**(4) (1996), 1051–1094.

---

[2]Available at http://www.math.metu.edu.tr/~dpierce/papers/differential

44. Wood, C. *The model theory of differential fields of characteristic $p \neq 0$*, Proc. Amer. Math. Soc. **40** (1973), 577–584.

45. Wood, C. *Prime model extensions for differential fields of characteristic $p \neq 0$*, J. Symbolic Logic **39** (1974), 469–477.

Differential Algebra and Related Topics, pp. 151–169
Proceedings of the International Workshop
Eds. L. Guo, P. J. Cassidy, W. F. Keigher & W. Y. Sit
© 2002 World Scientific Publishing Company

INVERSE DIFFERENTIAL GALOIS THEORY

ANDY R. MAGID

Department of Mathematics,
University of Oklahoma,
Norman OK 73019, USA
E-mail: amagid@ou.edu

Let F be a differential field with field of constants C, which we assume to be algebraically closed and of characteristic 0. Let G be an algebraic group over C. The *Inverse Galois Problem* for G over F asks whether there is a differential Galois extension $E \supset F$ whose differential Galois group $G(E/F)$ is isomorphic to G. More generally, if $E \supset F$ is a differential Galois extension and $\phi : G \to G(E/F)$ is an epimorphism of algebraic groups over C, the *Lifting Problem* for G and E over F asks whether there is a differential Galois extension $K \supset F$ containing E with $G(K/F)$ isomorphic to G such that the natural map $G(K/F) \to G(E/F)$ becomes isomorphic to ϕ. (The inverse problem is the lifting problem for the case $E = F$.) The easiest contexts in which to understand the problems are when G is connected algebraic and, for the lifting problem, when the kernel of ϕ is commutative unipotent or a torus, and this tutorial will focus on those.

1   Introduction

This tutorial assumes that the reader is familiar with the rudiments of differential Galois theory as well as the usual "polynomial" Galois theory. Of course the latter, strictly speaking, is a subset of the former, at least in the characteristic zero context which will be considered here. In addition to the tutorial of M. van der Put in this volume, which is the obvious initial reference, we point the reader to the definitive account of differential Galois theory in the forthcoming monograph by van der Put and M. Singer [8], as well as the author's lecture notes [5]. These treatments emphasize a special case of the theorem, due to Ellis Kolchin (as is much of the rest of the subject) [1], that in a differential Galois extension $E \supset F$ with (connected linear algebraic) Galois group G, E (as a field) can be regarded as the field of rational functions on the group G with scalars F, and that in this identification, the action of G as Galois group (as differential field automorphisms) becomes the translation action of G on rational functions. Actually Kolchin's theorem asserts, in the general case, that E is the function field of a principal homogeneous space for

---

[1]The author was recently (January, 2001) informed by A. Thaler of the National Science Foundation that he received research support as a junior investigator under Kolchin's NSF grant as a Ritt Instructor at Columbia. A quarter century elapsed between that support and the author's publications in the subject.

G over F. In this form, it subsumes the similar result from polynomial Galois theory, in which the isomorphism

$$\phi : E \otimes_F E \to C(G, E) \quad \phi(a \otimes b)(\sigma) = a\sigma(b)$$

($C(G, E)$ denotes the E valued functions on G) exhibits E as a principal homogeneous space for G over F.

The import of Kolchin's theorem, at least for the case when G is connected linear algebraic, is that the structure, as a *field* extension, of a differential Galois extension with group G is simply $E = F(G) \supset F$. To make this field extension a *differential* one, then, is a matter of finding an appropriate derivation D_E of E extending that of F, and making the extension a differential Galois one such that G is the differential Galois group. This requires, in part, that the natural translation action of G on $F(G)$ be as differential automorphisms, or that D_E be a G-invariant derivation of $F(G)$. The G-invariant derivations of $F(G)$ are $F \otimes \text{Lie}(G)$, so having a differential Galois extension with group G means there is a suitable derivation in $F \otimes \text{Lie}(G)$ making $F(G) \supset F$ a differential Galois extension.

So if one begins with a differential field F, and a connected, linear, algebraic group G, the task of finding a differential Galois extension of F with group G is equivalent to finding a suitable derivation in $F \otimes \text{Lie}(G)$. The apparent simplicity of this problem versus the corresponding problem in polynomial Galois theory is obvious: in polynomial Galois theory, finding the field extension E of the field F is everything, whereas in the differential context the field extension is already present. Nonetheless, the problem has proved challenging in the differential context as well.

1.1 *Picard-Vessiot extensions*

To properly state the inverse Galois problem, we begin by recalling the definition of Picard-Vessiot extensions, which are the differential Galois extensions with which we will deal. The notation and conventions established in this section will be followed throughout this tutorial.

F denotes a differential field of characteristic zero with derivation $D = D_F$ and algebraically closed field of constants $C = \{a \in F \mid D(a) = 0\}$.

$E \supset F$ is a *Picard-Vessiot*, or *Differential Galois* extension for an order n monic linear homogeneous differential operator L over F,

$$L = Y^{(n)} + a_{n-1}Y^{(n-1)} + \cdots + a_1 Y^{(1)} + a_0 Y \tag{1.1}$$

$a_i \in F$, Y a differential indeterminate, if

(a) E is a differential field extension of F generated over F by

$$V = \{y \in E \mid L(y) = 0\},$$

(b) The constants of E are those of F ("no new constants"), and

(c) $\dim_C(V) = n$ ("full set of solutions").

(See [5, Defn. 3.2, p.24].)

Let $G(E/F) = \operatorname{Aut}_F^{\operatorname{diff}}(E)$ be the group of differential field automorphisms of E over F. Because E is generated over F by $V = L^{-1}(0)$, the restriction map $G(E/F) \to GL(L^{-1}(0))$ is an injection (which turns out to have Zariski closed image). Note that $GL(V)$ is an algebraic group over the field of constants C of F and E.

With these definitions and notational conventions, we can state the relevant fundamental theorem of differential Galois theory:

Fundamental Theorem for Picard-Vessiot Extensions *Let $E \supset F$ be a Picard-Vessiot extension. Then $G = G(E/F)$ has a canonical structure of affine algebraic group over C and there is a one-one lattice inverting correspondence between differential subfields K, $E \supset K \supset F$, and Zariski closed subgroups H of G given by $K \mapsto G(E/K)$ and $H \mapsto E^H$. ($E^H = \{a \in E \mid h(a) = a \text{ for all } h \in H\}$.) If E is itself a Picard-Vessiot extension, then the restriction map $G \to G(K/F)$ is a surjection with kernel $G(E/K)$. If H is normal in G, then E^H is a Picard-Vessiot extension.*

The theorem can be derived directly, but for purposes of the inverse problem, it is best to see it as a consequence of Kolchin's theorem on the structure of differential Galois extensions as principal homogeneous spaces:

Principal Homogeneous Space Theorem *Let $E \supset F$ be a Picard-Vessiot extension with $G(E/F) = G$. Let $R \subseteq E$ be the differential integral closure of F in E. Then R is a finitely generated differential F-algebra with quotient field E differentially and G equivariantly isomorphic to $C[G]$ over the algebraic closure $\overline{F}$:*

$$\overline{F} \otimes_F R \cong \overline{F} \otimes_C C[G].$$

If G is connected, and R has an F-point, then R is G equivariantly and differentially isomorphic to $F \otimes_C C[G]$.

Here by "differential integral closure" we mean:

$$R = \{a \in E \mid \dim_C(\langle Ga \rangle) < \infty\}$$
$$= \{a \in E \mid L(a) = 0 \text{ for some operator } L \text{ as in Equation (1.1)}\}.$$

For a proof of the Principal Homogeneous Space Theorem, see [5, Thm. 5.12, p.67]; for a proof of the Fundamental Theorem from the Principal Homogeneous Space Theorem see [5, Thm. 6.5, p.77].

1.2 Statement of the inverse problem

Although differential Galois theory includes more general situations than the Picard-Vessiot case, we will restrict our attention to that case from this point, and use "differential Galois" as a synonym for "Picard-Vessiot". Before formulating the inverse problem, we note the direct problem:

Direct Problem: given L, find a Picard-Vessiot extension $E \supset F$ for L and determine the group $G = G(E/F)$.

Then the inverse problems are formulated as the converse:

Inverse Problem: given an algebraic group G over C, find an operator L and Picard-Vessiot extension $E \supset F$ for L such that $G(E/F)$ is isomorphic to G.

Lifting Problem: if $E \supset F$ is a differential Galois extension and $\phi : G \to G(E/F)$ is an epimorphism of algebraic groups over C, find a Picard-Vessiot extension $K \supset F$ containing E with $G(K/F)$ isomorphic to G such that the natural map $G(K/F) \to G(E/F)$ becomes isomorphic to ϕ.

The inverse problem is a natural one for differential Galois theory, just as it is for polynomial Galois theory. The most interesting case of the inverse problem for polynomial Galois theory, for its naturalness and its (to date) intractability, is the case of the base field $\mathbb{Q}$. Similarly, a natural case for the differential inverse Galois problem is the base field $\mathbb{C}(t)$ with derivation $\frac{d}{dt}$.

Ellis Kolchin [1], at the Moscow International Congress of Mathematicians in 1966, called attention to the Inverse Problem in this classical case of base field $\mathbb{C}(t)$, as well as the analogy with polynomial Galois theory. A little over a decade later, in 1979, Carol Tretkoff and Marvin Tretkoff [10] observed that the theorem from analysis that "every discrete group is a monodromy group" implied a solution of the Inverse Problem in the classical case. During the 60's and 70's, Jerald Kovacic [2], [3] had developed an algebraic methodology, more precisely Lie methods, for the Inverse Problem. Then in the mid-90's, Claudine Mitschi and Michael Singer [6] built on Kovacic's work to solve the Inverse Problem in the classical case for connected linear groups by algebraic methods. For an account of this and related work, see [7]. For the inverse problem for finite groups, see [9].

2   The derivation approach to the inverse problem

From now on, we assume G is a connected linear algebraic group over C, the algebraically closed field of constants of the differential field F. We seek a Picard-Vessiot extension $E \supset F$ with $G(E/F) \cong G$. We are going to assume that the differential integral closure of F in E has an F-point. Then by the Principal Homogeneous Space Theorem above, E is G-isomorphic to the quotient field $F(G)$ of $R = F[G] = F \otimes_C C[G]$ where the G-action on R is by translation. We only need to determine the structure of E as a differential field, namely the derivation on E. We state this as a proposition.

Proposition 2.1 *Let G be a connected linear algebraic group, let $R = F[G]$ and suppose D_R is a G-equivariant derivation of R extending D_F such that the quotient field $F(G)$ of R has the same constants as F. Then (using the extension of D_R to the quotient field), $F(G) \supset F$ is a Picard-Vessiot extension with $G(F(G)/F) \cong G$.*

To prove the proposition, one uses the fact that a differential field extension $E \supset F$ that is generated over F, has no new constants, has a group Γ of differential automorphisms fixing F such that the fixed field $E^{\Gamma} = F$, and is generated over F as a differential field by a finite dimensional Γ-stable vector space V over F is necessarily a Picard-Vessiot extension. Clearly $F(G)$ meets the latter two of these conditions, with $\Gamma = G$ and V any G-stable finite dimensional subspace of $C[G]$ generating $C[G]$, and the hypothesis on D guarantees the first condition. This makes $F(G) \supset F$ a Picard-Vessiot extension, and then one can show that $G(F(G)/F) = G$.

The definition of Picard-Vessiot extension referred to an operator L. In the present case, an operator for the extension is obtained by taking Wronskians:

$$L = \frac{w(Y, \alpha_1, \dots, \alpha_n)}{w(\alpha_1, \dots, \alpha_n)}$$

where $\alpha_1, \dots \alpha_n$ is a C basis of V and Y is a differential indeterminate. We mention this here in connection with Kolchin's posing of the Inverse Galois Problem: he actually asked not just for the extension, but for the operator L realizing G as the differential Galois group of a Picard-Vessiot extension of F for L.

We return to the notation of the proposition. The derivations of $F(G)$ invariant under the translation action of G are $F \otimes_C \mathrm{Lie}(G)$. Suppose that $D_1, \dots, D_n$ is a C-basis of $\mathrm{Lie}(G)$. Then the proposition tells us that we can solve the inverse Galois problem for G provided we can find $\phi_1, \dots, \phi_n \in F$

such that the derivation

$$D := D_F \otimes 1 + \sum_{i=1}^{n} \phi_i \otimes D_i$$

defined on $F(G)$ has only C as constants.

There are no elementary general conditions to tell precisely when a derivation D of a differential integral domain S-finitely generated over F as an F-algebra has the property that its field of quotients E is a no new constants extension of F. One powerful sufficient condition is that S have no non-trivial differential ideals [5, Cor. 1.18, p.11], which is important for proving the existence of Picard-Vessiot extensions, but is neither necessary nor particularly easy to verify.

Of course one can write the *no new constants* condition naively: namely, if $h = \frac{f}{g}$, $f, g \in S$ is a constant then

$$0 = D\left(\frac{f}{g}\right) = \frac{gD(f) - fD(g)}{g^2} \Rightarrow fD(g) = gD(f)$$

so the criteria should be

$$\forall f, g \in S, g \neq 0, fD(g) = gD(f) \Rightarrow \frac{f}{g} \in C.$$

While one can write this down for the case $S = F[G]$ and $D = D_F \otimes 1 + \sum_{i=1}^{n} \phi_i \otimes D_i$, this general formulation rapidly becomes unwieldy. For example we invite the reader to write down just the expression $fD(g)$ when $f = \sum_{i=1}^{n} \alpha_i \otimes f_i$ and $g = \sum_{i=1}^{n} \beta_i \otimes g_i$.

In case S is a unique factorization domain (UFD), then we can take f and g relatively prime, and then the equation $fD(g) = gD(f)$ implies that $f | D(f)$ and $g | D(g)$, say $D(f) = a_f f$ and $D(g) = a_g g$, and then the original equality implies that $a_f = a_g$. On the other hand, having f, g in $F[G]$ with $D(f) = af$ and $D(g) = ag$ certainly implies that that $\frac{f}{g}$ is a constant; for it to be a new constant we must have $\frac{f}{g} \notin C$, which is automatic if f and g are relatively prime. Thus, for the case of S a UFD, "no new constants in E" is equivalent to "no relatively prime f, g in S and a in S with $D(f) = af$ and $D(g) = ag$".

This sort of reasoning can be pushed a little further. Obviously if $D(f) = af$ with $a = 0$ then f is already a constant. So suppose we already know that the constants of S are C. If $D(f) = af$ and $f = f_1 f_2$ with f_1, f_2 relatively prime, then

$$af_1 f_2 = D(f_1 f_2) = f_1 D(f_2) + D(f_1)f_2$$

implies that $f_i | D(f_i)$, say $D(f_i) = a_i f_i$, and then that

$$a f_1 f_2 = (a_1 + a_2)(f_1 f_2);$$

and if $f = p^k$ is a power then $D(f) = af$ implies that

$$ap^k = D(p^k) = kp^{k-1} D(p).$$

If $a = 0$, then $f \in C$. Otherwise, we can divide the above by k and p^{k-1} to conclude that

$$D(p) = \frac{a}{k} p.$$

Finally, suppose that $D(f) = af$ and $D(g) = ag$ with f and g relatively prime. Suppose $f = p_1^{e_1} \dots p_k^{e_k}$ and $g = q_1^{d_1} \dots q_\ell^{d_\ell}$ with the p_i and the q_j irreducible and all relatively prime. Then for each i, j we have $D(p_i) = a_i p_i$, $D(q_j) = b_j q_j$ and $\sum e_i a_i = a = \sum d_j b_j$. This last analysis provides a criteria for new constants:

Proposition 2.2 *Suppose S is a differential UFD finitely generated as an algebra over F, and let E be its quotient field. Suppose that all constants of S are in C. Then E has a new constant if and only if there is a set $p_1, \dots, p_m$ of pairwise not associate irreducible elements of S satisfying*

(a) $p_i | D(p_i)$, $1 \le i \le m$, *and*

(b) $\left\{ \frac{D(p_i)}{p_i} \mid 1 \le i \le m \right\}$ *is linearly dependent over $\mathbb{Q}$.*

Proof. Suppose $\frac{f}{g}$ is a constant of E not in C with f, g in S relatively prime (and neither units). Then, in the notation preceding the proposition, the combined sets of irreducible factors of f and g provides a set of irreducibles as the proposition asserts. ($\sum e_i a_i = \sum d_j b_j$ implies that $\sum e_i a_i - \sum d_j b_j = 0$.) The case of f or g or both units is handled by a separate argument (see below).

Conversely, suppose there is a set of irreducible elements of S as in the proposition. Write the equation of linear independence over $\mathbb{Q}$, clear denominators to have an equation over $\mathbb{Z}$ and group terms with positive and negative coefficients. This yields an equation like $\sum e_i a_i = \sum d_j b_j$ from which the desired elements f and g can be constructed as in the discussion preceding the proposition.

To handle the case, deferred from above, where f and g are elements of S with one or both units and $\frac{f}{g}$ is a constant not in C, we can assume that f is the unit. (If g is the unit, we observe that $\frac{g}{f}$ is a new constant as well.) Since $\frac{f}{g} \notin C$, f and g are linearly independent over C. On the other hand, as above, $D(f) = af$ and $D(g) = ag$. Then for every $c, d \in C$, $D(cf + dg) = a(cf + dg)$.

We can select c, d so that $cf + dg$ is a non-unit (this uses that S is finitely generated as an F algebra) and not a multiple of g, unless both f and g are in F, in which case $\frac{f}{g} \in C$. Then $\frac{cf+dg}{g}$ is a constant not in C, and now with non-unit numerator. $\qquad\square$

To return to the case $S = F[G]$ that we want to use in the derivation-extension approach to the inverse problem, the above criteria are relevant when $F[G]$ is a UFD. Precise conditions for this are known (basically, G modulo its radical should be simply connected), but it is clearly satisfied when G is solvable: in that case G is a semi-direct product $U \rtimes T$ of a unipotent group and a torus, and $F[G] = F[U] \otimes F[T]$ is a tensor product of a polynomial ring with a Laurent polynomial ring, and hence a UFD. And indeed the derivation extension approach works in this case. We illustrate this now for the 2×2 upper triangular matrix group.

3 The inverse problem for a 2×2 upper triangular matrix group

Throughout this section, F denotes a differential field with field of constants $\mathbb{C}$, and B denotes the subgroup

$$\left\{ \begin{bmatrix} t & u \\ 0 & 1 \end{bmatrix} \mid t, u \in \mathbb{C}, t \neq 0 \right\}$$

of $GL_2(\mathbb{C})$. We also use the isomorphism $\mathbb{G}_a \rtimes \mathbb{G}_m \cong B$ as equality and write

$$(u, t) = \begin{bmatrix} t & u \\ 0 & 1 \end{bmatrix} = \begin{bmatrix} 1 & u \\ 0 & 1 \end{bmatrix} \cdot \begin{bmatrix} t & 0 \\ 0 & 1 \end{bmatrix}.$$

In this notation, we have the following:

(a) $(u, 1)(0, t) = (u, t)$,

(b) $(0, t)(u, 1)(0, t)^{-1} = (tu, 1)$,

(c) $(0, t)(u, 1) = (tu, t)$, and

(d) $(v, s)(u, t) = (v + su, st)$.

We define functions $x, y : B \to \mathbb{C}$ by coordinate projection:

$$x((u, t)) = u, \quad y((u, t)) = t.$$

Then $x, y \in \mathbb{C}[B]$ and in fact

$$\mathbb{C}[B] = \mathbb{C}[x, y, y^{-1}]$$

so that $F[B] = F[x, y, y^{-1}]$ as well.

We use the usual conventions for actions of a group G on functions f on G: for $\gamma \in G$, define $(\gamma \cdot f)(\alpha) = f(\alpha\gamma)$ and $(f \cdot \gamma)(\alpha) = f(\gamma\alpha)$. Then B acts on $\mathbb{C}[B]$ as follows:

$$(u, t) \cdot x = x + uy, \ (u, t) \cdot y = ty,$$

$$x \cdot (u, t) = u + tx, \ y \cdot (u, t) = ty.$$

We will determine $\text{Lie}(B)$ as the $\mathbb{C}$-derivations of $\mathbb{C}[B]$ invariant under the left action. We look for derivations D which carry x and y to linear expressions

$$D(x) = ax + by + c$$

and

$$D(y) = dx + ey + f$$

with $a, d, c, d, e, f \in \mathbb{C}$. Then

$$D((u, t) \cdot x) = (a + ud)x + (b + ue)y + c + ud$$

while

$$(u, t) \cdot D(x) = ax + (au + bt)y + c.$$

Comparing coefficients implies that $a = e$ and $b = d = f = 0$ so that $D(x) = ax + c$ and $D(y) = ay$. For a basis of $\text{Lie}(B)$, then, we can take D_1, D_2 defined by

$$D_1(x) = x, \ D_1(y) = y$$

using $a = 1$ and $c = 0$, and

$$D_2(x) = 1, \ D_2(y) = 0$$

using $a = 0$ and $c = 1$. For future reference, we note also that

$$[D_2, D_1] = D_2.$$

Using the basis $\{D_1, D_2\}$, a general element of $\text{Lie}(B)$ is then $aD_1 + bD_2$ $(a, b \in \mathbb{C})$ which sends x to $ax + b$ and which sends y to ay, and a general element of the derivations $F \otimes \text{Lie}(B)$ of $F[B]$ is $f_1 D_1 + f_2 D_2$ which sends x to $f_1 x + f_2$ and y to $f_1 y$.

Since $(u, t) \cdot y = ty = y \cdot (u, t)$, $\mathbb{C}[y, y^{-1}]$ is a B-stable subalgebra of $\mathbb{C}[B]$ (in fact $\mathbb{C}[y, y^{-1}]$ is the coordinate ring of $\mathbb{G}_m$ and the inclusion $\mathbb{C}[y, y^{-1}] \subset \mathbb{C}[x, y, y^{-1}]$ is the morphism on coordinate rings induced from the projection $B = \mathbb{G}_a \rtimes \mathbb{G}_m \to \mathbb{G}_m$). The derivation $D_F + \phi_1 D_1 + \phi_2 D_2$ also preserves $F[y, y^{-1}]$ and acts on it by sending y to $\phi_1 y$. We want to focus on such a derivation of this subalgebra, and we change notation slightly so that we have a $\phi \in F$ and consider the derivation $D = D_F + \phi D_1$ of $F[y, y^{-1}]$ which acts like D_F on F and sends y to ϕy (and is invariant under B). We would like to know if D has any new constants in the quotient field $F(y)$ of $F[y, y^{-1}]$, and we apply our previous analysis.

To do so, we look for elements $f \in F[y, y^{-1}]$ of the form

$$f = \sum_{j=-m}^{m} \alpha_j y^j$$

satisfying an equation

$$D(f) = af.$$

Since

$$D(\alpha y^j) = \alpha' y^j + \alpha j y^{j-1} D(y) = (\alpha' + j\alpha\phi)y^j,$$

we have that D preserves top degree and hence that $a \in F$. Then, for each j,

$$a\alpha_j = \alpha'_j + j\alpha_j\phi$$

or

$$\alpha'_j = (a - j\phi)\alpha_j.$$

Suppose that $b \neq 0$ is any element of F. Then

$$D(bf) = b'f + bD(f) = (b' + ba)f = \left(\frac{b'}{b} + a\right)(bf)$$

so that $bf | D(bf)$ as well (and of course f and bf have the same divisibility properties). If m is the top degree of f, we can take $b = \frac{1}{\alpha_m}$ and replace f by bf so that $\alpha_m = 1$. Then

$$\alpha'_m = (a - m\phi)\alpha_m$$

implies that

$$a = m\phi$$

so that

$$\alpha'_j = (m - j)\phi\alpha_j.$$

Suppose that $\alpha_j \neq 0$ for some $j < m$. Then

$$D\left(\frac{\alpha_j}{y^{m-j}}\right) = 0$$

and thus α_j / y^{m-j} is a new constant of $F(y)$.

In general, if there is an element $0 \neq \alpha \in F$ such that

$$\alpha' = r\phi\alpha,$$

where $0 \neq r = \frac{p}{q} \in \mathbb{Q}$ with $p, q \in \mathbb{Z}$, then

$$D\left(\frac{\alpha^q}{y^p}\right) = 0$$

so α^q/y^p is a new constant of $F(y)$.

Thus for $F(y) \supset F$ to be a no new constants extension, we must assume that

$$Y' = r\phi Y \qquad (3.1)$$

has no non-zero solutions in F for any $0 \neq r \in \mathbb{Q}$. Conversely, if there are no non-zero solutions in F to Equation (3.1) for any r, then the coefficients α_j, $j < m$ of f are all zero, so that $f = y^m$ (or more generally, before the multiplication by b, $f = \alpha_m y^m$). Thus the only elements of $F[y, y^{-1}]$ which divide their derivatives are F scalar multiples of powers of y, and it follows from the proposition in the preceding section that $F(y) \supset F$ as no new constants.

Before continuing to handle $F[x, y, y^{-1}]$, we note two observations:

(1) Provided we can find $\phi \in F$ for which Equation (3.1) has no solutions, we can solve the inverse problem for $\mathbb{G}_m$ using $F(y)$ with $y' = \phi y$.

(2) Suppose $K \supset F$ is any Picard–Vessiot extension with group $\mathbb{G}_m$. Then one can show that $K = F(y)$ where $y' = \phi y$ for some $\phi \in F$ ([5, Example 5.24, p.71]), and then ϕ has to be such that Equation (3.1) has no non–zero solutions by our reasoning here. It follows that every Picard–Vessiot extension of F with group $\mathbb{G}_m$ fits into the discussion above. This remark will be important when we turn to the lifting problem.

Now we want to return to the full ring $F[x, y, y^{-1}]$ and its derivation $D = D_F + \phi_1 D_1 + \phi_2 D_2$. We assume that $\phi = \phi_1$ has been chosen so that Equation (3.1) has no non-zero solutions for any $r \neq 0$, so that the only non-zero elements of $F[y, y^{-1}]$ which divide their derivatives are of the form ay^m. We want to select a $\psi = \phi_2$ so that $F(x, y) \supset F$ has no new constants. Because $F[x, y, y^{-1}]$ is a UFD, we again look for elements which divide their derivatives.

Assume f is such an element. We regard $F[x, y, y^{-1}]$ as $F[y, y^{-1}][x]$ and write

$$f = \sum_{i=0}^{n} a_i(y)x^i, \ a_i(y) \in F[y, y^{-1}], \ a_n \neq 0.$$

We assume $D(f) = bf$ for some $b \in F[x, y, y^{-1}]$.

Since

$$D(a(y)x^m) = a(y)'x^m + a(y)mx^{m-1}(\phi x + \psi)$$
$$= (a(y)' + a(y)m\phi)x^m + ma(y)\psi x^{m-1},$$

we have

$$D(f) = \sum_{i=0}^{n} b_i(y)x^i$$

where

$$b_n(y) = a_n(y)' + a_n(y)n\phi$$

and

$$b_m(y) = a_m(y)' + a_m(y)m\phi + (m+1)a_{m+1}\psi$$

for $m < n$. In particular, we have $\deg_x(D(f)) \le \deg_x(f)$. Since $\deg_x(bf) \ge \deg_x(f)$, we see that $\deg_x(b) = 0$ and $b \in F[y, y^{-1}]$. Then comparing top coefficients of bf and $D(f)$ gives

$$ba_n(y) = a_n(y)' + a_n(y)n\phi,$$

so $a_n(y)' = (n\phi - b)a_n(y)$ and $a_n|D(a_n)$. Since $a_n \ne 0$, and a_n divides its derivative in $F[y, y^{-1}]$, we know that $a_n = ay^m$ for some $0 \ne a \in F$ and $m \in \mathbb{Z}$. In particular, a_n is a unit of $F[y, y^{-1}]$. Thus we can multiply f by a_n^{-1} and assume $a_n = 1$ (and we still have that $f|D(f)$), and then taking $a_n = 1$ in the above equations implies that $b = n\phi$.

Now assume $n > 0$ and compare the coefficients of x^{n-1} in f and $D(f) = n\phi f$. We find that

$$a_{n-1}(y)' = \phi a_{n-1} + n\psi.$$

This suggests that we consider the differential equation

$$Y' = \phi Y + n\psi \qquad (3.2)$$

over $F(y)$: for if Equation (3.2) has no solutions in $F(y)$, then in particular it has none in $F[y, y^{-1}]$ and hence there is no element $a_{n-1}(y)$ as above. Thus if Equation (3.2) has no solutions in $F(y)$, then the assumption that $n > 0$ is in error, and it follows that $ay^m = f \in F[y, y^{-1}]$ (the original f, with general top coefficient) and then we conclude, as in the case of $F[y, y^{-1}]$, that since there are no relatively prime elements of $F[x, y, y^{-1}]$ which divide their derivatives, there are no new constants in $F(x, y) \supset F$.

It is, of course, conceivable that Equation (3.2) has a solution in $F(y)$ but that nonetheless $F(x, y) \supset F$ has no new constants. So suppose that

$$z' = \phi z + n\psi$$

for some $z \in F(y)$ and let

$$w = \frac{z}{n}.$$

Then $(x - z)' = \phi(x - z)$ in $F(x, y)$, so $\frac{x-w}{y} \in F(x, y)$ is a new constant.

To summarize:

I. If $\phi \in F$ is chosen so that the Equation (3.1) has no non-zero solutions in F for any $q \neq 0$, then $F(y) \supset F$ has no new constants and is a Picard-Vessiot extension for $\mathbb{G}_m$.

II. If in addition $\psi \in F$ is chosen so that the Equation (3.2) has no solutions in $F(y)$ (where $y' = \phi y$), then $F(x, y) \supset F$ has no new constants and is a Picard-Vessiot extension for B.

III. If for every $\psi \in F$ there is $n \in \mathbb{N}$ such that the Equation (3.2) has a solution in $F(y)$ (where $y' = \phi y$) then for every ψ, $F(x, y) \supset F$ has a new constant.

Points I. and II. show how it is possible to solve the inverse Galois problem for $\mathbb{G}_m$ and for B: find examples of fields F and functions ϕ and ψ for which the relevant differential equations have no solution. And point III., combined with the previous observation (2) above shows how the lifting problem could fail to have a solution: for that observation pointed out that any Picard-Vessiot extension $K \supset F$ with group $\mathbb{G}_m$ had to be of the form $K = F(y)$ with $D(y) = \phi y$ for some $\phi \in F$, so solving the extension problem for $B \to \mathbb{G}_m$ is the same as finding a derivation $D = D_F + \phi D_1 + \psi D_2$ of $F[x, y, y^{-1}]$ with no new constants in the quotient field.

For the case $F = \mathbb{C}(t)$, it is easy to find functions $\phi \in F$ such that Equation (3.1) has no non-zero solutions: one can take $\phi = 1$. Then one can find functions $\psi \in F$ such that Equation (3.2) has no solutions; for example, $\psi = \frac{1}{t-\alpha}$, $\alpha \in \mathbb{C}$ [5, Lemma, p. 95]. This solves the inverse problem for B over $\mathbb{C}(t)$. Examples of fields F and functions $\phi \in F$ for which Equation (3.1) has no non-zero solution but for every $\psi \in F$ there is $n \in \mathbb{N}$ such that the Equation (3.2) has a solution in $F(y)$ can presumably be constructed by the technique of "digging holes" (see [4, Exercise 26,27, p. 325]) in differentially closed fields, thereby exhibiting examples in which the lifting problem has no solution. Whether one can do this in finite extensions of $\mathbb{C}(t)$ is not clear.

4 Solvable groups

The inverse problem for (connected) solvable algebraic groups G divides naturally into the unipotent and torus cases, reflecting first the semi-direct product decomposition $G = U \cdot T$ of G into its unipotent radical U and maximal torus T, and then the (split) extension

$$1 \to U \to G \to G/U(\cong T) \to 1.$$

To solve the inverse problem for solvable G, therefore, we can look at the inverse problem for tori and for unipotent groups separately, and then at the lifting problem for unipotent groups.

4.1 Tori

Let $T = \mathbb{G}_m^{(n)}$ be a torus over C, and let t_i, $1 \le i \le n$, be the coordinate projection functions on T, so $C[T]$ is the Laurent polynomial ring $C[t_1^{\pm 1}, \dots, t_n^{\pm 1}]$. Note that $F[T] = F \otimes C[T]$ is a UFD. Let D_i, $1 \le i \le n$, be the derivation of $C[T]$ with $D_i(t_j) = \delta_{ij} t_j$; $D_1, \dots, D_n$ is a C basis of $\mathrm{Lie}(T)$. To exhibit T as a differential Galois group over F, we seek $\phi_1, \dots, \phi_n$ so that with the derivation

$$D = D_F \otimes 1 + \sum_{i=1}^{n} \phi_i \otimes D_i,$$

$F[T]$ has no new constants in its quotient field $F(T)$.

As we saw in the previous section, for the case $n = 1$, it was necessary and sufficient that the differential equation (3.1) has no solutions in F for any $0 \ne r \in \mathbb{Q}$. Therefore it will be necessary that the ϕ_i be such that the differential equations

$$Y' = r\phi_i Y$$

have no solutions in F for any $0 \ne r \in \mathbb{Q}$, $1 \le i \le n$.

When $n > 1$, we will also need of course that the ϕ_i be distinct. More generally, if there is a linear dependence relation

$$a_1\phi_1 + \cdots + a_n\phi_n = 0$$

with $a_i \in \mathbb{Z}$ and not all $a_i = 0$, then for

$$t = \prod_{i=1}^{n} t_i^{a_i},$$

the logarithmic derivative gives

$$\frac{t'}{t} = \sum_i a_i \frac{t_i'}{t_i} = \sum_i a_i \phi_i = 0$$

and hence a new constant t. So we will also need that the ϕ_i are linearly independent over $\mathbb{Q}$. If there is an element $\alpha \in F$ and rational numbers $a_i = \frac{m_i}{d}$, $1 \le i \le n$, not all zero, such that

$$\alpha' = \left(\sum_i a_i \phi_i \right) \alpha,$$

then

$$\left(\frac{\alpha^d}{\prod_i t_i^{m_i}} \right)' = 0$$

and hence $F(T)$ has a new constant. Hence to avoid new constants we must require that the differential equation

$$Y' = \left(\sum_i a_i \phi_i \right) Y \tag{4.1}$$

has no non-zero solutions in F for any $a_1, \ldots, a_n \in \mathbb{Q}$, not all $a_i = 0$. (Note that this implies that $\phi_1, \ldots, \phi_n$ are linearly independent over $\mathbb{Q}$). As we will now see, this necessary condition is sufficient.

To look for new constants in $F(T)$, we consider the equation $D(f) = af$ in $F[T]$. We adopt the following notation: for $m = (m_1, \ldots, m_n) \in \mathbb{Z}^n$, let $t^m = \prod_i t_i^{m_i}$ and let $\langle \phi, m \rangle = \sum_i m_i \phi_i$. For $\alpha \in F$, we then have

$$D(\alpha t^m) = (\alpha' + \langle \phi, m \rangle \alpha) t^m.$$

If

$$D(f) = af$$

where

$$f = \sum_m \alpha_m t^m, \alpha_m \in F,$$

then, comparing coefficients, we see that $a \in F$ and

$$a\alpha_m = \alpha_m' + \langle \phi, m \rangle \alpha_m,$$

so

$$\alpha_m' = (a - \langle \phi, m \rangle) \alpha_m.$$

If $\alpha_m \neq 0$, then

$$\left(\frac{\alpha_q}{\alpha_m}\right)' = (\langle\phi, m\rangle - \langle\phi, q\rangle)\left(\frac{\alpha_q}{\alpha_m}\right),$$

so with

$$\beta = \left(\frac{\alpha_q}{\alpha_m}\right)$$

we have

$$\beta' = \langle m - q, \phi\rangle\beta$$

and hence, if $q \neq m$ and $\alpha_q \neq 0$, a non-zero solution to Equation (4.1) with $a_i = m_i - q_i$ not all zero. In the absence of such solutions to Equation (4.1), then, f consists of at most a single term αt^m with $\alpha \neq 0$. But then f is a unit. It follows that there are no irreducible elements of $F[T]$ which divide their derivatives, and hence no new constants in $F(T)$. In summary, we have:

Proposition 4.2 Let $\phi_i \in F$, $1 \leq i \leq n$ be such that

$$Y' = (\sum_{i=1}^{n} a_i\phi_i)Y$$

has no non-zero solutions in F for any $a_1, \ldots, a_n \in \mathbb{Q}$, not all $a_i = 0$. Let

$$D = D_F \otimes 1 + \sum_{i=1}^{n} \phi_i \otimes D_i.$$

Then, with the derivation D, $F(T) \supset F$ is a Picard-Vessiot extension with Galois group T.

4.2 Unipotents

We handle the case of unipotent groups U by reduction to the case of the one-dimensional group $\mathbb{G}_a$, plus a solution to the lifting problem for certain group extensions with kernel $\mathbb{G}_a$. All the groups in question will have UFD coordinate rings, so we can use the method of looking for functions which divide their derivatives. We begin by remarking that, as for tori, this necessary condition for no new constants is sufficient in the unipotent case:

Proposition 4.3 Let U be a unipotent group over C and let D be a U-equivariant derivation of $F[U]$ extending D_F such that $F(U)$ has no new constants. Then there are no elements a, f in $F[U]$ with f irreducible such that $D(f) = af$.

Proof. Use induction on the dimension of U. Let $Z = \mathbb{G}_a$ be a one dimensional central subgroup of U with quotient $\overline{U}$ and let $u : U \to Z$ be a Z-equivariant retraction. Regard u as a C-valued function on U. Then $F[U] = F[\overline{U}][u]$ (polynomial ring), $D(u) = 1$, and since $F[\overline{U}] = F[U]^Z$, we have $D(F[\overline{U}]) \subseteq F[\overline{U}]$. Suppose we have a and f with $D(f) = af$. We write f as a polynomial in u with coefficients in $F[\overline{U}]$. One sees that $a, D(u) \in F[\overline{U}]$. Then if f has degree n in u, its leading coefficient b_n satisfies $D(b_n) = ab_n$, which by induction applied to $\overline{U}$ implies that b_n is a unit. Then f is replaced by $b_n^{-1} f$ so f is monic and its coefficient b_{n-1} of degree $n-1$ satisfies $D(u) = D(-\frac{b_{n-1}}{n})$. Thus $u - \frac{b_{n-1}}{n}$ is a constant and since $F(U)$ has no new constants, this implies that $u \in F[\overline{U}]$, which is a contradiction. $\qquad\square$

Now we state, without proof, the lifting theorem for $\mathbb{G}_a$:

Theorem 4.4 *Suppose that the algebraic group G over C contains a normal subgroup Z isomorphic to $\mathbb{G}_a$ and that there is a G-equivariant retraction $u : G \to Z$. Let $\overline{G} = G/Z$, and assume $F[\overline{G}]$ is a UFD. Let D_n be a basis of $\mathrm{Ker}(\mathrm{Lie}(G) \to \mathrm{Lie}(\overline{G})) = \mathrm{Lie}(Z)$. Let $\overline{D_i}$, $1 \le i \le n-1$, be a basis of $\mathrm{Lie}(\overline{G})$ such that there exist $f_i \in F$ with*

$$\overline{D} = D_F \otimes 1 + \sum_{i=1}^{n-1} f_i \otimes \overline{D_i}$$

and such that $\overline{D}(f) = af$ has no solutions in $F[\overline{G}]$ (f irreducible). Further suppose there are $D_i \in \mathrm{Lie}(G)$, $1 \le i \le n-1$, such that $D_i | F[\overline{G}] = \overline{D_i}$ and $[D_i, D_n] = c_i D_n$, $c_i \in C$. Then $F[G]$ is a UFD and an element $f_n \in F$ can be found so that if

$$D = D_F \otimes 1 + \sum_{i=1}^{n} f_i \otimes D_i ,$$

then $D(f) = af$ has no solutions in $F[G]$ (f irreducible), provided for any given element $h \in F$ and character χ of $\overline{G}$ there is $f_n \in F$ such that the differential equation $y' = f_n \chi + h$ has no solution $y \in F[\overline{G}]$.

For the proof, see [5, Thm. 7.6, p.93].

The theorem provides an inductive way to construct Picard-Vessiot extensions of F. In the case of a unipotent group U, there are no non-trivial characters of any quotient of U, so the equation $y' = f_n \chi + h$ becomes $y' = f_n + h$, and the problem is simply to find non-derivatives. As long as F contains plenty of such non-derivatives, then constructing extensions with unipotent Galois group is always possible:

Corollary 4.5 *Suppose there are elements ϕ_i, $i = 1, 2, 3, \ldots$, in F such that solutions of the equations $y' = \phi_i$ are algebraically independent over F. Then any unipotent group is a differential Galois group over F.*

Proof. Proceed by induction on the dimension of U, the inductive step being handled by the theorem. The theorem requires that $y' = f_n + h$ has no solution. If $y' = h$ has no solution in $F[\overline{G}]$, then take $f_n = 0$. If it has a solution, then take $f_n = \phi_m$, where m is minimal with respect to $y' = \phi_m$ having no solution in $F(\overline{G})$. $\square$

When $F = \mathbb{C}(t)$, the functions

$$\phi_i = \frac{1}{t - i}, \ i = 1, 2, 3, \ldots,$$

are such a family.

4.3   General solvable case

The theorem also implies a solution to the inverse Galois problem in the general solvable case:

Corollary 4.6 *Let $G = U \cdot T$ be a connected solvable algebraic group over C with unipotent radical U and maximal torus $T = \mathbb{G}_m^{(n)}$. Suppose that there are $\phi_1, \ldots, \phi_n \in F$ such that*

$$Y' = \left(\sum a_i \phi_i\right) Y$$

has no non-zero solutions in F when $a_1, \ldots, a_n \in \mathbb{Q}$ are not all zero. Suppose further that for any $g \in F$, there are countably many $\psi_m^g \in F$ such that the solutions of $Y' = gY + \psi_m^g$ are algebraically independent (for distinct m) over F. Then there is a derivation D of $F[G]$ such that $F(G) \supset F$ is a Picard-Vessiot extension with differential Galois group G.

Proof. The condition on the ϕ's guarantees a derivation of $F[T]$ such that $F(T) \supset F$ has no new constants. Now we apply the theorem repeatedly: at each stage, the problem is the equation $y' = f_n \chi + h$. When h is not a derivative in $F[\overline{G}]$, then we take $f_n = 0$. When $h = D(k)$, then we change variables from y to $w = (y - k)\chi^{-1}$, which changes the equation to $w' = f_n + gw$ where $g \in F$. Since $F[\overline{G}]$ is of finite transcendence degree over F, there is a ψ_m^g so that solutions of $Y' = gY + \psi_m^g$ are not in it, and then taking $f_n = \psi_m^g$ provides desired coefficient for the derivation required by the theorem. $\square$

For the case of $F = \mathbb{C}(t)$, the existence of the functions ψ_m^g, in fact ones of the form $\frac{1}{t-\alpha}$, are guaranteed by an analytic argument [5, Lemma, p.95].

Acknowledgments

Research supported by National Science Foundation grant DMS 0070748.

References

1. Kolchin, E. R. *Some problems in differential algebra*, Proc. Intl. Cong. Math., Moscow, 1966, reprinted in Bass, H. et al, *Selected Works of Ellis Kolchin with Commentary*, American Mathematical Society, Providence RI, 1999, 343–346.
2. Kovacic, J. *The inverse problem in the Galois theory of differential fields*, Ann. of Math. (2) **89** (1969), 583–608.
3. Kovacic, J., *On the inverse problem in the Galois theory of differential fields. II*, Ann. of Math. (2) **93** (1971), 269–284.
4. Lang, S. *Algebra*, Third Edition, Addison-Wesley, Reading MA, 1993.
5. Magid, A. *Lectures on Differential Galois Theory*, University Lecture Series **7**, American Mathematical Society, Providence RI, 1997 (second printing with corrections).
6. Mitschi, C. and Singer, M. *Connected linear groups as differential Galois groups*, J. Algebra **184** (1996), 333–361.
7. van der Put, M. *Recent work on differential Galois theory*, Séminaire Bourbaki 1997–1998, Astérique **252** (1998), 341–367.
8. van der Put, M. and Singer, M. *Differential Galois Theory* (to appear), see http://www4.ncsu.edu/~singer/ms_papers.html.
9. van der Put, M. and Ulmer, F. *Differential equations and finite groups*, J. Algebra **226** (2000), 920–966.
10. Tretkoff, C. and Tretkoff, M. *Solution of the inverse problem of differential Galois theory in the classical case*, Amer. J. Math. **101** (1979), 1327–1332.

Differential Algebra and Related Topics, pp. 171–189
Proceedings of the International Workshop
Eds. L. Guo, P. J. Cassidy, W. F. Keigher & W. Y. Sit
© 2002 World Scientific Publishing Company

DIFFERENTIAL GALOIS THEORY, UNIVERSAL RINGS AND UNIVERSAL GROUPS

MARIUS VAN DER PUT

*Department of Mathematics, University of Groningen,
P.O.Box 800, 9700 AV Groningen, The Netherlands
E-mail: mvdput@math.rug.nl*

After a more or less standard introduction to differential Galois theory, we consider the problem of determining the universal Picard-Vessiot ring of a class of differential equations and its differential Galois group for various differential fields K and classes of equations. As a special case, we sketch a new proof of a result presented in the inspiring paper by J. Martinet and J.-P. Ramis [6].

1 The basic concepts

In the sequel K will denote an ordinary differential field, i.e., a field with a given non-trivial differentiation denoted by $f \mapsto f'$, such that the field of constants $C := \{a \in K | \ a' = 0\}$ is an algebraically closed field of characteristic 0. A *linear differential equation in matrix form* is given by $y' = Ay$ where y is a vector of length m and A is a $m \times m$ matrix with coefficients in K.

A "basis-independent notion" is that of a *differential module* (M, ∂) *over* K. By definition, M is a finite dimensional vector space over K and a C-linear map $\partial : M \to M$ is satisfying $\partial(am) = a'm + a\partial m$ for all $a \in K$, $m \in M$. After choosing a basis $e_1, \ldots, e_m$ of M over K we associate to ∂ a matrix $A = (a_{i,j})$ by the formula $\partial e_i = -\sum_j a_{j,i}e_j$. The equation $y' = Ay$, where $y = (y_1, \ldots, y_m) \in K^m$ is now equivalent to the equation $\partial \sum_i y_i e_i = 0$.

The solution space of the differential equation $y' = Ay$ is a C-vector space of dimension $\leq m$. If this dimension is strictly less than m, then the equation does not have enough solutions with coordinates in K. An extension of K is needed in order to have enough solutions. This leads to the concept of a Picard-Vessiot ring R of $y' = Ay$ over K. The ring R is a *Picard-Vessiot ring* if the following properties hold:

- R is a K-algebra and there is given an extension $r \mapsto r'$ of the differentiation of K to the K-algebra R.

- The only ideals of R invariant under the differention of R are (0) and R.

- There exists an invertible matrix F with coefficients in R, such that $F' = AF$. (We call F a *fundamental matrix*.)

- R is generated over K by the coefficients of F and $\det(F)^{-1}$.

Theorem 1.1 *Let $y' = Ay$ be a matrix differential equation over K of size $m \times m$.*

(a) *A Picard-Vessiot ring exists and is unique up to isomorphism.*

(b) *"The" Picard-Vessiot ring R has no zero-divisors and its field of fractions $\mathrm{Qt}(R)$, called the Picard-Vessiot field, has C as its field of constants.*

(c) *The differential Galois group of $y' = Ay$ over K is defined as the group G of differential automorphisms of R/K (that is, the K-algebra automorphisms of R that commute with the differentiation of R). The solution space V of $y' = Ay$ is defined by $V = \{y \in R^m | \; y' = Ay\}$. This is a vector space of dimension m over C. The group G operates faithfully on V and the image of G in $\mathrm{GL}(V)$ is a linear algebraic subgroup.*

The following examples illustrate the above theorem.

Examples 1.2 Some Picard-Vessiot rings and their groups.

(a) Let $K = C(z)$, where C is an algebraically closed field of characteristic 0, the differentiation is $f \mapsto f' = \frac{df}{dz}$ and the equation is $y' = \frac{\alpha}{z}y$ with $\alpha \in C$. The Picard-Vessiot ring is $R = K[t]$ with $t' = \frac{\alpha}{z}t$. If α is not a rational number, then t is transcendental over K and the differential Galois group is $\mathbf{G}_m = C^*$, the multiplicative group. The action of an element σ in this group is given by $\sigma t = ct$ for some $c \in C^*$. If α is a rational number with denominator d, then t is algebraic over K. Its minimal equation over K is given by $t^d - f = 0$, where $f \in K$ is a nonzero solution of $f' = \frac{d\alpha}{z}f$. The differential Galois group is the cyclic group of order d and coincides with the ordinary Galois group of the field extension $K[t]/K$.

(b) Let K be as in the Example 1.2(a). Suppose that the inhomogeneous equation $y' = f$ with $f \in K$ has no solution in K. Then one can transform this inhomogeneous equation into the equation $x' = Ax$ with A the matrix $\begin{pmatrix} 0 & f \\ 0 & 0 \end{pmatrix}$. The Picard-Vessiot extension is $K[t]$ with t transcendental over K and $t' = f$. The differential Galois group is the additive group $\mathbf{G}_a = C$. An element σ in this group acts by $\sigma t = t + c$ for $c \in C$.

(c) The Airy equation $y'' = zy$ over the field K of Example 1.2(a) can also be written in matrix form $x' = Ax$, where A is the matrix $\begin{pmatrix} 0 & 1 \\ z & 0 \end{pmatrix}$. A rather non-trivial calculation shows that the Picard-Vessiot ring has the form $K[y_1, y_2, y_1', y_2']$, where the differentiation is given by $y_1'' = zy_1$ and $y_2'' = zy_2$, and where the only algebraic relation between the y_1, y_2, y_1', y_2' over K is $y_1y_2' - y_1'y_2 - 1 = 0$. Its differential Galois group is $\mathrm{SL}_2(C)$. A matrix $\sigma = \begin{pmatrix} a & b \\ c & d \end{pmatrix}$ in this group acts by $y_1 \mapsto ay_1 + by_2$, $y_2 \mapsto cy_1 + dy_2$ and similarly for its action on y_1' and y_2'.

Picard-Vessiot theory and differential Galois groups are nowadays well explained in many papers and books, e.g., [3,4,5,7].

2   Universal Picard-Vessiot rings

For a differential field K, we consider the collection of all differential modules over K. This is a category, denoted by Diff_K. The objects of this category are the differential modules. A morphism $f : (M_1, \partial_1) \to (M_2, \partial_2)$ is a K-linear map satisfying $\partial_2 \circ f = f \circ \partial_1$. Then $\text{Hom}((M_1, \partial_1), (M_2, \partial_2))$ is easily seen to be a finite dimensional vector space over C. The category Diff_K has far more structure. There are direct sums, kernels, cokernels, tensor products et cetera. The more precise description is that Diff_K is a neutral Tannakian category. The latter means that Diff_K is equivalent to the category of all finite dimensional representations of a certain *affine group scheme* G over C. For the benefit of the reader we have inserted some basic material on affine group schemes.

2.1   The formalism of affine group schemes

As above, C denotes an algebraically closed field of characteristic 0. Let B be a commutative C-algebra that is provided with

- a co-multiplication $m^* : B \to B \otimes_C B$ which is a morphism of C-algebras,
- a co-unit $e^* : B \to C$ that is a C-algebra morphism, and
- a co-inverse $i^* : B \to B$ that is a C-algebra, morphism.

satisfying certain "co-rules", namely:

(a) the maps $(m^* \otimes id) \circ m^*$ and $(id \otimes m^*) \circ m^*$ from B to $B \otimes_C B \otimes_C B$ coincide.

(b) The maps $(id \otimes e^*) \circ m^*$ and $(e^* \otimes id) \circ m^*$ from B to $C \otimes_C B = B$ are the identity.

(c) The maps $product \circ (id \otimes i^*) \circ m^*$ and $product \circ (i^* \otimes id) \circ m^*$ from B to B coincide with $B \xrightarrow{e^*} C \subset B$. Here $product : B \otimes_C B \to B$ is the product-map given by $product(b_1 \otimes b_2) = b_1 b_2$.

In fact B with the above structure is usually called a *Hopf algebra*. The affine group scheme associated to B above is the affine scheme $G := \text{Spec}(B)$. By definition, this is the set of prime ideals of B provided with a Zariski topology and a structure sheaf. We will not at all use these structures explicitly.

What we only need to know is that a morphism $f : \operatorname{Spec}(B_1) \to \operatorname{Spec}(B_2)$ between affine schemes over C is the "same thing" as a C-algebra homomorphism $f^* : B_2 \to B_1$. More precisely, a C-algebra homomorphism $h : B_2 \to B_1$ induces a map $\tilde{h} : \operatorname{Spec}(B_1) \to \operatorname{Spec}(B_2)$ given by $\tilde{h}(\underline{P}) = h^{-1}(\underline{P})$ for any $\underline{P} \in \operatorname{Spec}(B_1)$. The map $\tilde{h}$ is continuous with respect to the Zariski topologies on $\operatorname{Spec}(B_1)$ and $\operatorname{Spec}(B_2)$. Moreover h induces a morphism between the two structure sheaves of C-algebras. In other words, $\tilde{h}$ is a morphism between affine schemes over C. The above statement asserts that every morphism $f : \operatorname{Spec}(B_1) \to \operatorname{Spec}(B_2)$ of affine schemes over C is equal to $\tilde{h}$ for a unique C-algebra homomorphism $h : B_2 \to B_1$. In particular, the multiplication $m : G \times G \to G$ on the affine group $G = \operatorname{Spec}(B)$ is defined by $m^* : B \to B \otimes_C B$. Similarly for the unit element (neutral element) e of G and the inverse $i : G \to G$. The co-rules are formally equivalent to the group laws for G.

We note that in case B is a finitely generated algebra over C, then G is (by definition) a linear algebraic group over C and can be embedded as a Zariski closed subgroup into some $\operatorname{GL}_n(C)$.

For a general affine group scheme $G = \operatorname{Spec}(B)$, there is a collection of subalgebras $\{B_i\}_{i \in I}$ such that

- Each B_i is a finitely generated algebra over C.
- For all j, $m^* B_j \subset B_j \otimes B_j$ and $i^*(B_j) \subset B_j$.
- For any i and j, there is a k with $B_i \cup B_j \subset B_k$.
- $\bigcup_{i \in I} B_i = B$.

Thus every B_i is a Hopf algebra for the induced structures and $G_i = \operatorname{Spec}(B_i)$ is a linear algebraic group since B_i is a finitely generated C-algebra. Now G can be seen as the projective limit of the linear algebraic groups G_i. In other words, an affine group scheme is the projective limit of linear algebraic groups. The converse also holds.

An affine group scheme $G = \operatorname{Spec}(B)$ admits a translation into a certain covariant functor $\mathcal{G}$. This functor is defined on the category of all commutative C-algebras and has values in the category of groups. For any C-algebra A we define the set $\mathcal{G}(A)$ as $\operatorname{Hom}_C(B, A)$, which is the set of the C-algebra homomorphisms from B to A. For $f_1, f_2 \in \operatorname{Hom}_C(B, A)$ we define a product f by

$$f : B \xrightarrow{m^*} B \otimes_C B \xrightarrow{f_1 \otimes f_2} A \otimes_C A \xrightarrow{product} A.$$

This makes $\mathcal{G}(A)$ into a group. We find that $\mathcal{G}$ is a functor from the category of the commutative C-algebras to the category of groups. Moreover $\mathcal{G}$ is representable (in fact by B). These somewhat abstract considerations are useful because the converse is also true:

Let $\mathcal{F}$ be a representable covariant functor from the category of all commutative C-algebras to the category of groups. Then $\mathcal{F}$ is isomorphic to the functor $\mathcal{G}$ associated to a uniquely determined (up to canonical isomorphism) affine group scheme G.

For $i = 1, 2, 3$, let G_i be affine group schemes and let $\mathcal{G}_i$ be their corresponding functors. Then an exact sequence of group schemes $1 \to G_1 \to G_2 \to G_3 \to 1$ is equivalent to an exact sequence $1 \to \mathcal{G}_1(A) \to \mathcal{G}_2(A) \to \mathcal{G}_3(A) \to 1$ of groups for every commutative C-algebra A which depends functorially on A.

The Lie algebra $\mathrm{Lie}(G)$ of an affine group scheme $G = \mathrm{Spec}(B)$ consists of the C-linear maps $D : B \to C$ satisfying $D(b_1 b_2) = e^*(b_1)D(b_2) + D(b_1)e^*(b_2)$.

If G is the projective limit of linear algebraic groups G_i, then $\mathrm{Lie}(G)$ is the projective limit of the $\mathrm{Lie}(G_i)$. In the sequel we will consider representations of $\mathrm{Lie}(G)$ on finite dimensional vector spaces W over C. In general this is a homomorphism $h : \mathrm{Lie}(G) \to \mathrm{End}(W)$ of Lie algebras over C. In our context we will require that h factors over some $\mathrm{Lie}(G_i)$, that is, for some homomorphism h_i, we have $\mathrm{Lie}(G) \overset{h}{\to} \mathrm{End}(W) = \mathrm{Lie}(G) \to \mathrm{Lie}(G_i) \overset{h_i}{\to} \mathrm{End}(W)$.

For an affine group scheme $G = \mathrm{Spec}(B)$ over C we define a G-module V (or a *finite dimensional representation of G*) to be a finite dimensional vector space V over C provided with a C-linear action of G. The latter is given by a C-linear map $\tau : V \to B \otimes V$ which has the properties:

(i) $(e^* \otimes id) \circ \tau : V \to B \otimes V \to C \otimes V = V$ is the identity map.

(ii) The maps

$$(m^* \otimes id) \circ \tau : V \to B \otimes V \to B \otimes B \otimes V$$

and

$$(id \otimes \tau) \circ \tau : V \to B \otimes V \to B \otimes B \otimes V$$

coincide.

A morphism $f : (V_1, \tau_1) \to (V_2, \tau_2)$ between two G-modules is a C-linear map such that $\tau_2 \circ f = (id \otimes f) \circ \tau_1$. The category of all G-modules is denoted by Repr_G.

The notion of a G-module V is equivalent to a morphism $G \to \mathrm{GL}(V)$ of affine group schemes over C.

One could define a *neutral Tannakian category over C* as a category $\mathcal{C}$ (with a lot of structure) which is equivalent to the category Repr_G for some affine group scheme G. The actual, rather long and involved, definition of "neutral Tannakian category over C" in [1] can be read as a criterion for recognizing whether a given category $\mathcal{C}$ (with a lot of structure) is a neutral Tannakian category. The affine group scheme G such that $\mathcal{C}$ is equivalent to Repr_G is uniquely determined by $\mathcal{C}$.

Let H be any group. The objects of the category Repr_H are the finite dimensional complex representations of H. The morphisms of Repr_H are the homomorphisms of complex representations. It can be shown that Repr_H is a neutral Tannakian category. Therefore Repr_H is isomorphic to Repr_T, where T is an affine group scheme over $\mathbf{C}$. This group scheme will be called the *algebraic hull of H*. In Section 3 we will see in detail the structure of the algebraic hull of $\mathbf{Z}$.

2.2 Classes of differential modules

It can be shown that for any differential field K (such that its field of constants is algebraically closed and has characteristic 0), the category Diff_K is a neutral Tannakian category. Consider a subcategory $\mathcal{C} \subset \mathrm{Diff}_K$ such that:

(i) For any two objects A_1, A_2 in $\mathcal{C}$, $\mathrm{Hom}_{\mathcal{C}}(A_1, A_2)$ is the same as $\mathrm{Hom}(A_1, A_2)$ for the category Diff_K (that is, $\mathcal{C}$ is a *full subcategory of* Diff_K), and

(ii) $\mathcal{C}$ is closed under all "operations of linear algebra", i.e., kernels, cokernels, duals, direct sums, subobjects and tensor products.

Then $\mathcal{C}$ is a neutral Tannakian category and determines a certain affine group scheme G. This G will be called the *(universal) differential Galois group* of $\mathcal{C}$. A very special case occurs when we consider a single differential module M over K. Now we can form the full subcategory $\{\{M\}\}$ of Diff_K, whose objects are the direct sums of all subquotients of all $M \otimes M \otimes \cdots \otimes M \otimes M^* \otimes \cdots \otimes M^*$ (where M^* stands for the dual of M). It turns out that the differential Galois group of the category $\{\{M\}\}$ is the same as the ordinary differential Galois group of the differential module M.

The main theme of this paper is to give examples of differential fields K and subcategories $\mathcal{C}$ of Diff_K, satisfying properties (i)–(ii) above, for which the differential Galois group can be explicitly computed. Moreover we will try to find an explicit universal Picard-Vessiot ring for $\mathcal{C}$, which we now define. The *universal Picard-Vessiot ring* of $\mathcal{C}$ is a K-algebra R such that

- There is given an extension $r \mapsto r'$ of the differentiation of K to the K-algebra R.

- The only ideals of R invariant under the differention of R are (0) and R.

- For every differential module in $\mathcal{C}$, presented as a matrix differential equation $y' = Ay$, there exists an invertible matrix F with coefficients in R, such that $F' = AF$.

- R is generated over K by the coefficients of the fundamental matrices F and $\det(F)^{-1}$ for all equations $y' = Ay$ in $\mathcal{C}$.

It is easily seen that R exists and is unique up to isomorphism. Indeed, R can be obtained as the direct limit of the Picard-Vessiot rings for all the equations $y' = Ay$ occurring in $\mathcal{C}$. Moreover, the differential Galois group G of $\mathcal{C}$ can be seen to be the projective limit of the ordinary differential Galois groups of the equations $y' = Ay$ occurring in $\mathcal{C}$. Furthermore, this G has the interpretation as the group of the K-linear automorphisms of R which commute with the differentiation of R. This needs a more detailed explication.

Let R denote the universal Picard-Vessiot ring of $\mathcal{C}$. As in Theorem 1.1, we would like to define the affine group scheme G of the category $\mathcal{C}$ as the group of the K-algebra automorphisms of R, commuting with the differentiation on R. In this way, G is just a group. In order to give G the structure of an affine group scheme, we have to define a suitable covariant functor $\mathcal{F}$ from the category of the commutative C-algebras to the category of groups. We define this covariant functor $\mathcal{F}$ on the category of the commutative C-algebras with values in the category of all groups as follows: For any commutative C-algebra A, consider the $(A \otimes_C K)$-algebra $A \otimes_C R$ with the extension of the differentiation given by $(a \otimes r)' = a \otimes r'$. Then $\mathcal{F}(A)$ is by definition the group of the $(A \otimes_C K)$-linear automorphisms of $A \otimes_C R$, which commute with the differentiation. It is clear that $\mathcal{F}$ is a covariant functor from the category of the commutative C-algebras to the category of all groups. We can give R a presentation with generators and relations and produce expressions for the derivatives of the generators. Using this, it can be shown that $\mathcal{F}$ is representable. Thus $\mathcal{F}$ is equivalent with an affine group scheme. The latter can be seen to be the universal differential Galois group G of $\mathcal{C}$.

In the sequel we will for convenience identify the affine group scheme G with the group $\mathcal{F}(C)$. The Lie algebra $\mathrm{Lie}(G)$ can be identified with the Lie algebra over C consisting of all the K-linear derivations $D : R \to R$ commuting with the differentiation on R.

As we said before, our aim is to produce examples of differential fields K and categories $\mathcal{C}$ where both the universal differential Galois and the universal Picard-Vessiot ring are explicit. For ordinary Galois theory our problem is analoguous to producing a field K and a class $\mathcal{C}$ of finite Galois extensions of

K such that the corresponding "universal field U" and its (pro-finite) Galois group are explicitly known. The two obvious examples in ordinary Galois theory are:

(a) C is the collection of all finite separable extensions, U is the separable algebraic closure of K and the group is the Galois group $\mathrm{Gal}(U/K)$. For K a finite field or a local field, there are explicit answers.

(b) K is, say, a number field and C is the collection of all finite abelian extensions of K. The field U is the maximal (infinite) abelian extension of K. Class field theory describes both U and its profinite Galois group.

3   Regular singular equations

The differential field $\widehat{K}$ is the field of all formal complex Laurent series $\mathbf{C}((z))$, provided with the differentiation $f \mapsto f' = \frac{df}{dz}$. A matrix differential equation $y' = Ay$ over $\widehat{K}$ is said to be *regular singular* if the equation is equivalent to $y' = \frac{B}{z}y$ for some matrix B with coefficients in $\mathbf{C}[[z]]$. In this context "equivalent" means that there is an invertible matrix T with coefficients in $\widehat{K}$ such that $T^{-1}AT - T^{-1}T' = \frac{B}{z}$. One can prove that a regular singular equation $y' = Ay$ is also equivalent to an equation $y' = \frac{D}{z}y$ where D is a constant matrix. The matrix $e^{2\pi i D}$ is called *the formal monodromy* of the equation $y' = Ay$. We will make this more precise by working with differential modules over $\widehat{K}$. Furthermore, we prefer in this context to use the derivation $\delta := z\frac{d}{dz}$ instead of derivation $\frac{d}{dz}$.

A differential $\widehat{K}$-module (M, δ_M) (and with this choice of δ one has $\delta_M(fm) = \delta(f)m + f\delta_M(m)$ for $f \in \widehat{K}$, $m \in M$) is called *regular singular* if there exists a $\mathbf{C}[[z]]$-lattice $\Lambda \subset M$ which is invariant under δ_M. We recall that a $\mathbf{C}[[z]]$-lattice $\Lambda \subset M$ is a $\mathbf{C}[[z]]$-submodule of M having the form $\mathbf{C}[[z]]m_1 + \cdots + \mathbf{C}[[z]]m_s$, where $m_1, \ldots, m_s$ is a basis of M over $\widehat{K} = \mathbf{C}((z))$. It is easily seen that this definition is compatible with the above definition for matrix differential equations. Let C denote the full subcategory of $\mathrm{Diff}_{\widehat{K}}$ whose objects are the regular singular differential modules. It is rather clear that C has also the required property (ii) of Section 2.2. We start now by describing the universal Picard-Vessiot ring $R_{regsing}$ for C. This ring is written as $R_{regsing} = \widehat{K}[\{z^a\}_{a \in \mathbf{C}}, \ell]$, that is, a $\widehat{K}$-algebra generated by the symbols $z^a, a \in \mathbf{C}$ and ℓ. The only relations are:

(a) For $a \in \mathbf{Z}$ the symbol z^a is equal to the element $z^a \in \widehat{K}$.

(b) For all $a, b \in \mathbf{C}$, $z^a \cdot z^b = z^{a+b}$.

We extend the differentiation δ on $\widehat{K}$ to R by defining $\delta z^a = az^a$ and $\delta \ell = 1$. It can be shown that the only ideals of $R_{regsing}$, invariant under δ, are (0) and $R_{regsing}$. Furthermore, z^a and ℓ are solutions of the regular singular equations $z\frac{d}{dz}y = ay$ and $z\frac{d}{dz}\ell = 1$. Finally, using the fact that every regular singular matrix equation is equivalent to $z\frac{d}{dz}y = Dy$ with D a constant matrix, we find that $R_{regsing}$ is indeed the universal Picard-Vessiot ring for $\mathcal{C}$.

For any regular singular differential module (M, δ_M) we define its solution space $V := \ker(\delta_M, R_{regsing} \otimes M)$. ¿From the above it follows that V is a finite dimensional vector space over $\mathbf{C}$ and that the canonical map $R_{regsing} \otimes_{\mathbf{C}} V \to R_{regsing} \otimes_{\widehat{K}} M$ is an isomorphism.

Let $G_{regsing}$ denote the group of the $\widehat{K}$-linear automorphisms of $R_{regsing}$ commuting with the differentiation of $R_{regsing}$. The action of $G_{regsing}$ on $R_{regsing} \otimes_{\widehat{K}} M$ commutes with δ_M on $R_{regsing} \otimes_{\widehat{K}} M$. In particular, the solution space V is invariant under the action of $G_{regsing}$ and we obtain a homomorphism $G_{regsing} \to \mathrm{GL}(V)$. The image in $\mathrm{GL}(V)$ is the differential Galois group of M.

The *formal monodromy* γ is defined to be the $\widehat{K}$-algebra automorphism of $R_{regsing}$ given by the formulas: $\gamma z^a = e^{2\pi i a}z^a$ and $\gamma \ell = \ell + 2\pi i$. Clearly γ commutes with the differentiation of $R_{regsing}$. In particular γ induces an invertible $\mathbf{C}$-linear map $\gamma_M \in \mathrm{GL}(V)$, which will be called the *formal monodromy of M*. This construction yields a functor $\mathcal{M} : \mathcal{C} \to \mathrm{Repr}_{\mathbf{Z}}$. The latter category is that of the representations of $\mathbf{Z}$ on finite dimensional vector spaces over $\mathbf{C}$. Any representation $\rho : \mathbf{Z} \to \mathrm{GL}(V)$, where V is a finite dimensional vector space over $\mathbf{C}$, is of course determined by the single element $\rho(1) \in \mathrm{GL}(V)$. The morphisms of $\mathrm{Repr}_{\mathbf{Z}}$ are defined in the obvious way. In $\mathrm{Repr}_{\mathbf{Z}}$, the "constructions of linear algebra" can be performed. In particular, for two representations $\rho_i : \mathbf{Z} \to \mathrm{GL}(V_i)$, $i = 1, 2$ we can define the tensor product $\rho := \rho_1 \otimes \rho_2 : \mathbf{Z} \to \mathrm{GL}(V_1 \otimes_{\mathbf{C}} V_2)$ by the formula $\rho(n)(v_1 \otimes v_2) = (\rho_1(n)v_1) \otimes (\rho_2(n)v_2)$. As stated before in Section 2.1, $\mathrm{Repr}_{\mathbf{Z}}$ is a neutral Tannakian category over $\mathbf{C}$. The algebraic hull $\overline{\mathbf{Z}}$ is the affine group scheme over $\mathbf{C}$ such that $\mathrm{Repr}_{\mathbf{Z}}$ is equivalent to the category $\mathrm{Repr}_{\overline{\mathbf{Z}}}$. An important statement is:

Theorem 3.1 $\mathcal{M} : \mathcal{C} \to \mathrm{Repr}_{\mathbf{Z}}$ *is an equivalence of categories (preserving all operations of linear algebra, in particular the tensor product).*

In other words, regular singular differential modules are classified by their formal monodromy. We conclude that the differential Galois group $G_{regsing}$ of $\mathcal{C}$ is the algebraic hull of $\mathbf{Z}$. On the other hand $G_{regsing}$ is also the group of the $\widehat{K}$-linear differential automorphisms of $R_{regsing}$ (seen as an affine group scheme). This will lead to a concrete description of $\overline{\mathbf{Z}}$.

Every automorphism has uniquely the form $\sigma_{h,c}$, where $c \in \mathbf{C}$, and $h \in$ $\mathrm{Hom}(\mathbf{C}/\mathbf{Z}, \mathbf{C}^*)$, and is given explicitly by $\sigma_{h,c}(z^a) = h(a)z^a$ and $\sigma_{h,c}\ell = \ell + c$. Therefore $G_{regsing} \cong \mathrm{Hom}(\mathbf{C}/\mathbf{Z}, \mathbf{C}^*) \times \mathbf{G}_a$. The first group is the projective limit of the groups $\mathrm{Hom}(F, \mathbf{C}^*)$, taken over the finitely generated subgroups F of $\mathbf{C}/\mathbf{Z}$. Each $\mathrm{Hom}(F, \mathbf{C}^*)$ is in an obvious way a linear algebraic group (in fact a finite extension of a torus). This gives $\mathrm{Hom}(\mathbf{C}/\mathbf{Z}, \mathbf{C}^*)$ the structure of an affine group scheme.

We can also define a Hopf algebra B such that $\overline{\mathbf{Z}} \cong G_{regsing} = \mathrm{Spec}(B)$, namely $B = \mathbf{C}[\{s(q)\}_{q \in \mathbf{C}/\mathbf{Z}}, t]$, where the only relations between the generators are $s(q_1 + q_2) = s(q_1) \cdot s(q_2)$. The map $m^* : B \to B \otimes B$ is given by $m^*s(q) = s(q) \otimes s(q)$ and $m^*t = (1 \otimes t) + (t \otimes 1)$. Consider the subalgebra $B_0 = \mathbf{C}[\{s(q)\}_{q \in \mathbf{Q}/\mathbf{Z}}]$. There is an obvious surjective map $\overline{\mathbf{Z}} = \mathrm{Spec}(B) \to \mathrm{Spec}(B_0)$. Every prime ideal p of B_0 turns out to be the kernel of a $\mathbf{C}$-linear homomorphism $h : B_0 \to \mathbf{C}$. These homomorphisms are in bijective correspondence with the elements of $\widehat{\mathbf{Z}}$, the profinite completion of the group $\mathbf{Z}$. Therefore the affine group scheme $\mathrm{Spec}(B_0)$ (or better its group of $\mathbf{C}$-valued points) can be identified with $\widehat{\mathbf{Z}}$. Moreover $\mathrm{Spec}(B_0)$ can be seen to be G/G^o, where G is the affine group scheme $\overline{\mathbf{Z}}$ and G^o is the component of the identity of G.

4 Formal differential equations

The differential field is again $\widehat{K} = \mathbf{C}((z))$ and we will use $\delta = z\frac{d}{dz}$ as differentiation on $\widehat{K}$. For the category $\mathcal{C}$ we take now $\mathrm{Diff}_{\widehat{K}}$. The formal classification of differential modules over $\widehat{K}$ leads to an explicit determination of the universal Picard-Vessiot ring. The result is:

Theorem 4.1 *The universal Picard-Vessiot ring of* $\mathrm{Diff}_{\widehat{K}}$ *is*

$$R_{formal} := \widehat{K}[\{z^a\}_{a \in \mathbf{C}}, \ell, \{e(q)\}_{q \in \mathcal{Q}}]$$

where $\mathcal{Q} = \cup_{m \geq 1} z^{-1/m} \mathbf{C}[z^{-1/m}]$. *The* $\widehat{K}$-*algebra* R_{formal} *is generated by the symbols* $z^a, \ell, e(q)$. *The relations are:*

(a) *For* $a \in \mathbf{Z}$ *the symbol* z^a *is equal to* $z^a \in \widehat{K}$.

(b) *For all* $a, b \in \mathbf{C}$, $z^{a+b} = z^a \cdot z^b$.

(c) *For all* $q_1, q_2 \in \mathcal{Q}$, $e(q_1 + q_2) = e(q_1) \cdot e(q_2)$,

The differentiation on R_{formal} *extends* δ *on* $\widehat{K}$ *and is given by* $\delta z^a = az^a$, $\delta \ell = 1$ *and* $\delta e(q) = qe(q)$.

The *interpretation of the symbols on a suitable sector at the point* $z = 0$ is: z^a stands for the holomorphic function with the same name, ℓ stands for $\log z$, and $e(q)$ stands for $\exp(\int q \frac{dz}{z})$. We can reformulate Theorem 4.1 by stating that all solutions y of a matrix differential equation $y' = Ay$ over $\widehat{K}$ can be written as vectors with coefficients which are combinations of the symbols $z^a, \ell, e(q)$ and formal Laurent series. Moreover these symbols (or their interpretations) have no more algebraic relations than the ones given in Theorem 4.1.

The universal differential Galois group G_{formal}, is equal to the affine group scheme of the $\widehat{K}$-linear automorphisms of R_{formal} commuting with the differentiation on R_{formal}. There is a split exact sequence of affine group schemes $1 \to \text{Hom}(\mathcal{Q}, \mathbf{C}^*) \to G_{formal} \to \overline{\langle \gamma \rangle} \to 1$. Again γ denotes the *formal monodromy* and is defined by $\gamma(z^a) = e^{2\pi i a} z^a$, $\gamma\ell = \ell + 2\pi i$, and $\gamma e(q) = e(\gamma q)$. We note that γ acts in an obvious way on the algebraic closure of $\widehat{K}$ (which is a subfield of R_{formal}) and thus has a natural action on $\mathcal{Q}$. Furthermore, $\langle \gamma \rangle$ denotes the group generated by γ and $\overline{\langle \gamma \rangle}$ is the algebraic hull of $\langle \gamma \rangle$. The affine group scheme $\text{Hom}(\mathcal{Q}, \mathbf{C}^*)$ is the projective limit of the linear algebraic groups $\text{Hom}(F, \mathbf{C}^*)$, taken over all finitely generated subgroups F of $\mathcal{Q}$. The affine group scheme $\text{Hom}(\mathcal{Q}, \mathbf{C}^*)$ is an infinite dimensional torus and and is known as *the exponential torus*.

The morphism $G_{formal} \to \overline{\langle \gamma \rangle} = G_{regsing}$ is obtained by restricting any $\sigma \in G_{formal}$ to the subalgebra $R_{regsing} = \widehat{K}[\{z^a\}_{a \in C}, \ell]$ of R_{formal}. We have seen in Section 3 that the affine group scheme of the $\widehat{K}$-automorphisms commuting with the differentiation of $R_{regsing}$ is $\overline{\langle \gamma \rangle}$. The subgroup $\text{Hom}(\mathcal{Q}, \mathbf{C}^*)$ of G_{formal} consists of the $\sigma \in G_{formal}$, such that $\sigma(z^a) = z^a$ for all $a \in \mathbf{C}$ and $\sigma\ell = \ell$. Such $\sigma \in G_{formal}$ is then given by $\sigma e(q) = h(q) \cdot e(q)$ where h is a homomorphism of $\mathcal{Q}$ to $\mathbf{C}^*$. Finally the affine group scheme G_{formal} is not commutative since γ does not commute with the elements of the normal subgroup scheme $\text{Hom}(\mathcal{Q}, \mathbf{C}^*)$.

In the sequel we attack the most interesting case where the differential field is $K = \mathbf{C}(\{z\})$, the field of the convergent Laurent series provided with the differentiation $f \mapsto z\frac{df}{dz}$. The essential tool we will be using here is the *multisummation* of divergent solutions of a differential equation $y' = Ay$, where the matrix A has coefficients in K.

5 Multisummation and Stokes maps

We keep the notation $\widehat{K} = \mathbf{C}((z))$ and let $K = \mathbf{C}(\{z\})$ denote its subfield consisting of the convergent Laurent series. For both fields we will use the differentiation $f \mapsto \delta(f) := z\frac{df}{dz}$.

Consider a matrix differential equation $\delta y = Ay$, where A is a matrix of size $m \times m$ with coefficients in K. This can also be seen as an equation over $\widehat{K}$. In particular there is a solution space $V \subset R^m_{formal}$ consisting of the elements $y \in R^m_{formal}$ satisfying $\delta(y) = Ay$. The space V has dimension m over $\mathbf{C}$. A real number $d \in \mathbf{R}$ is seen as the direction e^{id} at $z = 0$. In this context we do not want to identify d and $d + k \cdot 2\pi$ for $k \in \mathbf{Z}$, since we will be dealing with multivalued meromorphic functions which are branched at $z = 0$.

Theorem 5.1 [6, Theorem 5] *Consider the matrix differential equation $\delta(y) = Ay$ where A is a matrix of size $m \times m$ with coefficients in K. Let $V \subset R^m_{formal}$ denote its solution space.*

(a) *There are finitely many exceptional directions in $\mathbf{R}/2\pi\mathbf{Z}$ for the equation. These are called the singular directions. The set of singular directions depends only on the formal equivalence class (that is, taken over $\widehat{K}$) of the equation $\delta(y) = Ay$ (see Lemma 5.3 for an explicit description). The other directions will be called regular.*

(b) *Let d be a regular direction. Consider the sector S defined by the opening $(d-r, d+r)$. Here $r > 0$ is a fixed constant only depending on the formal equivalence class of the equation $\delta(y) = Ay$. The multisummation S_d in the direction d is a uniquely defined $\mathbf{C}$-linear bijection $S_d : V \to W(S)$, where $W(S)$ is the solution space of the equation $\delta(y) = Ay$ taken in the space of the meromorphic functions living on the sector S.*

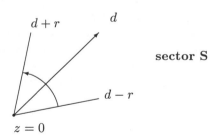

We sketch the general idea of the multisummation map. In the construction of the map S_d the symbols $z^a, \ell, e(q)$ are mapped to their interpretations $z^a, \log(z), exp(\int q \frac{dz}{z})$ on the given sector S. This interpretation depends on $d \in \mathbf{R}$ and not just on d modulo 2π. In the expression for the elements of V, finitely many divergent power series $\widehat{f}_1, \ldots, \widehat{f}_s \in \widehat{K}$ occur. They satisfy certain linear differential equations over K, derived from the original equation $\delta(y) = Ay$. Then S_d maps the $\widehat{f}_i$ to meromorphic functions F_i, $i = 1, \ldots, s$

living on the sector S, having $\widehat{f}_i$, $i = 1, \ldots, s$ as asymptotic expansion and such that for each i the meromorphic function F_i and the formal series $\widehat{f}_i$ satisfy the same linear differential equation over K. The construction S_d commutes with differentiation and with all operations of linear algebra. The multisummation theorem is the cumulative work of many mathematicians, for example, W. Balser, B.L.J. Braaksma, J. Ecalle, B. Malgrange, J. Martinet, J.-P. Ramis, Y. Sibuya *et al.*

In Lemma 5.3 we will describe the set of singular directions explicitly. Let d be a singular direction. Then d^+, d^- denote directions close enough to d and such that $d^- < d < d^+$. Let S^-, S^+ denote the sectors with openings $(d^- - r, d^- + r)$ and $(d^+ - r, d^+ + r)$ corresponding to the directions d^-, d^+.

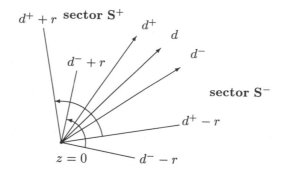

The Stokes map St_d is defined by

$$St_d : V \xrightarrow{S_{d^-}} W(S^-) \xrightarrow{\text{analytic cont.}} W(S^+) \xrightarrow{S_{d^+}^{-1}} V.$$

One of the main results on Stokes maps is:

Theorem 5.2 (J. Martinet and J.-P. Ramis, [6, Theorem 6]) *The differential Galois group of $\delta(y) = Ay$ over K is the smallest algebraic subgroup of $\mathrm{GL}(V)$ containing the differential Galois group of $\delta(y) = Ay$ over $\widehat{K}$ and the collection of the Stokes maps St_d (taken over all directions $d \in [0, 2\pi)$).*

For the formulation of the second main result, Theorem 5.4, on the Stokes maps we will need some additional notation and terminology. The solution space $V \subset R_{formal}^m$ can be decomposed as a finite direct sum of vector spaces $\{V_q | q \in \mathcal{Q}\}$. The $\widehat{K}$-algebra R_{formal} can be written as the direct sum $R_{formal} = \oplus_{q \in \mathcal{Q}} R_{regsing} e(q)$, where we recall that $R_{regsing} = \widehat{K}[\{z^a\}_{a \in \mathbb{C}}, \ell]$. Let $V_q := V \cap (R_{regsing} e(q))^m$.

For $q \in Q$, consider the multivariant function $F_q(z) := \exp(\int q \frac{dz}{z})$. A direction $d \in \mathbf{R}$ will be called *singular for q* if the function $r \in \mathbf{R}_{>0} \mapsto F_q(re^{id})$ (with the interpretation of q given by d) is of *maximal decrease* for $r \to 0$. For two elements $q_1, q_2 \in Q$ such that both spaces V_{q_1}, V_{q_2} are non-zero, we say that $d \in \mathbf{R}$ is a *singular direction for the pair q_1, q_2* if d is singular for the direction $q_1 - q_2$.

Lemma 5.3 *The collection of the singular directions of $\delta(y) = Ay$ is equal to the union of all singular directions of all pairs q_1, q_2 such that both spaces V_{q_1}, V_{q_2} are non-zero.*

Theorem 5.4 (J. Martinet, J.-P. Ramis *et al.*, [6], Theorem 10)

(a) *For a (singular) direction d of $\delta(y) = Ay$, the Stokes map*
$St_d : V = V_{q_1} \oplus \cdots \oplus V_{q_s} \to V$ *has the form*

$$St_d = id + \sum_{d \text{ singular for } q_i - q_j} A_{i,j},$$

where $A_{i,j} \in \mathrm{Hom}_C(V_i, V_j)$. In particular, St_d is a unipotent map.

(b) *Suppose that there are given:*

(1) *a "standard" formal differential equation $\delta(y) = By$ over $\widehat{K}$ which has a solution space V with decompostion $V = V_{q_1} \oplus \cdots \oplus V_{q_s}$ and*
(2) *for every singular direction $d \in [0, 2\pi)$ a map $F_d : V \to V$ of the form given in part (a).*

Then there exists a unique (up to equivalence over K) differential equation $\delta(y) = Ay$ over K which is formally equivalent (that is, over $\widehat{K}$) with $\delta(y) = By$ and such that $St_d = F_d$ for every singular direction $d \in [0, 2\pi)$.

The first part of this theorem gives the special form which a Stokes map has. The second part says that all possible Stokes maps (of the prescribed form) actually occur for a differential equation. Moreover the meromorphic differential equation is determined (up to isomorphism over K) by its formal equivalence class and the Stokes maps.

6 Meromorphic differential equations

In this section, K and $\widehat{K}$ have the same meaning as in the Sections 4 and 5. Our first claim is an explicit expression for the universal Picard-Vessiot ring R_{conv} of Diff_K. For this purpose we define a K-algebra $\mathcal{D}$ with $K \subset \mathcal{D} \subset \widehat{K}$ as follows: $f \in \widehat{K}$ belongs to $\mathcal{D}$ if and only if f satisfies some linear scalar differential equation $f^{(n)} + a_{n-1}f^{(n-1)} + \cdots + a_1 f^{(1)} + a_0 f = 0$ with all coefficients $a_i \in K$. This condition on f can be restated as follows: f belongs to $\mathcal{D}$ if and only if the K-linear subspace of $\widehat{K}$ generated by all the derivatives of f is finite dimensional. It follows easily that $\mathcal{D}$ is an algebra over K stable under differentiation. The following example shows that $\mathcal{D}$ *is not a field*.

Example 6.1 The differential equation $y^{(2)} = z^{-3}y$ (here we have used the ordinary differentiation $\frac{d}{dz}$) has a solution $f = \sum_{n\geq 2} a_n z^n \in \widehat{K}$ given by $a_2 = 1$ and $a_{n+1} = n(n-1)a_n$ for $n \geq 2$. Clearly f is a divergent power series and by definition $f \in \mathcal{D}$. Suppose that also $f^{-1} \in \mathcal{D}$. Then also $u := \frac{f'}{f}$ lies in $\mathcal{D}$ and there is a finite dimensional K-vector space W with $K \subset W \subset \widehat{K}$ which is invariant under differentiation and contains u. We note that $u' + u^2 = z^{-3}$ and consequently $u^2 \in W$. Suppose that $u^n \in W$. Then $(u^n)' = nu^{n-1}u' = nu^{n-1}(-u^2 + z^{-3}) \in W$ and thus $u^{n+1} \in W$. Since all the powers of u belong to W, the element u must be algebraic over K. It is known that K is algebraically closed in $\widehat{K}$ and thus $u \in K$. The element u can be written as $\frac{2}{z} + b_0 + b_1 z + \cdots$ and since $f' = uf$ one finds $f = z^2 \cdot \exp(b_0 z + b_1 \frac{z^2}{2} + \cdots)$. The latter is a convergent power series and we have obtained a contradiction.

Lemma 6.2 $R_{conv} := \mathcal{D}[\{z^a\}_{a\in\mathbf{C}}, \ell, \{e(q)\}_{q\in\mathcal{Q}}]$ *is the universal Picard-Vessiot ring for* Diff_K.

Proof. The algebra R_{formal} contains R_{conv} and R_{conv} is generated, as a K-algebra, by all solutions of linear scalar differential equations over K. From this the lemma follows. $\qquad\square$

The universal differential Galois group for Diff_K is denoted by G_{conv}. The inclusion $R_{conv} \subset R_{formal}$ induces an injective morphism of affine group schemes $G_{formal} \to G_{conv}$. One defines the normal affine subgroup scheme $N \subset G_{conv}$ to consist of the σ such that $\sigma(z^a) = z^a$ for all $a \in \mathbf{C}$, $\sigma e(q) = e(q)$ for all $q \in \mathcal{Q}$, and $\sigma\ell = \ell$. Moreover St denotes the affine group scheme of G_{conv} generated by $\{St_d| d \in \mathbf{R}\}$. This needs a little explanation.

For a given differential equation $\delta(y) = Ay$ over K and any direction $d \in \mathbf{R}$, the map St_d, defined in Section 5, acts as a K-algebra automorphism on the Picard-Vessiot ring R_A of the equation and commutes with the differentiation of R_A. If d happens to be a regular direction for the equation $\delta(y) = Ay$ then St_d is the identity. The universal Picard-Vessiot ring R_{conv} is the direct limit of the Picard-Vessiot rings R_A, taken over all equations $\delta(y) = Ay$ over K. The invariance of St_d under all operations of linear algebra shows that the maps St_d on the various algebras R_A glue to an K-linear automorphism

St_d of R_{conv} commuting with the differentiation of R_{conv}. Thus $St_d \in G_{conv}$. Moreover $St_d \in N$, since by construction St_d leaves all z^a, $e(q)$, ℓ invariant. Therefore $St \subset N$. It can be shown, by producing rather involved examples of linear differential equations, that actually $St \neq N$.

Lemma 6.3 $G_{conv}/N \cong G_{formal}$ and G_{conv} is the semi-direct product of G_{formal} and N.

The problem of determining the structure of G_{conv} is now reduced to finding the structure of the affine group scheme N and the action by conjugation of G_{formal} on N. Instead of considering the connected affine group scheme N one studies the "easier" object $\mathrm{Lie}(N)$, the Lie algebra of N.

This Lie algebra can also be expressed in terms of the universal Picard-Vessiot ring R_{conv}. First of all $\mathrm{Lie}(G_{conv})$ can be identified with the Lie algebra over C consisting of all K-linear derivations $D : R_{conv} \to R_{conv}$ which commute with the differentiation on R_{conv}. In the same way, $\mathrm{Lie}(N)$ can be identified with the Lie algebra over C consisting of all K-linear derivations $D : R_{conv} \to R_{conv}$, commuting with the differentiation of R_{conv} and such that $D(z^a) = 0$ for all $a \in \mathbf{C}$, $D(\ell) = 0$, and $D(e(q)) = 0$ for all $q \in \mathcal{Q}$.

From the construction of R_{conv} it follows that there exists a collection $\{W_i\}_{i \in I}$ consisting of finite dimensional K-linear subspaces of R_{conv} such that:

(a) Each W_i is invariant under the action of G_{conv}.

(b) For all $i, j \in I$, there is a $k \in I$ with $W_i + W_j \subset W_k$.

(c) $R_{conv} = \bigcup_{i \in I} W_i$.

Indeed, the universal Picard-Vessiot ring R_{conv} is the filtered union of the Picard-Vessiot rings R_M, taken over all differential modules M over K. Each R_M is invariant under the action of G_{conv}. Actually, the restriction of the action of G_{conv} to R_M coincides with the action of the differential Galois group G_M of the differential module M on R_M. It is not difficult to show that any finite subset of R_M lies in a finite dimensional K-linear subspace W of R_M which is invariant under G_M.

The action of any St_d on R_{conv} is *locally unipotent*, that is, the restriction of St_d to each W_i is unipotent. Then $\Delta_d = \log St_d$, where $\log St_d$ is defined as usual by $\sum_{n \geq 1} \frac{(-1)^{n+1}}{n} (St_d - 1)^n$, is a well defined *locally nilpotent* K-linear map from R_{conv} to itself. The fact that St_d belongs to N implies that Δ_d belongs to $\mathrm{Lie}(N)$. The action of Δ_d on R_{conv} is determined by its restriction $\Delta_d : \mathcal{D} \to R_{conv}$ since Δ_d is 0 on the elements z^a, ℓ, $e(q)$. The algebra R_{conv} has a direct sum decomposition $R_{conv} = \oplus_{q \in \mathcal{Q}} R_{conv,\,q}$ where $R_{conv,\,q} :=$

$\mathcal{D}[\{z^a\}_{a\in\mathbf{C}},\ell]e(q)$. This allows us to decompose $\Delta_d : \mathcal{D} \to R_{conv}$ as direct sum $\sum_{q\in Q} \Delta_{d,q}$ by the formula $\Delta_d(f) = \sum_{q\in Q} \Delta_{d,q}(f)$, where $\Delta_{d,q}(f) \in R_{conv, q}$ for each $q \in Q$. We note that $\Delta_{d,q} = 0$ if d is not a singular direction for q. The map $\Delta_{d,q} : \mathcal{D} \to R_{conv}$ has a unique extension to an element in $\text{Lie}(N)$. The elements of $S := \{\Delta_{d,q}|\ d \text{ singular direction for } q\}$ are sometimes called *alien derivations*. The group $G_{formal} \subset G_{conv}$ acts on $\text{Lie}(N)$ by conjugation. For a homomorphism $h : Q \to \mathbf{C}^*$ one writes τ_h for the element of this group given by the formulas τ_h leaves z^a, ℓ invariant and $\tau e(q) = h(q) \cdot e(q)$. Let γ denote, as before, the formal monodromy. According to the structure of G_{formal} explained in Section 4, it suffices to know the action by conjugation of the τ_h and γ on $\text{Lie}(N)$. For the elements $\Delta_{d,q}$ one has the explicit formulas:

$$\gamma\Delta_{d,q}\gamma^{-1} = \Delta_{d-2\pi,\gamma(q)},$$
$$\tau_h\Delta_{d,q}\tau_h^{-1} = h(q) \cdot \Delta_{d,q}.$$

We would like to state that S, the set of the alien derivations, generates the Lie algebra $\text{Lie}(N)$ and that these elements are independent. This is close to being correct. The fact that the $\Delta_{d,q}$ acts locally nilpotent on R_{conv}, however, complicates the final statement. In order to be more precise we have to go through some general constructions with Lie algebras.

6.1 A construction with free Lie algebras

We follow some classical constructions, see [2]. Let S be any set. Let W denote a vector space over $\mathbf{C}$ with basis S. By $W^{\otimes m}$ we denote the m-fold tensor product $W \otimes_\mathbf{C} \cdots \otimes_\mathbf{C} W$ (note that this is *not* the symmetric tensor product). Then $F\{S\} := \mathbf{C} \oplus \sum_{m\geq 1}^{\oplus} W^{\otimes m}$ is the *free associative algebra on the set S*. It comes equipped with a map $i : S \to F\{S\}$. The universal property of $(i, F\{S\})$ reads: For any associative $\mathbf{C}$-algebra B and any map $\phi : S \to B$, there is a unique $\mathbf{C}$-algebra homomorphism $\phi' : F\{S\} \to B$ with $\phi' \circ i = \phi$.

The algebra $F\{S\}$ is also a Lie algebra with respect to the Lie brackets $[\ ,\ ]$ defined by $[A, B] = AB - BA$ for all $A, B \in F\{S\}$. The *free Lie algebra on the set S* is denoted by $\text{Lie}\{S\}$ and is defined as the Lie subalgebra of $F\{S\}$ generated by $W \subset F\{S\}$. This Lie algebra is equipped with an obvious map $i : S \to \text{Lie}\{S\}$ and the pair $(i, \text{Lie}\{S\})$ has the following universal property: For any complex Lie algebra L and any map $\phi : S \to L$ there is a unique homomorphism $\phi' : \text{Lie}\{S\} \to L$ of complex Lie algebras such that $\phi' \circ i = \phi$.

Furthermore, for any associative complex algebra B and any homomorphism $\psi : \text{Lie}\{S\} \to B$ of complex Lie algebras (where B is given its canonical structure as complex Lie algebra) there is a unique homomorphism

$\psi' : F\{S\} \to B$ of complex algebras such that the restriction of ψ' to $\text{Lie}\{S\}$ coincides with ψ.

Consider now a finite dimensional complex vector space W and an action of $\text{Lie}\{S\}$ on W. This amounts to a homomorphism $\psi : \text{Lie}\{S\} \to \text{End}(W)$ of complex Lie algebras or to a $\mathbf{C}$-algebra homomorphism $\psi' : F\{S\} \to \text{End}(W)$. Here we are only interested in those ψ such that:

(1) $\psi(s) = \psi'(s)$ is nilpotent for all $s \in S$.

(2) there are only finitely many $s \in S$ with $\psi(s) \neq 0$.

For any ψ satisfying (1) and (2) one considers the ideal $\ker \psi$ in the Lie algebra $\text{Lie}\{S\}$ and its quotient Lie algebra $\text{Lie}\{S\}/\ker \psi$. We define now a sort of completion $\widehat{\text{Lie}\{S\}}$ of $\text{Lie}\{S\}$ as the projective limit of $\text{Lie}\{S\}/\ker \psi$, taken over all ψ satisfying (1) and (2). We can now formulate the description of J. Martinet and J.-P. Ramis for the structure of N and its Lie algebra $\text{Lie}(N)$.

Theorem 6.4 *Let S denote again the set of alien derivations*

$$\{\Delta_{d,q}\mid d \in \mathbf{R},\ q \in \mathcal{Q},\ d \text{ is singular for } q\}.$$

The affine group scheme N is connected and there exists an isomorphism of complex Lie algebras $\psi : \widehat{\text{Lie}\{S\}} \to \text{Lie}(N)$, which respects the action of G_{formal} on both Lie algebras.

Remark 6.5 Theorem 6.4 and a proof of the theorem are indicated in [6], Theorems 17 and 18. Unfortunately this publication was not followed by a more detailed exposition. A first "complete" proof, along the lines of [6] will be provided in [9].

Remark 6.6 The proof may start by defining $\psi_0 : \text{Lie}\{S\} \to \text{Lie}(N)$ so that ψ_0 maps a generator $\Delta_{d,q} \in S$ to the same object but now seen as an element of $\text{Lie}(N)$. The $\Delta_{d,q} \in \text{Lie}(N)$ are locally nilpotent. This implies that ψ_0 induces a morphism ψ from the Lie algebra $\widehat{\text{Lie}\{S\}}$ to $\text{Lie}(N)$. Proving that ψ is actually an isomorphism is based upon Theorem 5.4 and requires some more technical tools.

References

1. Deligne, P., Milne, J.S. *Tannakian categories.* Lect. Notes in Math. **900**, Springer-Verlag, Berlin 1982.
2. Jacobson, N. *Lie Algebras*, Dover Publications, Inc. New York, 1962.

3. Kaplansky, I. *An Introduction to Differential Algebra*, 2nd ed., Hermann Paris, 1976.

4. Kolchin, E. R. *Differential Algebra and Algebraic Groups*, Academic Press, New York, 1973.

5. Magid, A. *Lectures on Differential Galois Theory*, AMS University Lecture Series 7, 1994.

6. Martinet, J., Ramis, J.-P. *Elementary acceleration and multisummability*, Annales de l'Institut Henri Poincaré, Physique Théorétique **54**(4), (1991), 331–401.

7. van der Put, M. *Galois theory of differential equations, algebraic groups and Lie algebras.* J. Symbolic Comput. **28**, (1999), 441–472.

8. van der Put, M. *Recent Work on Differential Galois Theory.* Séminaire Bourbaki, Juin 1998.

9. van der Put, M., Singer, M. F. *Differential Galois Theory*, Book in preparation.

Differential Algebra and Related Topics, pp. 191–218
Proceedings of the International Workshop
Eds. L. Guo, P. J. Cassidy, W. F. Keigher & W. Y. Sit
© 2002 World Scientific Publishing Company

CYCLIC VECTORS

R.C. CHURCHILL

Department of Mathematics,
Hunter College and Graduate Center, CUNY,
695 Park Avenue, New York, NY 10021, USA
E-mail: rchurchi@math.hunter.cuny.edu

JERALD J. KOVACIC

Department of Mathematics,
The City College of New York, CUNY
Convent Avenue at 138th Street, New York, NY 10031, USA
E-mail: jkovacic@member.ams.org

A differential structure on an n-dimensional vector space V over a field K is a pair (δ, D), where δ is a derivation on K and $D : V \to V$ is an additive mapping satisfying the Leibniz (product) rule. A cyclic vector is a vector v such that $(v, Dv, \ldots, D^{n-1}v)$ is a basis of V. In this note we offer two existence proofs for cyclic vectors under very mild hypotheses on K. The first proof is in the form of an algorithm; the second includes a simple description of the collection of all cyclic vectors. A MAPLE implementation of the algorithm for the case $(K, \delta) = (\mathbb{Q}(x), \frac{d}{dx})$ is detailed in the Appendix.

1 Introduction

Let K be a field and V a K-space (*i.e.*, a vector space over K). By a *differential structure* on V we mean a pair (δ, D), where $\delta : k \mapsto k'$ is a derivation on K, *i.e.*, K is a differential field, and $D : V \to V$ is an additive group homomorphism satisfying the *Leibniz rule*

$$D(kv) = k'v + kDv \qquad (k \in K, \ v \in V).$$

The triple (V, δ, D) is called a *differential vector space* or *differential system*. In this paper we restrict our attention to finite dimensional differential systems, $\dim V = n$.

A vector $v \in V$ is *cyclic* if $(v, Dv, \ldots, D^{n-1}v)$ is a basis for V. The Cyclic Vector Theorem asserts the existence of such a vector under quite mild assumptions on K.

The earliest proof known to these authors is found in [23], but is restricted to fields of meromorphic functions. The argument is by contradiction. The second proof appears in [10]. That treatment is constructive, in fact almost an algorithm, but is specific to the field of rational functions.

Many other proofs have since appeared, perhaps the most influential being that of Deligne [12]. His non-constructive approach apparently caused some consternation: Adjamagbo [1], in the abstract, remarks that the *"reductio ad absurdum* [argument of Deligne] ... caused much ink to flow".

Dabèche [11] has a constructive proof using special properties of the field of meromorphic functions (poles, evaluation, etc.). Other constructive arguments can be found in [6], [8], and [13] (using methods attributed to Ramis [25]).

Differential algebraists first learned of the Cyclic Vector Theorem, and of Loewy's proof, from [21]. A generalization to partial differential fields was presented in [28]. It is amusing to note that the "New York school" knew of Loewy but not of Cope. The "French school" knew of Cope but not of Loewy. And neither knew of the work of the other!

Using entirely different ideas, Katz [19] found cyclic vectors *locally* on $\mathrm{Spec}(R)$ for quite general differential rings. However, for the Cyclic Vector Theorem itself his proof requires strong hypothesis (see Remark 4 on page 68): K must be a differential field in which $(n-1)!$ is invertible and which is finitely generated over an algebraically closed field of constants (for example $K = \mathbb{C}(x)$).

Hilali [16] computes several invariants of a differential system avoiding use of the cyclic vector theorem, which he claims is "très coûteuse en pratique" (middle of page 405).

Adjamagbo [1] gives an algorithm for finding cyclic vectors over differential division rings. He uses ideas from the theory of modules over the non-commutative ring of linear differential operators (generalizations of Weyl algebras); in particular, left and right Euclidean division.

Adjamagbo [1] also sketches a novel (albeit non-trivial) proof of the Cyclic Vector Theorem in Remark 2, page 546. He refers to a paper in preparation that explains the connection between cyclic vectors and primitive elements. With those results, the Cyclic Vector Theorem follows from the existence of a primitive element [20, Proposition 9, p. 103]. The paper was published later as [2]; it relies on theorems of [26] and [27].

Barkatou [4] solves a related problem and, in Remark 3, p. 192, claims that the same method can be used to find a cyclic vector. §5 of that paper deals with the cyclic vector algorithm in more detail. Working over $\mathbb{C}(x)$ he states that it is always possible to find a cyclic vector whose components are polynomials of degree less than n, but does not give a specific reference. He also states that it is "well-known" that the probability that a random vector is cyclic is 1. His algorithm, at the bottom of page 194, is:

Algorithm 1.1 To find a cyclic vector:

Step 1. Choose a vector with random polynomial coefficients.

Step 2. If the vector is cyclic, we are done.

Step 3. Otherwise, go to Step 1.

Theoretically this "algorithm" could fail to terminate; in practice it generally provides a cyclic vector in just a few steps. The reason is easy to explain: such vectors form a non-empty open (and therefore dense) subset of affine space in the Kolchin topology (see Theorem 7.3).

Algorithm 1.1 seems to be the basis for the implementation in MAPLE. The DETools package (attributed to Mark van Hoeij) appears to first try $(1, 0, \ldots, 0)$, and then $(p_1, \ldots, p_n)$ where the p_i are random polynomials, with coefficients from the set $\{-1, 0, 1\}$, of degree one higher than in the previous trial.

In this note we offer two proofs of the Cyclic Vector Theorem with no restriction on the characteristic. The first (found in §3) is similar in spirit to that given in [12], but is effective and to these authors seems considerably more elementary than any of the constructive proofs referenced above. When K contains the real field the method enables one to produce a cyclic vector which can be viewed as an arbitrarily small perturbation $v + \sum_{j=2}^n \epsilon_j v_j$ of any preassigned $v \in V$.

We discuss the details of the algorithm, and its complexity, in §4. The Appendix contains a MAPLE implementation for the case $(K, \delta) = (\mathbb{Q}(x), \frac{d}{dx})$.

The second proof (found in §7) is influenced by [10], but is here cast in a more general setting. In §5 we remark on the hypotheses of our algorithm; in §6 we show they cannot be weakened.

We view a differential structure as a coordinate-free formulation of a first order (system of) homogeneous linear differential equation(s); proving the existence of a cyclic vector for the associated adjoint system is then seen to be equivalent to converting the given equation to an n-th order homogeneous form. The ideas are reviewed in §2 (or see [29], pp. 131–2).

The explicit construction of cyclic vectors is important for calculating invariants of differential systems at singular points. For example, Varadarajan [29, p. 152–3] explains how cyclic vectors can be used to compute principal levels ("Katz invariants") of connections. Levelt [22] makes essential use of cyclic vectors to find normal forms for differential operators. Additional applications can be found in [5], [7], [14], [16] and [17]. Hilali [15] contrasts the efficiency of particular algorithms with and without the use of cyclic vectors.

2   Linear Differential Equations

Throughout this paper we assume (V, δ, D) is a finite dimensional differential system. Thus $\delta : k \mapsto k'$ is a derivation on K, V is K-space of finite dimension $n > 0$, and $D : V \to V$ is a differential structure. We follow custom and when $v \in V$ write $D(v)$ as Dv unless confusion might otherwise result. The space (Lie algebra) of $n \times n$ matrices with entries in K is denoted by $\mathrm{gl}(n, K)$.

Fix a basis $\mathbf{e} = (e_1, \ldots, e_n)$ of V. We may use $\mathbf{e}$ to associate a matrix $A = (a_{ij}) \in \mathrm{gl}(n, K)$ with D by means of the formula

$$De_j = \sum_{i=1}^{n} a_{ij} e_i, \qquad (2.1)$$

and the correspondence $D \mapsto A$ is easily seen to be bijective. We refer to A as the *defining* $\mathbf{e}$-*matrix* of D. If for any $v = \sum_j v_j e_j \in V$ we let $v_{\mathbf{e}}$ (resp. $v'_{\mathbf{e}}$) denote the column vector with j-th entry v_j (resp. v'_j) we then have

$$(Dv)_{\mathbf{e}} = v'_{\mathbf{e}} + A v_{\mathbf{e}}. \qquad (2.2)$$

This last equality explains how Algorithm 1.1 in the Introduction can be implemented: choose any non-zero $v \in V$ and form the matrix

$$\begin{pmatrix} v_1 & (Dv)_1 & \cdots & (D^{n-1}v)_1 \\ \vdots & \vdots & & \vdots \\ v_n & (Dv)_n & \cdots & (D^{n-1}v)_n \end{pmatrix}; \qquad (2.3)$$

if the determinant does not vanish the vector v is cyclic; otherwise choose a different non-zero v and try again; etc.

Vectors annihilated by D are said to be *horizontal* (with respect to D). Notice from (2.2) that a vector $v \in V$ is horizontal if and only if $v_{\mathbf{e}}$ is a solution of the first order (system of) homogeneous linear equation(s)

$$Y' + AY = 0. \qquad (2.4)$$

This is the *defining* $\mathbf{e}$-*equation* for D. We now see the relationship between differential structures and homogeneous linear differential equations mentioned in the introduction.

To understand the significance of cyclic vectors in the differential equation context first observe that to each differential structure D on V there is a naturally associated differential structure D^* on the dual space V^* defined by

$$(D^* u^*)v = \delta(u^* v) - u^*(Dv) \quad \text{for } u^* \in V^* \text{ and } v \in V. \qquad (2.5)$$

It is easy to check that D^*u^* is K-linear and that D^* is a differential structure on V^*. The matrix associated with the basis e^* of V^* dual to e is $-A^\tau$, where the τ denotes the transpose, and as a consequence the corresponding differential equation is the (classical) adjoint equation

$$Y' - A^\tau Y = 0 \tag{2.6}$$

of (2.4). The natural identification $V^{**} \sim V$ induces an identification $D^{**} \sim D$; in particular, by regarding e as the dual basis of e^* we may view (2.4) as the adjoint equation of (2.6).

Suppose the basis e^* of V^* corresponds to a cyclic vector $v^* \in V^*$ for D^*, i.e., suppose $e^* = (v^*, D^*v^*, \ldots, D^{*n-1}v^*)$. Then $-A^\tau \in gl(n, K)$ has the form

$$-A^\tau = \begin{pmatrix} 0 & \cdots\cdots & 0 & q_n \\ 1 & 0 & \vdots & q_{n-1} \\ 0 & 1 & 0 & \vdots & \vdots \\ \vdots & & \ddots & \ddots & \vdots & \vdots \\ & & & 1 & 0 & q_2 \\ 0 & \cdots\cdots & 0 & 1 & q_1 \end{pmatrix}, \tag{2.7}$$

and the defining e-matrix of D is therefore

$$A = \begin{pmatrix} 0 & -1 & 0 & \cdots\cdots & 0 \\ \vdots & 0 & -1 & & \vdots \\ & & 0 & \ddots & \\ \vdots & & & \ddots & -1 & 0 \\ 0 & \cdots\cdots\cdots & & 0 & -1 \\ -q_n & -q_{n-1} & \cdots\cdots & -q_2 & -q_1 \end{pmatrix}. \tag{2.8}$$

Conversely, when a defining basis matrix of D has this form one sees directly from the companion matrix structures of A and $-A^\tau$ that the initial element of the dual basis is cyclic for D^*.

The relevance of (2.8) is completely standard: when A has this form a column vector $v = (v_1, \ldots, v_n)^\tau$ is a solution of the associated linear homogeneous system $Y' + AY = 0$ if and only if $y := v_1$ is a solution of the n-th-order homogeneous linear differential equation

$$y^{(n)} - q_1 y^{(n-1)} - \cdots - q_{n-1} y' - q_n y = 0. \tag{2.9}$$

We conclude that the existence of a cyclic vector for D^* is equivalent to the existence of a defining equation for D which can be expressed in n-th order form.

There is an alternate way to view this equivalence. Specifically, suppose $\mathbf{e} = (e_1, \ldots, e_n)$ and $\mathbf{f} = (f_1, \ldots, f_n)$ are bases of V, and A and $B \in \mathrm{gl}(n, K)$ are the defining $\mathbf{e}$ and $\mathbf{f}$-matrices of a differential structure D. Let $T = (t_{ij}) \in \mathrm{GL}(n, K)$ be the transition matrix between the given bases, *i.e.*, $e_j = \sum_i t_{ij} f_i$ for $j = 1, \ldots, n$, and write T' for the matrix $(t'_{ij}) \in \mathrm{gl}(n, K)$. Then from the Leibniz rule one sees that

$$A = T^{-1}BT + T^{-1}T'. \qquad (2.10)$$

The mapping of $\mathrm{gl}(n, K) \times GL(n, K)$ into $\mathrm{gl}(n, K)$ defined by

$$(B, T) \mapsto T^{-1}BT + T^{-1}T'$$

is a right action of $GL(n, K)$ on $\mathrm{gl}(n, K)$, often referred to as the action by *gauge transformations*. From (2.10) and the preceding discussion, we see that by viewing B as a defining matrix for a differential structure D the existence of a cyclic vector for D^* is equivalent to the existence of an element A in the $\mathrm{GL}(n, K)$-orbit of B having the form displayed in (2.8).

In view of the duality discussed immediately following (2.6) we can work with either D^* or D for purposes of constructing cyclic vectors. We choose D to ease notation.

3   The Algorithm

Here $V = (V, \delta, D)$ is a differential K-space of finite dimension $n \geq 2$. (Any non-zero vector is cyclic when $n = 1$.) The field of constants of K (the kernel of δ), which always contains the prime field, is denoted K_C.

Hypotheses 3.1 We assume that:

1) the degree of the extension $K_C \subset K$ is at least n (possibly infinite); and

2) K_C contains at least n non-zero elements.

These hypotheses have been formulated so as to emphasize the finite nature of the algorithm. In fact 2) is a simple consequence of 1), as will be seen in Proposition 5.1(d). Assuming Hypotheses 3.1, we describe a simple algorithm for constructing a cyclic vector for D.

Suppose $v \in V$, $v \neq 0$, is *not* a cyclic vector, in which case

$$v \wedge Dv \wedge \cdots \wedge D^{m-1}v \neq 0, \qquad (3.2)$$

where $1 \leq m < n$, and

$$D^m v = \sum_{k=0}^{m-1} a_k D^k v \qquad (3.3)$$

for some $a_0, \ldots, a_{m-1} \in K$. Choose any vector $u \in V$ not in the span of

$\{v, Dv, \ldots, D^{m-1}v\}$. We will indicate how one can select elements $\lambda_0 \in K_C$ and $k_0 \in K$ such that for

$$\bar{v} := v + \lambda_0 k_0 u,$$

one has

$$\bar{v} \wedge D\bar{v} \wedge \cdots \wedge D^{m-1}\bar{v} \wedge D^m\bar{v} \neq 0, \tag{3.4}$$

i.e., the differential vector space spanned by $\{\bar{v}, D\bar{v}, \ldots, D^{n-1}\bar{v}\}$ is strictly larger than that spanned by $\{v, Dv, \ldots, D^{n-1}v\}$.

To this end, set

$$v_0 := v, \quad v_i := D^i v, \qquad i = 1, \ldots, m-1,$$

and extend this K-linearly independent set to a basis of V by first adjoining u and then, if necessary, $e_1, \ldots, e_{n-m-1}$. For $0 \le k \le m$, write

$$D^k u = \sum_{i=0}^{m-1} \alpha_{ik} v_i + \beta_k u + \sum_{j=1}^{n-m-1} \gamma_{jk} e_j, \tag{3.5}$$

where $\alpha_{ik}, \beta_k, \gamma_{jk} \in K$. (In particular, $\alpha_{i0} = \gamma_{j0} = 0$ and $\beta_0 = 1$.)

Define linear differential operators by the formulae

$$L_r(y) := \sum_{k=0}^{r} \binom{r}{k} \beta_k y^{(r-k)} = y^{(r)} + r\beta_1 y^{(r-1)} + \cdots + \beta_r y, \tag{3.6}$$

for $r = 0 \ldots, m$. Note that $L_0(y) = y$. Next define

$$L(y) := L_m(y) - \sum_{r=0}^{m-1} a_r L_r(y) = y^{(m)} + c_{m-1} y^{(m-1)} + \cdots + c_0 y, \tag{3.7}$$

where

$$c_i := \binom{m}{m-i} \beta_{m-i} - \sum_{r=i}^{m-1} \binom{r}{r-i} a_r \beta_{r-i}, \quad i = 0, \ldots, m-1.$$

Since the operator L in (3.7) is of order less than n, any collection of n distinct elements of K linearly independent over K_C will include a non-solution of $L(y) = 0$. This collection can be prespecified, i.e., is independent of the differential system.

We choose a set $S_K \subset K$ containing at least n elements linearly independent over K_C. We also choose a set $S_{K_C} \subset K_C$ containing at least n non-zero elements.

Proposition 3.8 Suppose $k_0 \in S_K$ is not a solution of (3.7). Then there exists a $\lambda_0 \in S_{K_C}$ for which $\bar{v} := v + \lambda_0 k_0 u$ satisfies $\bar{v} \wedge D\bar{v} \wedge \cdots \wedge D^m\bar{v} \neq 0$.

Proof. Let λ be an indeterminate over K. Extend the derivation δ on K to $K(\lambda)$ by defining $\lambda' = \delta\lambda = 0$. Formally this is done by first defining δ on $K[\lambda] = K_C[\lambda] \otimes_{K_C} K$ by the formula

$$\delta(P \otimes k) = P \otimes k'$$

and then extending via the quotient rule.

The tensor product $\widehat{V} = K(\lambda) \otimes_K V$ has the natural structure of a differential $K(\lambda)$-space, obtained by defining

$$D(Q \otimes w) = \delta Q \otimes w + Q \otimes Dw$$

for any $w \in V$. To simplify notation we write this as

$$D(Qw) = Q'w + QDw.$$

Induction on the integer $r \geq 1$ gives the Leibniz rule

$$D^r(Qw) = \sum_{k=0}^{r} \binom{r}{k} Q^{(r-k)} D^k w.$$

Let $\widehat{v} := v + \lambda k_0 u \ (= 1 \otimes v + k_0 \lambda \otimes u) \in \widehat{V}$. Then, for $0 \leq r < m$, we have

$$D^r \widehat{v} = D^r v + \lambda \sum_{k=0}^{r} \binom{r}{k} k_0^{(r-k)} D^k u$$

$$= v_r + \lambda \sum_{k=0}^{r} \binom{r}{k} k_0^{(r-k)} \left(\sum_{i=0}^{m-1} \alpha_{ik} v_i + \beta_k u + \sum_{j=1}^{n-m-1} \gamma_{jk} e_j \right)$$

$$= v_r + \lambda \sum_{i=0}^{m-1} \sum_{k=0}^{r} \binom{r}{k} k_0^{(r-k)} \alpha_{ik} v_i + \lambda L_r(k_0) u$$

$$+ \lambda \sum_{j=1}^{n-m-1} \sum_{k=0}^{r} \binom{r}{k} k_0^{(r-k)} \gamma_{jk} e_j$$

$$= v_r + \lambda \sum_{i=0}^{m-1} \theta_{ir} v_i + \lambda L_r(k_0) u + \lambda \sum_{j=1}^{n-m-1} \tau_{jr} e_j,$$

for some $\theta_{ir}, \tau_{jr} \in K$. A similar calculation, using (3.3), gives

$$D^m \widehat{v} = \sum_{i=0}^{m-1} a_i v_i + \lambda \sum_{i=0}^{m-1} \sum_{k=0}^{m} \binom{m}{k} \alpha_{ik} k_0^{(m-k)} v_i$$

$$+ \lambda \sum_{k=0}^{m} \binom{m}{k} \beta_k k_0^{(m-k)} u + \lambda \sum_{j=1}^{n-m-1} \sum_{k=0}^{m} \binom{m}{k} \gamma_{jk} k_0^{(m-k)} e_j$$

$$= \sum_{i=0}^{m-1} a_i v_i + \lambda \sum_{i=0}^{m-1} \theta_{im} v_i + \lambda L_m(k_0)u + \lambda \sum_{j=1}^{n-m-1} \tau_{jm} e_j \, .$$

To summarize:

$$\widehat{v} = \quad v_0 \quad + \quad 0 \quad + \lambda L_0(k_0)u + \quad 0$$

$$D\widehat{v} = \quad v_1 \quad + \lambda \sum_{i=0}^{m-1} \theta_{i1} v_i + \lambda L_1(k_0)u + \lambda \sum_{j=1}^{n-m-1} \tau_{j1} e_j$$

$$\vdots \qquad \vdots \qquad \vdots \qquad \vdots \qquad \vdots$$

$$D^r \widehat{v} = \quad v_r \quad + \lambda \sum_{i=0}^{m-1} \theta_{ir} v_i + \lambda L_r(k_0)u + \lambda \sum_{j=1}^{n-m-1} \tau_{jr} e_j \qquad (3.9)$$

$$\vdots \qquad \vdots \qquad \vdots \qquad \vdots \qquad \vdots$$

$$D^m \widehat{v} = \sum_{i=0}^{m-1} a_i v_i + \lambda \sum_{i=0}^{m-1} \theta_{im} v_i + \lambda L_m(k_0)u + \lambda \sum_{j=1}^{n-m-1} \tau_{jm} e_j.$$

By the four "columns" of this array we mean the columns, delineated by the + signs, appearing to the right of the equality signs, e.g., the first column has initial entry v_0; the second has initial entry 0; the third has initial entry $\lambda L_0(k_0)u = \lambda k_0 u$; and the fourth has initial entry 0.

Now consider the vector

$$\widehat{w} := \widehat{v} \wedge D\widehat{v} \wedge \cdots \wedge D^m \widehat{v} \in \bigwedge^{m+1} \widehat{V} \, .$$

We initially view $\widehat{w}$ as a polynomial in λ with "coefficients" in the K-space $\bigwedge^{m+1} V$. As such it has degree at most $m+1$ and the "constant term" is 0, as can be seen by substituting $\lambda = 0$ in the formulae above. Accordingly we write

$$\widehat{w} = \lambda w_1 + \lambda^2 w_2 \cdots + \lambda^{m+1} w_{m+1} \, , \qquad (3.10)$$

where $w_i \in \bigwedge^{m+1} V$ for $i = 1, \ldots, m+1$.

We claim that the coefficient of $v_0 \wedge v_1 \wedge \cdots \wedge v_{m-1} \wedge u$ for w_1 is $L(k_0) \neq 0$. Assuming this is the case it follows that the coefficient of $v_0 \wedge v_1 \wedge \cdots \wedge v_{m-1} \wedge u$ for $\widehat{w}$ is a non-zero polynomial in $K[\lambda]$ whose degree is at most $m+1$ and whose constant term is 0. Since $S_{K_C} \subset K_C$ has at least $n > m$ non-zero elements there exists an element $\lambda_0 \in S_{K_C}$ which is not a zero of that polynomial, and the vector $\overline{v} := v + \lambda_0 k_0 u \in V$ will then satisfy the conclusion of the proposition.

To verify the claim, first note that, in computing this wedge product, the vector u can only result from a term in the third column of (3.9). But that term simultaneously contributes one λ, and since w_1 is the coefficient of λ in (3.10) we conclude that the remaining factors must be the terms from the first column in the m remaining rows. The coefficient in question is therefore determined from the calculation

$$\sum_{r=0}^{m-1} v_0 \wedge \cdots \wedge v_{r-1} \wedge L_r(k_0)u \wedge v_{r+1} \wedge \cdots \wedge v_{m-1} \wedge \sum_{j=0}^{m-1} a_j v_j$$

$$+ \; v_0 \wedge \cdots \wedge v_{m-1} \wedge L_m(k_0)u$$

$$= \sum_{r=0}^{m-1} a_r L_r(k_0)v_0 \wedge \cdots \wedge v_{r-1} \wedge u \wedge v_{r+1} \wedge \cdots \wedge v_{m-1} \wedge v_r$$

$$+ \; L_m(k_0)v_0 \wedge \cdots \wedge v_{m-1} \wedge u$$

$$= \Big(L_m(k_0) - \sum_{r=0}^{m-1} a_r L_r(k_0)\Big)v_0 \wedge \cdots \wedge v_{m-1} \wedge u$$

$$= L(k_0)\, v_0 \wedge \cdots \wedge v_{m-1} \wedge u \,,$$

and the claim follows. $\qquad\qquad\qquad\qquad\qquad\qquad\qquad\qquad\qquad\qquad\square$

Theorem 3.11 (The Cyclic Vector Theorem) *Suppose K_C contains at least n non-zero elements and that the extension $K_C \subset K$ is infinite or, if finite, of degree at least n. Then every n-dimensional differential K-space (V, δ, D) admits a cyclic vector.*

When K has characteristic 0 the degree n alternative on $K_C \subset K$ is impossible (see Proposition 5.1(a)).

Proof. First note that our hypotheses imply that K contains at least n elements linearly independent over constants. Therefore, for any $m < n$ the linear differential equation (3.7) admits a non-solution and Proposition 3.8 applies. Starting with an arbitrary non-zero vector $v \in V$ at most $n - 1$ applications of the proposition produce a cyclic vector. $\qquad\qquad\qquad\square$

4   Remarks on the Algorithm

We fix a set $S_K \subset K$ containing at least n elements linearly independent over K_C, and a set $S_{K_C} \subset K_C$ containing at least n distinct non-zero elements. For example, if $K = \mathbb{Q}(x)$ it is quite natural to choose $S_K = \{1, x, x^2, \ldots, x^{n-1}\}$

and $S_{K_C} = S_{\mathbb{Q}} = \{1, 2, 3, \ldots, n\}$. We review the steps of the algorithm for purposes of reference. A MAPLE implementation of the case $(K, \delta) = (\mathbb{Q}(x), \frac{d}{dx})$ is presented in the Appendix.

Algorithm 4.1 To find a cyclic vector:

Step 1. Choose any non-zero vector $v \in V$.

Step 2. Compute the matrix M with columns $v, Dv, \ldots, D^{n-1}v$ as in (2.3).

Step 3. Compute the rank m of M, find $a_0, \ldots, a_{m-1} \in K$ as in (3.3), and a supplement to the subspace spanned by $v, Dv, \ldots, D^{m-1}v$.

Step 4. If $m = n$ the algorithm terminates and v is a cyclic vector.

Step 5. Choose any vector u not in the span of $\{v, Dv, \ldots, D^{m-1}v\}$.

Step 6. Compute the β_k as in (3.5).

Step 7. Find $k_0 \in S_K$ so that $L(k_0) \neq 0$ as in (3.7).

Step 8. Find $\lambda_0 \in S_{K_C}$ such that for $\overline{v} := v + \lambda_0 k_0 u$ the rank of the matrix in (2.3) is bigger than m.

Step 9. Go to Step 2.

In practice a differential structure $D : V \to V$ assumes the form of an $n \times n$ matrix A with coefficients in K. In particular, a basis $\mathbf{e}$ identifying V with K^n is implicit, and Dv, for any vector $v \in V$, is computed as in (2.2).

For Step 1 one might begin with a vector having a simple form, e.g. $(1, 0, \ldots, 0)$. When working over $K = \mathbb{Q}(x)$ this would increase the likelihood of constructing a cyclic vector not involving high powers of x or large coefficients. Alternatively, and more generally, one could choose a non-zero vector at random to increase the likelihood that the algorithm terminates upon first reaching Step 4. It is worth noting that when this random vector approach is used, replacing Step 5 with "Go to Step 1" results in the algorithm of the introduction.

To calculate the matrix M of Step 2 requires $O(n^2)$ differentiations and $O(n^3)$ multiplications in the field K. Assuming that differentiation is not too much more expensive than multiplication, we see that this step is $O(n^3)$.

Step 3 can be accomplished by applying Gaussian elimination as described in [9], Algorithm 2.3.6, p. 60, or see the procedure CVRank of the Appendix. This step requires $O(n^3)$ multiplications and divisions (see, for example, the discussion on pages 47-50 of [9]).

Each time Step 4 is reached the value of m increases; this step is involved at most $n - 1$ times.

For Step 6 we first compute the matrix with columns $u, Du, \ldots, D^m u$; this was seen in Step 2 to be $O(n^3)$. We then express the answer in the basis $v, Dv, \ldots, D^{m-1}, u, e_1, \ldots, e_{n-m-1}$. This requires matrix inversion and is $O(n^3)$.

In Step 7 the set S_K is known in advance and as a result all the derivatives of its elements are also known in advance. For $S_K = S_{\mathbb{Q}(x)} = \{1, x, x^2, \ldots, x^{n-1}\}$ this simply requires finding the smallest index i so that $c_i \neq 0$. The step is $O(n)$.

Step 8 involves computing the matrix of (2.3) and then its rank. The operations are $O(n^3)$, but may require repetition for each $\lambda_0 \in S_{K_C}$. The step is therefore $O(n^4)$; it is the most expensive of the algorithm.

An alternative approach would be to compute the polynomial for λ described in the proof of Proposition 3.8. Doing so would take $O(n^3)$ multiplications in $K[\lambda]$, and therefore at least $O(n^4)$ multiplications in K.

In Step 9 we loop back to Step 2. However the computations in Steps 2 and 3 repeat those done in Step 9. In the implementation in the Appendix we actually loop back to Step 4.

Since each step is at worst $O(n^4)$, and since we iterate up to n times the entire algorithm is $O(n^5)$. Of course this only counts multiplications and divisions; the actual complexity depends on the ground field K and the cost of multiplication and division within that field.

A more naive implementation might simply be to test all vectors having sums of products of elements of S_K and S_{K_C} as coefficients. There are n^3 such possibilities, and as a result the implementation would be $O(n^6)$.

The Appendix contains an implementation of the above algorithm. We also present a procedure that computes a companion matrix. The input is a matrix $A \in gl(n, \mathbb{Q}(x))$; the output consists of a matrix $T \in GL(n, \mathbb{Q}(x))$ and a matrix $C \in gl(n, \mathbb{Q}(x))$ such that $C = T^{-1}AT + T^{-1}T'$ has the companion matrix form

$$\begin{pmatrix} 0 & \cdots\cdots\cdots & 0 & q_n \\ 1 & 0 & \vdots & q_{n-1} \\ 0 & 1 & 0 & \vdots & \vdots \\ \vdots & & \ddots & \ddots & \vdots & \vdots \\ \vdots & & & 1 & 0 & q_2 \\ 0 & \cdots\cdots & 0 & 1 & q_1 \end{pmatrix} ; \tag{4.2}$$

a cyclic vector is provided by the first column of T.

We have also included the details of a command for converting a first order system $Y' + AY = 0$ of n homogeneous linear differential equations to

n-th order form. The input is a matrix $A \in gl(n, \mathbb{Q}(x))$; the output consists of a matrix $T \in GL(n, \mathbb{Q}(x))$ and a matrix $B \in gl(n, \mathbb{Q}(x))$ such that $B = T^{-1}AT + T^{-1}T'$ has the companion matrix form

$$
\begin{pmatrix}
0 & -1 & 0 \ldots\ldots & & 0 \\
\vdots & 0 & -1 & & \vdots \\
\vdots & & 0 & \ddots & \\
\vdots & & & \ddots & -1 & 0 \\
0 & & \ldots\ldots\ldots & & 0 & -1 \\
-q_n & -q_{n-1} & \ldots\ldots & -q_2 & -q_1
\end{pmatrix}. \tag{4.3}
$$

The corresponding n-th order form is

$$
y^{(n)} - q_1 y^{(n-1)} - \cdots - q_{n-1} y' - q_n y = 0. \tag{4.4}
$$

5   Remarks on the Hypotheses

These are based on a few elementary (and completely standard) observations on a differential field $K = (K, \delta)$ with field of constants K_C.

Proposition 5.1 (a) *When the characteristic of K is 0 any $k \in K \setminus K_C$ is transcendental over K_C, i.e., the subfield K_C is algebraically closed in K.*

(b) *The only derivation on a perfect field of positive characteristic is the trivial derivation. In particular, a finite field admits only the trivial derivation.*

(c) *When the characteristic of K is $p > 0$ any $k \in K \backslash K_C$ satisfies $k^p \in K_C$. In particular, the field extension $K_C \subset K$ is (algebraic and) purely inseparable.*

(d) *$K \neq K_C \Rightarrow K_C$ is infinite.*

It is immediate from (d) that Hypothesis 2) of §3 is redundant.

Proof. (a): Suppose k is algebraic over K_C with irreducible monic polynomial $P \in K_C[X]$. Then

$$
0 = (P(k))' = \frac{dP}{dX}(k)\, k',
$$

and since $dP/dX(k) \neq 0$ (the degree of dP/dX is too small) we conclude that $k' = 0$.

(b): When K is a perfect field of characteristic p each element of K has the form k^p for some $k \in K$. Therefore $(k^p)' = pk^{p-1}k' = 0$.

(c): $(k^p)' = 0$.

(d): For the characteristic 0 case this is trivial: K_C contains $\mathbb{Q}$.

When K has characteristic $p > 0$ the Frobenius mapping $k \mapsto k^p$ of K is an injection which by (c) has image in K_C. When K_C is finite the same then holds for K, which by the final assertion of (b) implies $K = K_C$. $\quad\square$

Corollary 5.2 *We have the following equivalences for Hypotheses 3.1.*

(a) *When K has characteristic 0 Hypotheses 3.1 are equivalent to $K \neq K_C$, i.e., to the assumption that the derivation on K is non-trivial.*

(b) *When K has characteristic $p > 0$ and $n \leq p$ Hypotheses 3.1 are equivalent to $K \neq K_C$. In this case K is a purely inseparable extension of K_C of degree at least p.*

(c) *When K has characteristic $p > 0$ and $n > p$ Hypotheses 3.1 are equivalent to the condition that K be a purely inseparable extension of K_C of degree at least n (possibly infinite).*

In all three cases the field of constants K_C must be infinite.

Proof. Hypothesis 1) obviously implies that $K \neq K_C$. Part (a) is immediate from Proposition 5.1(a). Parts (b) and (c) come from Proposition 5.1(c) and (d). $\quad\square$

6 Counterexamples

Here we investigate the necessity of Hypotheses 3.1. First we show, by example, that we cannot do with less than n elements of K linearly independent over K_C nor less than n non-zero elements of K_C.

Take $K = \mathbb{C}(x)$ ($x' = 1$) and let V be an n-dimensional K-space ($n \geq 2$) with basis $(e_1, \ldots, e_n)$. Now consider the differential structure on V given by

$$
De_i = \begin{cases} e_{i+1} & \text{if } i = 1, \ldots, n-2, \\ 0 & \text{if } i = n-1, \\ x^{-n}e_1 & \text{if } i = n. \end{cases}
$$

Following the notation of §3, we let

$$
v_0 = v = e_1,
$$
$$
v_i = D^i v = e_{i+1} \quad \text{for } i = 1, \ldots, n-2,
$$
$$
u = e_n.
$$

Then $D^{n-1}v = 0$, *i.e.*, the a_k of (3.3) are all 0. Next we compute the β_k of (3.5) (we do not care about the α_{ik}) and find that $\beta_0 = 1$ and $\beta_1 = \cdots = \beta_n = 0$. Using (3.7) and (3.6) we have $L(y) = y^{(n-1)} = 0$. Evidently $L(1) = L(x) = \cdots = L(x^{n-2}) = 0$, but $L(x^{n-1}) \neq 0$. This shows that the set $\{1, x, \ldots, x^{n-2}\}$, which consists of $n-1$ elements of K linearly independent over $K_C = \mathbb{Q}$, does not contain a *non*-solution of $L(y) = 0$. I.e. Hypothesis (1) is essential.

We choose a non-solution of $L(y) = 0$, namely $y = x^{n-1}$, and set

$$\bar{v} = v + \lambda x^{n-1} u,$$

where λ is an indeterminate constant, as in the proof of Proposition 3.8. We claim that for each $r = 0, \ldots, n-2$ we have

$$D^r \bar{v} = \sum_{i=1}^{r} \kappa_{ri} D^{i-1} \bar{v} + v_r + (-1)^r \lambda r! \binom{\lambda - n + r}{r} x^{n-r-1} u,$$

where $\kappa_{ri} \in K$. The case $r = 0$ is immediate from the definition of $\bar{v}$. Proceeding by induction, assume the formula holds for some $0 \leq r < n-2$ and apply D. The first two terms retain the desired form; the last becomes

$$(-1)^r \lambda r! \binom{\lambda - n + r}{r} \left((n - r - 1)x^{n-r-2} u + x^{-r-1} v\right).$$

Split this in two by writing $v = \bar{v} - \lambda x^{n-1} u$. The resulting term involving $\bar{v}$ combines with the predecessors, the remaining term is

$$(-1)^r \lambda r! \binom{\lambda - n + r}{r} (n - r - 1 - \lambda) x^{n-r-2} u,$$

and the claim follows.

Applying D to the formula for $r = n-2$ gives

$$D^{n-1} \bar{v} = \sum_{i=1}^{n-1} \kappa_{n-1,i} D^{i-1} \bar{v} + (-1)^{n-1} \lambda(\lambda - 1) \cdots (\lambda - n + 1) u,$$

and therefore

$$\bar{v} \wedge D\bar{v} \wedge \cdots \wedge D^{n-1} \bar{v}$$
$$= (-1)^{n-1} \lambda(\lambda - 1) \cdots (\lambda - n + 1) \bar{v} \wedge D\bar{v} \wedge \cdots \wedge D^{n-2} \bar{v} \wedge u.$$

We conclude that allowing $n+1$ possibilities for λ (counting 0) is also an essential requirement.

We now turn to examples in which our hypotheses fail, first considering the case $K = K_C$ (see (a) and (b) of Corollary 5.2). In that instance any differential structure D on a finite-dimensional K-space V is a K-endomorphism of V, and as such admits a unique minimal monic polynomial $P \in K[X]$. When $\deg(P) = m$ the collection $v, Dv, \ldots, D^m v$ is linearly dependent for any vector $v \in V$. Therefore V admits a cyclic vector if and only if the degree of P is $n = \dim(V)$.

In particular, when some defining basis matrix of D is a scalar multiple of the identity and $\dim(V) \geq 2$ there can be no cyclic vector.

On the other hand, V can always be written as a direct sum of cyclic subspaces (see, e.g., [24, Theorem 8, Chapter XI, page 390]). For a more extensive discussion of this case see [3].

To complete the discussion it remains to consider the characteristic $p > 0$ case with $n := \dim(V) > p$ and $K_C \subset K$ purely inseparable of degree less than n (see Corollary 5.2(c)). Here we offer a example illustrating that cyclic vectors need not exist. (Alternatively, see [19, page 68].) Specifically, consider the field $K := \mathbb{F}_p(x)$ where x is transcendental over $\mathbb{F}_p$ and $x' = 1$. By Proposition 5.1(c) the element x has degree p over K_C and as a result the degree of the extension $K_C \subset K$ is $p < n$.

Note that $(x^m)^{(p)} = 0$ for all integers m, and from $k \in K \Rightarrow k' \in \mathbb{F}_p$ that $\mathbb{F}_p$ is a differential subfield of K. By Proposition 5.1(b) the restricted derivation must be trivial, and it follows that $k^{(p)} = 0$ for all $k \in K$.

Now suppose $V = (V, D)$ is a differential K-space of dimension $n > p$ and that the identity matrix is a defining basis matrix for D. The corresponding basis elements $e_1, \ldots, e_n$ then satisfy $De_j = e_j$ and for any $v = \sum_j k_j e_j \in V$ we have

$$D^p v = \sum_j \left(k_j^{(p)} e_j + k_j D^p e_j \right) = v.$$

We conclude there is no cyclic vector for D.

7 An Alternate Approach

Here we sketch a proof of the Cyclic Vector Theorem in the spirit of that of [10]. We begin with a purely algebraic lemma (*i.e.*, derivations are not involved).

Lemma 7.1 *Suppose that $A \subset K$ is a set containing at least $n + 1$ distinct elements. Let $X_1, \ldots, X_r$ be indeterminates over K and let $Q(X_1, \ldots, X_r) \in K[X_1, \ldots, X_r]$, $Q \neq 0$, with $\deg Q \leq n$. Then there exist $a_1, \ldots, a_r \in A$ such that $Q(a_1, \ldots, a_r) \neq 0$.*

Proof. We use induction on the number s of indeterminates actually appearing in Q. If $s = 0$, *i.e.*, $Q \in K$, then $a_1, \ldots, a_r$ may be chosen arbitrarily.

Suppose that Q involves $s > 0$ variables and that the result is proved for polynomials involving fewer variables. Reordering the variables, if necessary, we may assume that Q only involves $X_1, \ldots, X_s$ and that the degree of Q in X_s is $d > 0$. Write

$$Q = Q_d X_s^d + Q_{d-1} X_s^{d-1} + \cdots + Q_0, \quad Q_i \in K[X_1, \ldots, X_{s-1}], \quad Q_d \neq 0.$$

By induction there exist $a_1, \ldots, a_{s-1} \in A$ such that $Q_d(a_1, \ldots, a_{s-1}) \neq 0$. Choose $a_{s+1}, \ldots, a_r \in A$ arbitrarily and define

$$\overline{Q} = Q(a_1, \ldots, a_{s-1}, X_s, a_{s+1}, \ldots, a_r) \in K[X_s].$$

Then $\overline{Q}$ is a non-zero polynomial in one variable, X_s, with $\deg \overline{Q} \leq n$. Such a polynomial has at most n roots, so there exists $a_s \in A$ with $\overline{Q}(a_s) \neq 0$. $\square$

Now let $y_1, \ldots, y_m$ be differential indeterminates over K. We let $K\{y_1, \ldots, y_m\}$ and $K\langle y_1, \ldots, y_n \rangle$ denote the differential ring of differential polynomials and the differential field of rational functions respectively. (For basic notions of differential algebra see [20] or [18].)

Proposition 7.2 *Suppose that $A \subset K$ is a set containing at least $n + 1$ elements. Let $P \in K\{y_1, \ldots, y_m\}$ be a non-zero differential polynomial of order $r - 1 \leq n - 1$ and degree $s \leq n$. Then for any $b_1, \ldots, b_r \in K$, linearly independent over K_C, there are elements $a_{ij} \in A$, $1 \leq i \leq m$, $1 \leq j \leq r$, such that for $k_i = a_{i1}b_1 + \cdots + a_{ir}b_r$ one has $P(k_1, \ldots, k_m) \neq 0$.*

Proof. Let X_{ij}, $1 \leq i \leq m$, $1 \leq j \leq r$, be indeterminates (not differential) over K. Because the $y_i^{(j-1)}$ are algebraically independent over K we may define a (non-differential) K-algebra homomorphism $\phi : K[y_i^{(j-1)}] \to K[X_{ij}]$ by

$$\phi : y_i^{(j-1)} \mapsto \sum_{k=1}^{r} b_k^{(j-1)} X_{ik}.$$

Since the $b_1, \ldots, b_r$ are linearly independent over K_C the Wronskian $\det(b_i^{(j-1)})$ does not vanish; it follows that ϕ is invertible, *i.e.*, is an isomorphism. The degree of $Q := \phi P$ is $s \leq n$. By Lemma 7.1 there are elements $a_{ij} \in A$, $1 \leq i \leq m$, $1 \leq j \leq r$, such that $Q(a_{ij}) \neq 0$. The desired result then follows from the observation that $P(k_1, \ldots, k_m) = Q(a_{ij}) \neq 0$ when the k_j are defined as in the statement of the proposition. $\square$

For the remainder of the section we assume Hypotheses 3.1. The next result explains the success of the "algorithm" of the Introduction. By the Kolchin topology we mean the topology on affine space K^n having the zero sets of differential polynomials as closed sets. We suppose that K satisfies Hypotheses 3.1.

Theorem 7.3 (The Cyclic Vector Theorem) *For any basis $(e_1, \ldots, e_n)$ of V the set of points $(k_1, \ldots, k_n) \in K^n$ for which $\sum_j k_j e_j$ is a cyclic vector is a non-empty Kolchin open (and therefore dense) subset of differential affine n-space K^n. In particular, V admits a cyclic vector for D.*

This formulation of the Cyclic Vector Theorem justifies the comments in Remarque 2.5, p. 133 of [2].

Proof. Let $y_1, \ldots, y_n$ be differential indeterminates over K. Then $W := K\langle y_1, \ldots, y_n \rangle \otimes_K V$ is a differential $K\langle y_1, \ldots, y_n \rangle$-space with differential structure $(\delta, \widehat{D})$, where

$$\widehat{D}(f \otimes v) = f' \otimes v + f \otimes Dv.$$

Choose any basis $\mathbf{e} = (e_1, \ldots, e_n)$ of V. We will show that the vector

$$w = \sum_{j=1}^{n} y_j \otimes e_j$$

is cyclic for $\widehat{D}$.

Define $p_{ij} \in K\langle y_1 \ldots, y_n \rangle$ by $\widehat{D}^{j-1} w = \sum_i p_{ij} \otimes e_i$. We claim that $p_{ij} = y_j^{(i-1)} + q_{ij}$, where q_{ij} is a linear differential polynomial in $y_1, \ldots, y_n$ of order strictly less than $i - 1$. The proof is a trivial induction on i: for $i = 1$ one has $p_{1j} = y_j + 0$; for $i > 1$ one uses $D^i w = D(D^{i-1})w$.

Recall (see (2.3)) that w is cyclic for $\widehat{D}$ if and only if $P = \det(p_{ij}) \neq 0$. To establish this nonvanishing first note that this differential polynomial has order less than n and degree at most n. Next observe that the matrix (p_{ij}) has a single entry involving $y_n^{(n-1)}$, namely p_{nn}, and that the coefficient of $y_n^{(n-1)}$ in the determinant is the minor $\det(p_{ij})_{1 \leq i,j \leq n}$. Arguing by induction we see that the coefficient of $y_1 y_2' \cdots y_n^{(n-1)}$ in P is 1, hence $P \neq 0$, and $w = \sum_j y_j \otimes e_j$ is therefore cyclic for $\widehat{D}$. The argument shows, in addition, that P has order $n - 1$ and degree n.

By Proposition 7.2 we can find $k_1, \ldots, k_n \in K$ such that $P(k_1, \ldots, k_n) \neq 0$. The result follows. $\qquad\square$

As before, we can prespecify the set S_K of elements of K linearly independent over K_C and the set A of the proposition.

Suppose $K = \mathbb{Q}(x)$, $S_K = S_{\mathbb{Q}(x)} = \{1, x, x^2, \ldots, x^{n-1}\}$, and $A = \{0, 1, \ldots, n\}$. Then the k_i are polynomials in x of degree less than n whose coefficients are non-negative integers up to n. There are at most $(n+1)^2$ ways to choose each k_i.

This suggests an easy to implement algorithm that, unlike the "algorithm" of the introduction, is guaranteed to terminate: simply try all possibilities. However this algorithm is $O(n^6)$. Indeed, there are $O(n^2)$ possibilities for each k_i (and there are n of them), and the test for cyclicity, e.g. computing the determinant of (2.3), is $O(n^3)$.

Acknowledgment

We would like to thank an anonymous referee for numerous suggestions which significantly improved the original manuscript, and for providing a substantial number of references.

Appendix - A MAPLE implementation

CVvector - find a cyclic vector

Calling Sequence:
    v := **CVvector**(A)
    v := **CVvector**(A,v0)

Parameters:
    A - square matrix defining a differential structure
    v0 - (optional) initial trial vector

Synopsis:
    CVvector returns a cyclic vector for the differential structure.
    v0, if present, is used as the starting vector.

Examples:
```
> with(linalg);
> A := matrix([[0,0,1/x^3],[1,0,0],[0,0,0]]);
```

$$A: = \begin{pmatrix} 0 & 0 & \frac{1}{x^3} \\ 1 & 0 & 0 \\ 0 & 0 & 0 \end{pmatrix}$$

```
> v := CVvector(A);
```

$$v := [1,0,3x^2]$$

```
> v := CVvector(A,[0,1,0]);
```

$$v := [x, 1, 1]$$

```
CVvector := proc (A, v0)
local n, v, M, a, B, m, m1, beta, c, i, r, u;
n := rowdim(A);

# Step 1 - choose any v in K^n
if nargs = 2 then
     v := v0;
else
     v := vector(n,0);
     v[1] := 1;
fi;

# Step 2 - compute the matrix with cols v, Dv, ...
M := CVdv (n, A, v, n);

# Step 3 - compute the rank, also a's and a supplement
m := CVrank(n, M, 'a', 'B');

# Step 4 - check if cyclic vector found,  come here from Step 8
while (m <> n) do

# Step 5 - compute u.  a's and e's have been computed by CVrank
u := col(B, m+1);

# Step 6 - compute the beta's
M := CVdv (n, A, u, m+1);
beta := row(evalm(inverse(B) &* M), m+1);

# Step 7 - find k_0 = x^ e so L(k_0) <> 0
for i from 0 to m-1 do
     c := binomial(m, m-i)*beta[m-i+1] -
          sum(binomial(r, r-i)*a[r+1]*beta[r-i+1], r=i..m-1);
     if c <> 0 then break; fi;
od;
if i <> 0 then u := evalm(x^i * u); fi;
```

```
# Step 8 - find lambda_0 = 1,2,...
for i from 0 to m do
     v  := evalm(v + u);
     M  := CVdv (n, A, v, n);
     m1 := CVrank(n, M, 'a', 'B');
     if m1 > m then break; fi;
od;
m := m1;

od;    # goto Step 4

RETURN(evalm(v));
end;
```

CVcompanion - compute a companion matrix

Calling Sequence:
    C := **CVcompanion** (A)
    C := **CVcompanion** (A, v0)
    C := **CVcompanion** (A, 'T')
    C := **CVcompanion** (A, v0, 'T')

Parameters:
    A - square matrix defining a differential structure
    v0 - (optional) initial trial vector
    T - (optional) used to return the transition matrix

Synopsis:
    CVcompanion returns a companion matrix C of the form (4.2)
    for the differential structure.
    v0, if present, is used as the starting vector,
    T, if present, is used to return the transition matrix.
    The formula is: $C := T\hat{}(-1) * A * T + T\hat{}(-1) * T'$, where
    T is the matrix with columns v, Dv, $D\hat{}2v$, ... and v is
    a cyclic vector.

Examples:
```
> with(linalg);
> A := matrix([[-x,1-x^2,x-x^3],[1,0,1],[0,1,x+1]]);
```

$$A := \begin{pmatrix} -x & 1-x^2 & x-x^3 \\ 1 & 0 & 1 \\ 0 & 1 & x+1 \end{pmatrix}$$

```
> C := CVcompanion (A, 'T');
```

$$C := \begin{pmatrix} 0 & 0 & x^2 \\ 1 & 0 & x \\ 0 & 1 & 1 \end{pmatrix}$$

```
> evalm(T);
```

$$\begin{pmatrix} 1 & -x & 0 \\ 0 & 1 & -x \\ 0 & 0 & 1 \end{pmatrix}$$

```
CVcompanion := proc(A, v0, T)
local v, n, lT, lTi;

if nargs > 1 and type(args[2],{'vector','list'}) then
     v := CVvector(A, v0);
else
     v := CVvector(A);
fi;

n := rowdim(A);
lT := CVdv(n, A, v, n);

if nargs = 3 then
     T := evalm(lT);
elif nargs = 2 and not type(args[2],{'vector','list'}) then
     v0 := evalm(lT);
fi;
lTi := inverse(lT);
RETURN (evalm(lTi &* A &* lT + lTi &* map(diff, lT, x)));
end;
```

CVscalar - convert matrix equation to scalar

Calling Sequence:
>   C := **CVscalar** (A)
>   C := **CVscalar** (A, v0)
>   C := **CVscalar** (A, 'T')
>   C := **CVscalar** (A, v0, 'T')

Parameters:
>   A - a matrix
>   v0 - (optional) initial trial vector
>   T - (optional) used to return the transition matrix

Synopsis:
>   **CVscalar** returns a companion matrix C of the form (4.4)
>   for the scalar equation equivalent to $Y' + AY = 0$.
>   The equation is $(D@@n)(y) + q[1]*(D@@(n-1))(y) + \cdots + q[n]*y$,
>   where $q[i] = C[n,n-i]$.
>   v0, if present, is used as the starting vector when **CVcompanion**
>   is called,
>   T, if present, is used to return the transition matrix.

Examples:
```
> with(linalg);
> A := matrix([[-x,-1,-1],[x^2,x-1,2*x-2],[0,1,1-x]]);
```

$$A := \begin{pmatrix} -x & -1 & -1 \\ x^2 & x-1 & 2x-2 \\ 0 & 1 & 1-x \end{pmatrix}$$

```
> C := CVscalar(A);
```

$$C := \begin{pmatrix} 0 & -1 & 0 \\ 0 & 0 & -1 \\ 0 & 0 & -x \end{pmatrix}$$

```
> q := row(C,3);
```

$$q := [0, 0, -x]$$

```
> # The corresponding linear differential equation is
> (D@@3)(y) - x * (D@@2)(y) = 0;
```

$$D^{(3)}(y) - xD^{(2)}(y) = 0$$

```
CVscalar := proc(A, v0, T)
local At, lT, C;
At := evalm(-1 * transpose(A));

if nargs > 1 and type(args[2],{'vector','list'}) then
    C := CVcompanion(At, v0, 'lT');
else
    C := CVcompanion(At, 'lT');
fi;

if nargs = 3 then
    T := evalm(transpose(inverse(lT)));
elif nargs = 2 and not type(args[2],{'vector','list'}) then
    v0 := evalm(transpose(inverse(lT)));
fi;

RETURN(evalm(-1 * transpose(C)));
end;
```

CVdv - internal subroutine to compute matrix v, Dv, ...

Calling Sequence:
    M := CVdv(n, A, v, m)

Parameters:
    n - the dimension of the vector space
    A - the n by n matrix defining the differential structure
    v - a vector
    m - number of columns to be computed (may be <> n)

Synopsis:
    CVdv returns the matrix with columns
    v, Dv, D^2v, ... D^(m-1)v
    This is used for Step 2 of CVvector.

```
CVdv := proc(n, A, v, m)
local i, j, k, w;
w := matrix(n,m);

for i from 1 to n do w[i,1] := v[i]; od;

for j from 2 to m do
    for i from 1 to n do
        w[i,j] := diff(w[i,j-1],x)
            + sum(A[i,k]* w[k,j-1],k=1..n);
    od;
od;
RETURN(evalm(w));
end;
```

CVrank - internal subroutine to compute rank

Calling Sequence:
   r := CVrank (n, M, a, S)

Parameters:
   n - the dimension of the vector space
   M - the n by n matrix whose rank is sought,
   usually M = CVdv (n, V, v, n);
   a - returns a vector with
   a[1] M[i,1] + $\cdots$ + a[r] M[i,r] = M[i,r+1]
   S - returns an invertible matrix s whose first r columns
   are M[i,1],...,M[i,r]

Synopsis:
   CVrank returns the rank of the matrix M and other data.
   The algorithm is as follows:
   Let B = Id. Perform row reduction on M, and corresponding column
   operations on B. Thus the product BM is constant (= original M). On
   return, the first r columns of M are the identity and the last n-r rows
   are 0. It follows that the first r columns of B are the same
   as those of M. This is used for Step 3 of CVvector.

```
CVrank := proc(n, M, a, S)
local  B, i, j, k, d;
B := band([1], n);
```

```
for i from 1 to n do
     for j from i to n do
          if M[j,i] <> 0 then break; fi;
     od;
     if j = n+1 then
          a := evalm(col(M, i));
          S := evalm(B);
          break;
     fi;
     if i <> j then
          M := swaprow(M, i, j);
          B := swapcol(B, j, i);
     fi;
     d := M[i,i];
     if d <> 1 then
          M := mulrow(M, i, 1/d);
          B := mulcol(B, i, d);
     fi;
     for k from 1 to n do
          if k <> i then
          d := M[k,i];
          if d <> 0 then
          M := addrow(M, i, k, -d);
          B := addcol(B, k, i, d);
          fi;
          fi;
     od;
od;
RETURN(i-1);
end;
```

References

1. Adjamagbo, K. *Sur l'effectivité du lemme du vecteur cyclique*, C. R. Acad. Sci. Paris Sér. I Math. **306**, no. 13 (1988), 543–546.
2. Adjamagbo, K., Rigal, L. *Sur les vecteurs cycliques et les éléments primitifs différentiels*, *Lois d'algèbres et variétés algébriques (Colmar, 1991)*, 117–133, Travaux en Cours, 50, Hermann, Paris, 1996.

3. Augot, D., Camion, P. *Forme de Frobenius et vecteurs cycliques*, C. R. Acad. Sci. Paris Sér. I Math. **318**(2) (1994), 183–188.

4. Barkatou, M. A. *An algorithm for computing a companion block diagonal form for a system of linear differential equations*, Appl. Algebra Engrg. Comm. Comput. **4**(3) (1993), 185–195.

5. Beauzamy, B. *Sous-espaces invariants pour les contractions de classe C_1. et vecteurs cycliques dans $C_0(\mathbf{Z})$*, J. Operator Theory **7** (1) (1982), 125–137.

6. Bertrand, D. *Systèmes différentielles et équations différentielles*, Séminaire d'arithmétique, Exposé V (1983), Saint-Etienne.

7. Bertrand, D. *Exposants des systèmes différentiels, vecteurs cycliques et majorations de multiplicités, Équations différentielles dans le champ complexe, Vol. I (Strasbourg, 1985)*, Publ. Inst. Rech. Math. Av., Univ. Louis Pasteur, Strasbourg, 1988, 61–85.

8. Bertrand, D. *Constructions effectives de vecteurs cycliques pour un D-module, Study group on ultrametric analysis, 12th year, 1984/85, No. 1*, Exp. No. 11, 7 pp., Secrétariat Math., Paris, 1985.

9. Cohen, H. *A Course in Computational Algebraic Number Theory*, Graduate Texts in Mathematics **138**, Springer-Verlag, Berlin, 1993.

10. Cope, F. T. *Formal solutions of irregular linear differential equations, Part II*, Amer. J. Math. **58** (1936), 130–140.

11. Dabèche, A. *Formes canoniques rationnelles d'un système différential à point singulier irrégulier, Équations différentielles et systèmes de Pfaff dans le champ complexe (Sem., Inst. Rech. Math. Avancée, Strasbourg, 1975)*, Lecture Notes in Math. **712**, Springer-Verlag, Berlin, 1979. 20–32.

12. Deligne, P. *Équations différentielles à points singuliers réguliers*, Lecture Notes in Mathematics **163**, Springer-Verlag, Berlin-New York, 1970.

13. Ekong, S. D. *Sur l'analyse algébrique. I.* Publications du Département de Mathématiques. Nouvelle Série, A. Vol 6, Publ. Dép. Math. Nouvelle Sér. A, Univ. Claude-Bernard, Lyon, 1985, 19–42.

14. Ekong, S. D. *Sur l'analyse algébrique. II.* Publications du Département de Mathématiques. Nouvelle Série, B. Vol 3, Publ. Dép. Math. Nouvelle Sér. B, Univ. Claude-Bernard, Lyon, 1988, 77–98.

15. Hilali, A. *Characterization of a linear differential system with a regular singularity, Computer algebra (London)*, Lecture Notes in Comput. Sci., **162**, Springer-Verlag, Berlin, 1983, 68–77.

16. Hilali, A. *Calcul des invariants de Malgrange et de Gérard-Levelt d'un système différentiel linéaire en un point singulier irrégulier*, J. Differential Equations **69** (3) (1987), 401–421.

17. Hilali, A., Wazner, A. *Un algorithme de calcul de l'invariant de Katz d'un système différentiel linéaire,* Ann. Inst. Fourier (Grenoble) **36**(3) (1986), 67–81.

18. Kaplansky, I. *An Introduction to Differential Algebra,* Second Edition. Actualités Sci. Ind., No. 1251, Publications de l'Inst. de Mathématique de l'Université de Nancago, No. V. Hermann, Paris, 1976.

19. Katz, N. M. *A simple algorithm for cyclic vectors,* Amer. J. Math. **109** (1) (1987), 65–70.

20. Kolchin, E. R. *Differential Algebra and Algebraic Groups,* Pure and Applied Mathematics, Vol. 54, Academic Press, New York-London, 1973.

21. Kovacic, J. *Loewy similarity and Picard-Vessiot extensions,* unpublished, 1970.

22. Levelt, A. H. M. *Jordan decomposition for a class of singular differential operators,* Ark. Mat. **13** (1975), 1–27.

23. Loewy, A. *Über einen Fundamentalsatz für Matrizen oder Lineare Homogene Differentialsysteme,* Sitzungsberichte der Heidelberger Akad. der Wiss., Math.-naturwiss. Klasse, Band 9A, 5. Abhandlung, 1918.

24. MacLane, S., Birkhoff, G. *Algebra,* Amer. Math. Soc., Chelsea, Providence, RI, 1999.

25. Ramis, J-P. *Théorèmes d'indices Gevrey pour les équations différentielles ordinaires,* Mem. Amer. Math. Soc. **48**, no. 296 (1984).

26. Seidenberg, A. *Some basic theorems in differential algebra (characteristic p, arbitrary),* Trans. Amer. Math. Soc. **73** (1952), 174–190.

27. Seidenberg, A. *An elimination theory for differential algebra,* Univ. California Publ. Math. (N.S.) **3** (1956), 31–66.

28. Sit, W. *On finite dimensional differential vector spaces,* unpublished, 1970.

29. Varadarajan, V. S. *Meromorphic Differential Equations,* Expositiones Math. **9**(2) (1991), 97–188.

Differential Algebra and Related Topics, pp. 219–255
Proceedings of the International Workshop
Eds. L. Guo, P. J. Cassidy, W. F. Keigher & W. Y. Sit

DIFFERENTIAL ALGEBRAIC TECHNIQUES IN HAMILTONIAN DYNAMICS

RICHARD C. CHURCHILL

Department of Mathematics,
Hunter College and Graduate Center, CUNY
695 Park Avenue,
New York, NY 10021, USA
E-mail: rchurchi@math.hunter.cuny.edu

We explain and illustrate how differential algebraic techniques can be used to establish the non-integrability of complex analytic Hamiltonian systems. Readers are assumed familiar with differential Galois theory, but not with Hamilton's equations; an introductory treatment of these entities, including a discussion of the significance of integrability, is therefore included. The examples we consider are handled as straightforward applications of a single result (Corollary 6.21), thereby allowing for a unified treatment of the work of several authors. Verification of the hypotheses is a matter of elementary calculation.

Over the last decade techniques from Differential Algebra have become increasingly important for establishing the non-integrability of complex Hamiltonian systems. This predominantly expository paper is an attempt to convey the flavor and significance of these methods to individuals with no background in Hamiltonian Dynamics, but who have a working knowledge of Differential Galois Theory.

Functions are always assumed C^∞ in the real case; meromorphic in the complex case.

1 Integrals of Ordinary Differential Equations

Let U be a non-empty open subset of $\mathbf{R}^m$. By adhering to the usual coordinate system we can view an autonomous (i.e., "time independent") ordinary differential equation on U as a function $F : U \to \mathbf{R}^m$, and a solution as a mapping $x = x(t)$ of a non-empty open interval $(a, b) \subset \mathbf{R}$ into the "phase space" U satisfying $x'(t) = F(x(t))$ for all $t \in (a, b)$. To indicate the equation we write

$$x' = F(x), \tag{1.1}$$

and when $m > 1$ occasionally refer to (1.1) as a *vector field* or *system (of [ordinary] differential equations)* on U.

A constant solution of (1.1) will be called an *equilibrium solution*.

We take the existence and uniqueness of (maximal) solutions (i.e., the existence and uniqueness of the associated "flow") for granted.

We recall the effect of a "change of variables" on (1.1). Specifically, suppose $V \subset \mathbf{R}^m$ is open and $\eta : V \to U$ is a surjection with non-singular derivative $D\eta(y) : \mathbf{R}^m \to \mathbf{R}^m$ (equivalently, non-singular $m \times m$ Jacobian matrix $\frac{d\eta}{dy}(y)$) at each $y \in V$. Assume in addition that $y = y(t)$ is a curve in V with the property that $x = x(t) := \eta(y(t))$ is a solution of (1.1). Identifying elements of $\mathbf{R}^m$ with column vectors we then see from the chain-rule

$$F(\eta(y(t))) = F(x(t)) = \dot{x}(t) = \frac{dx}{dt}(t) = \frac{d\eta}{dy}(y(t))\frac{dy}{dt}(t) = \frac{d\eta}{dy}(y(t))\dot{y}(t)\,,$$

and as a consequence that $y = y(t)$ is a solution of the differential equation

$$\dot{y} = \left(\frac{d\eta}{dy}(y)\right)^{-1} F(\eta(y))\,. \tag{1.2}$$

This is the *transformed equation* of equation (1.1) under the *change of variables* $x = \eta(y)$.

A remark on our notational conventions is in order. Vectors in $\mathbf{R}^m$ are identified, by means of the usual basis, with column vectors, but notational distinctions are made only in the case of tangents: for a C^1-function $x : (a,b) \subset \mathbf{R} \to \mathbf{R}^m$, the symbol $x'(t)$ denotes the tangent vector at $x(t)$, whereas $\dot{x}(t)$ denotes the corresponding column vector (i.e., $m \times 1$ matrix). For example, (1.1) reflects a (geometric) vector identity, whereas (1.2) reflects a column vector identity. The analogous conventions for holomorphic $x : U \subset \mathbf{C} \to \mathbf{C}^m$ will also be used.

A function $f : U \to \mathbf{R}$ is an *integral* of (1.1) if, for all t where defined, we have

$$\frac{d}{dt}f(x(t)) = 0 \tag{1.3}$$

whenever $x = x(t)$ is a solution of that equation. The geometrical meaning is: each solution curve $x = x(t)$ of (1.1) is contained entirely within a level surface of f. The existence of integrals allows for reduction in dimensions; certainly a desirable prospect when confronting a specific problem. Note from the chain-rule and (1.1) that (1.3) may be written

$$0 = \frac{d}{dt}f(x(t)) = Df(x(t))x'(t) = \langle \nabla f(x(t)), F(x(t))\rangle\,, \tag{1.4}$$

where $\langle\,,\,\rangle$ is the usual inner product and $\nabla f(x)$ is the usual gradient (vector) of f at x. Conclusion: $f : U \to \mathbf{R}$ *is an integral of* (1.1) *if and only if*

$$\langle \nabla f, F\rangle \equiv 0\,. \tag{1.5}$$

Integrals $f_1, \ldots, f_k : U \to \mathbf{R}$ of (1.1) are (*functionally*) *independent* when the gradients $\nabla f_j(x)$ are linearly independent over $\mathbf{R}$ for almost all $x \in U$.

Example 1.6 (The Lorenz Equation) The name refers to the system

$$\dot{x}_1 = -sx_1 + sx_2,$$
$$\dot{x}_2 = rx_1 - x_2 - x_1 x_3,$$
$$\dot{x}_3 = -bx_3 + x_1 x_2$$

on $\mathbf{R}^3$ in which s, r and b are real parameters. It was introduced in 1963, with $s = 10$, $b = 8/3$ and r the lone parameter, as a model for atmospheric convection in the study of long-range weather forecasting [22]. It became the object of intense study in dynamics as a result of the fortuitous discovery of a "strange attractor" within the flow corresponding to $r = 28$ (for our purposes: a geometrically bizarre collection of solutions with locally unpredictable ["chaotic"] behavior). Far less attention has been paid to the fact that there are parameter values corresponding to well-behaved solutions: for any $(0, b, r)$ the system admits the integral $f(x) = x_1$, the flows on the planes $x_1 = constant$ are linear, and as a result the equations are easily solved explicitly; for $(s, b, r) = (1/2, 1, 0)$ the system admits the integral $f(x) = \frac{x_2^2 + x_3^2}{(x_1^2 - x_3)^2}$, and this can be used to solve the associated equations in terms of elliptic functions (see, e.g., pages 344-7 of [30]).

Example 1.7 (The Kepler or Two-Body Problem) The terminology refers to the system on $U := (\mathbf{R}^3 \setminus \{0\}) \times \mathbf{R}^3 \subset \mathbf{R}^3 \times \mathbf{R}^3 \simeq \mathbf{R}^6$ given by

$$\dot{x}_1 = x_4,$$
$$\dot{x}_2 = x_5,$$
$$\dot{x}_3 = x_6,$$
$$\dot{x}_4 = -\frac{1}{(x_1^2 + x_2^2 + x_3^2)^{3/2}} \cdot x_1,$$
$$\dot{x}_5 = -\frac{1}{(x_1^2 + x_2^2 + x_3^2)^{3/2}} \cdot x_2,$$
$$\dot{x}_6 = -\frac{1}{(x_1^2 + x_2^2 + x_3^2)^{3/2}} \cdot x_3.$$

By introducing the vector variables $q = (x_1, x_2, x_3)$ and $p = (x_4, x_5, x_6)$, the equations assume the compact form

$$\begin{cases} q' = p, \\ p' = -q/|q|^3, \end{cases} \tag{1.8}$$

which we note is equivalent to the second order (vector) equation

$$q'' = -q/|q|^3. \tag{1.9}$$

A solution $q = q(t)$ of (1.9) describes the motion in $\mathbf{R}^3$ of a particle of mass 1 subject (only) to an inverse-square (normalized) gravitational force centered at the origin.

The vector field (1.8) has an abundance of integrals. In particular, the *angular momentum* function $\Omega = (\omega_1, \omega_2, \omega_3) : (q, p) \mapsto q \times p$ (= the usual cross product) is a "vector valued" integral in the sense that along any solution $(q, p) = (q(t), p(t))$ one has

$$\tfrac{d}{dt}\Omega(q, p) = q \times p' + q' \times p = -\tfrac{1}{|q|^3}(q \times q) + p \times p = 0\,,$$

and the associated component functions ω_j therefore provide integrals which one easily checks are functionally independent. A second vector-valued integral is given by the *Runge-Lenz* (or *eccentricity*) *vector* $E = (E_1, E_2, E_3)$: $(q, p) \mapsto \tfrac{q}{|q|} - p \times (q \times p)$; the component functions E_j provide additional integrals.

One can use the angular momentum integral in this example to illustrate the idea of reducing dimensions. Specifically, the calculation displayed in the previous paragraph implies that a solution $q = q(t)$ of (1.9) in $\mathbf{R}^3$ lies in the plane orthogonal to the vector $q(0) \times q'(0)$ or, if this vector vanishes, on the line in $\mathbf{R}^3$ spanned by $q(0)$. But now observe that equations (1.8) and (1.9) admit an $SO(3)$ symmetry: when $T : \mathbf{R}^3 \to \mathbf{R}^3$ is an orthogonal transformation and $q = q(t)$ satisfies (1.9) the same is true of $\hat{q} = \hat{q}(t) := Tq(t)$. All solutions of (1.9) can thereby be determined from solutions contained in the plane

$$\mathbf{R}^2 \simeq \{\, q = (q_1, q_2, q_3) \in \mathbf{R}^3 \ : \ q_3 = 0 \,\}\,,$$

and so it suffices to study (1.8) as a system defined not on $U \subset \mathbf{R}^6$ but rather on $\hat{U} := (\mathbf{R}^2 \backslash \{0\}) \times \mathbf{R}^2 \subset \mathbf{R}^2 \times \mathbf{R}^2 \simeq \mathbf{R}^4$. For this lower dimensional system, the angular momentum and Runge-Lenz vector (integrals) are replaced by $\Omega : (q, p) \mapsto \langle q, Jp \rangle$ and $E : (q, p) \mapsto q/|q| - \Omega(q, p)Jp$ respectively, where $J : \mathbf{R}^2 \to \mathbf{R}^2$ is clockwise rotation by $\pi/2$ radians.

Example 1.10 (Spinning Tops) Let μ be a constant, and for $j = 1, 2, 3$, let $\mathcal{I}_j$ and ℓ_j be non-zero constants, all generally assumed positive. Then the system

$$\dot{x}_1 = \mathcal{I}_3^{-1}x_2 x_6 - \mathcal{I}_2^{-1}x_3 x_5,$$
$$\dot{x}_2 = \mathcal{I}_1^{-1}x_3 x_4 - \mathcal{I}_3^{-1}x_1 x_6,$$
$$\dot{x}_3 = \mathcal{I}_2^{-1}x_1 x_5 - \mathcal{I}_1^{-1}x_2 x_4,$$
$$\dot{x}_4 = (\mathcal{I}_3^{-1} - \mathcal{I}_2^{-1})x_5 x_6 + \mu(\ell_3 x_2 - \ell_2 x_3),$$

$$\dot{x}_5 = (\mathcal{I}_1^{-1} - \mathcal{I}_3^{-1})x_4 x_6 + \mu(\ell_1 x_3 - \ell_3 x_1),$$
$$\dot{x}_6 = (\mathcal{I}_2^{-1} - \mathcal{I}_1^{-1})x_4 x_5 + \mu(\ell_2 x_1 - \ell_1 x_2)$$

constitutes the *Euler-Poisson equations* for a spinning top; it is more commonly written

$$\gamma' = \gamma \times \Omega,$$
$$\mathcal{M}' = \mathcal{M} \times \Omega + \mu \cdot (\gamma \times \ell),$$

where

$$\gamma = (x_1, x_2, x_3), \qquad \Omega = (\mathcal{I}_1^{-1} x_4, \mathcal{I}_2^{-1} x_5, \mathcal{I}_3^{-1} x_6),$$
$$\mathcal{M} = (x_4, x_5, x_6), \qquad \ell = (\ell_1, \ell_2, \ell_3).$$

As the reader is invited to check, three independent integrals are provided by

$$f_1(x) = \tfrac{1}{2}(\mathcal{I}_1^{-1} x_4^2 + \mathcal{I}_2^{-1} x_5^2 + \mathcal{I}_3^{-1} x_6^2) + \mu(\ell_1 x_1 + \ell_2 x_2 + \ell_3 x_3),$$
$$f_2(x) = x_1 x_4 + x_2 x_5 + x_3 x_6,$$
$$f_3(x) = x_1^2 + x_2^2 + x_3^2.$$

These enable us to reduce the dimensions of the problem from six to three. Indeed, for purposes of application (which we will not discuss) one generally restricts the equations to the cotangent bundle $T^* S^2 \simeq f_2^{-1}(\{c\}) \cap f_3^{-1}(\{1\})$ of S^2, where the real constant c can be arbitrary.

For a general discussion of such "rigid body problems" see, e.g., [5] or Chapters 3 and 5 of [14].

2 Linearized Equations

Suppose

$$x' = F(x) \tag{2.1}$$

is an autonomous differential equation defined on a nonempty open subset U of $\mathbf{R}^m$ and $x_j = x_j(t)$ are solutions for $j = 1, 2$. Then for $h = h(t) := x_2(t) - x_1(t)$, Taylor's theorem gives

$$h' = x_2' - x_1' = F(x_2) - F(x_1) = F(x_1 + h) - F(x_1)$$
$$= F(x_1) + DF(x_1)h + \cdots - F(x_1) = DF(x_1)h + \cdots,$$

and so "up to first order" the "variation" $h = h(t)$ of the solution $x_1 = x_1(t)$ satisfies the <u>linear</u> differential equation

$$h' = DF(x)h, \qquad x = x(t). \tag{2.2}$$

This system is the *linearized* or *variational equation* of (2.1) along the given solution. In calculations one writes $F = (F_1, \ldots, F_m)$ and identifies $DF(x)$ with the Jacobian matrix $\frac{dF}{dx}(x) := (\frac{\partial F_i}{\partial x_j}(x))$. The classical notations for h and h_j are δx and δx_j.

It is important to note that for any solution $x = x(t)$ of (2.1) we have, by straightforward differentiation,

$$x''(t) = DF(x(t))\, x'(t). \tag{2.3}$$

Conclusion: $x' = F(x(t))$ *is a solution of the linearized equation along* $x = x(t)$.

Example 2.4 The system

$$\begin{aligned}
\dot{x}_1 &= \quad x_2, \\
\dot{x}_2 &= -x_1 + x_2 x_3, \\
\dot{x}_3 &= \quad x_3^2
\end{aligned}$$

admits $x = x(t) = (\sin t, \cos t, 0)$ as a solution, and the relevant Jacobian matrix for $F(x) = (F_1(x), F_2(x), F_3(x)) = (x_2, -x_1 + x_2 x_3, x_3^2)$ is

$$\frac{dF}{dx}(x) = \begin{pmatrix} 0 & 1 & 0 \\ -1 & x_3 & x_2 \\ 0 & 0 & 2x_3 \end{pmatrix}.$$

The variational equation along this solution is therefore

$$\begin{pmatrix} \dot{h}_1 \\ \dot{h}_2 \\ \dot{h}_3 \end{pmatrix} = \begin{pmatrix} 0 & 1 & 0 \\ -1 & 0 & \cos t \\ 0 & 0 & 0 \end{pmatrix} \begin{pmatrix} h_1 \\ h_2 \\ h_3 \end{pmatrix}$$

i.e.,

$$\begin{aligned}
\dot{h}_1 &= \quad h_2, \\
\dot{h}_2 &= -h_1 + (\cos t) h_3, \\
\dot{h}_3 &= \quad 0\,.
\end{aligned}$$

Notice that $h = h(t) := \dot{x}(t) = (\cos t, -\sin t, 0)^\tau$ is a solution. (The superscript τ always denotes transposition.)

3   Hamiltonian Systems – The Classical Formulation

In this section $n \geq 1$ is an integer and the usual coordinate functions $x_1, \ldots, x_{2n}$ on $\mathbf{R}^{2n}$ are occasionally written $q_1, \ldots, q_n, p_1, \ldots, p_n$; the q_j ("positions") and p_j ("momenta") are said to be *conjugate variables*.

A *Hamiltonian system* on a nonempty open set $U \subset \mathbf{R}^{2n}$ is a system of (ordinary) differential equations of the form

$$\dot{q}_j = \frac{\partial H}{\partial p_j}(q, p), \quad \dot{p}_j = -\frac{\partial H}{\partial q_j}(q, p), \quad j = 1, \ldots, n, \tag{3.1}$$

where $H : U \to \mathbf{R}$. H is the corresponding *Hamiltonian function* and n is the (number of) *degrees of freedom* of the system. Equations (3.1) are also known as *Hamilton's equations for H*.

Example 3.2 Let $U = \mathbf{R}^{2n}$, let $\omega_1, \ldots, \omega_n$ be constants, and define $H : \mathbf{R}^{2n} \to \mathbf{R}$ by

$$H(q, p) = \sum_{j=1}^{n} \frac{\omega_j}{2}(q_j^2 + p_j^2).$$

The corresponding Hamiltonian system is

$$\dot{q}_j = \omega_j p_j, \quad \dot{p}_j = -\omega_j q_j, \quad j = 1, \ldots, n,$$

but would more likely be written

$$\ddot{q}_j = -\omega_j^2 q_j, \quad j = 1, \ldots, n.$$

This is the classical *linear harmonic oscillator*; the ω_j are the *frequencies*.

Example 3.3 Let $k > 0$ and $r > 0$ be integers and set $n = kr$. In addition, let $m_1, \ldots, m_r$ denote positive constants and let $1/m$ denote the $n \times n$ block diagonal matrix with j-th block $(1/m_j)I_k$, where I_k is the $k \times k$ identity matrix, $j = 1, \ldots, r$. If $W \subset \mathbf{R}^n$ is open and $H : W \times \mathbf{R}^n \to \mathbf{R}$ has the form

$$H(q, p) = \tfrac{1}{2}\langle (1/m)p, p \rangle + V(q), \tag{3.4}$$

where $\langle \, , \, \rangle$ is the usual inner product and $V : W \to \mathbf{R}$, then (in vector notation) the associated Hamiltonian system is

$$q_j' = (1/m_j)p_j, \quad p_j' = -V_{q_j}, \quad j = 1, \ldots, r,$$

where

$$V_{q_j}(q) = \left(\frac{\partial V}{\partial q_{j_1}}(q), \ldots, \frac{\partial V}{\partial q_{j_k}}(q) \right).$$

Since $m_j q_j' = p_j$ implies $m_j q_j'' = p_j'$ these equations may also be written

$$m_j q_j'' = -V_{q_j}, \quad j = 1, \ldots, r.$$

But these are precisely Newton's equations for r particles, of masses $m_1, \ldots, m_r$ respectively, moving in $\mathbf{R}^k$ subject to a force with potential function V. Hamilton's equations thus include Newton's equations for particle motion in conservative force fields.

A Hamiltonian function as in (3.4) is said to have *classical (mechanical) form* or *(classical) kinetic plus potential (energy) form*. The example explains why Hamiltonians are sometimes called *energy functions*.

Example 3.5 The Kepler Problem, i.e., equation (1.8) of Example 1.7, is a Hamiltonian system with Hamiltonian $H(q,p) = \frac{1}{2}|p|^2 - 1/|q|$. Note this is a special case of the previous example: take $r = 1 = m_1$, $k = n = 3$ or 2, and $V(q) = -1/|q|$. For $n = 3$ we interpret the equations and Hamiltonian as being defined on the open set $U \subset \mathbf{R}^6$ of Example 1.7; for $n = 2$ we replace U by the open set $\hat{U} \subset \mathbf{R}^4$ of that example.

Example 3.6 The two-degree of freedom Hamiltonian system governed by

$$H(q_1, q_2, p_1, p_2) = \frac{1}{2}p_1^2 + \frac{1}{2}p_2^2 + \frac{A}{2}q_1^2 + \frac{B}{2}q_2^2 + \frac{1}{3}q_1^3 + \lambda q_1 q_2^2$$

is the *Hénon-Heiles Hamiltonian*. It was originally introduced (1964) in a form equivalent to the case $A = B = -\lambda = 1$, which models galactic problems admitting an axially symmetric potential function [17]. The research literature on the general case is extensive.

When f is a function $U \to \mathbf{R}$, it proves convenient to write the coordinate vector of the gradient $\nabla f(x)$ (relative to the usual basis) as

$$f_x = f_x(x) := \begin{pmatrix} \frac{\partial f}{\partial x_1}(x) \\ \vdots \\ \frac{\partial f}{\partial x_{2n}}(x) \end{pmatrix}. \tag{3.7}$$

In terms of the usual coordinate functions $x = (x_1, \ldots, x_n)$, Hamilton's equations (3.1) may then be written

$$\dot{x} = JH_x(x), \tag{3.8}$$

where

$$J := \begin{pmatrix} 0 & I \\ -I & 0 \end{pmatrix} \tag{3.9}$$

with 0 and I denoting the $n \times n$ zero and identity matrices respectively. J is the *canonical matrix*.

The transformed equations of a Hamiltonian system under a change of variables will generally not be Hamiltonian; to maintain that property one must restrict to transformations of a special type. To introduce these, call a $2n \times 2n$ (real or complex) matrix M *symplectic* if

$$M^\tau J M = J, \qquad (3.10)$$

where the τ denotes transposition and J is the canonical matrix (3.9), and call a change of variables $x = \eta(y)$ *canonical* if the Jacobian matrix $\frac{d\eta}{dy}(y)$ is symplectic at each point y of the open domain $V \subset \mathbf{R}^{2n}$ of η. From $M^\tau J M = J$ and the observation that $det(J) = 1$ one sees that a symplectic matrix is non-singular; taking inverses in (3.10) and noting that $J^\tau = J^{-1} = -J$ one then has $M^{-1}J(M^\tau)^{-1} = J$, which shows that M^{-1} is also symplectic; it proves convenient to express this last equality in the form

$$M^{-1}J = JM^\tau. \qquad (3.11)$$

When $\eta : V \to U$ is canonical we now see from (1.2) and (3.11) that the transformed equations of (3.8) are given by

$$
\begin{aligned}
\dot{y} &= \left(\tfrac{d\eta}{dy}(y)\right)^{-1} J H_x(\eta(y)) \\
&= \left(\tfrac{d\eta}{dy}(y)\right)^{-1} J \left(\tfrac{dH}{dx}(\eta(y))\right)^\tau \\
&= J \left(\tfrac{d\eta}{dy}(y)\right)^\tau \left(\tfrac{dH}{dx}(\eta(y))\right)^\tau \\
&= J \left(\tfrac{dH}{dx}(\eta(y)) \tfrac{d\eta}{dy}(y)\right)^\tau \\
&= J \left(\tfrac{d(H \circ \eta)}{dy}(y)\right)^\tau \\
&= J(H \circ \eta)_y(y).
\end{aligned}
$$

In other words, when $x = \eta(y)$ is a canonical change-of-variables, the system (3.8) with Hamiltonian H transforms to the Hamiltonian system with Hamiltonian $H \circ \eta$.

Example 3.12 Consider a Hamiltonian function

$$H(q,p) = \tfrac{1}{2}\langle p,p \rangle + V(q)$$

of classical form defined on an open set $U \simeq \hat{U} \times \mathbf{R}^n \subset \mathbf{R}^{2n}$, where $\hat{U}$ is open in $\mathbf{R}^n$. Now suppose $\hat{V} \subset \mathbf{R}^n$ is open and $\beta : \hat{V} \to \hat{U}$ has non-singular derivative $D\beta(x) : \mathbf{R}^n \to \mathbf{R}^n$ at each $x \in \hat{V}$. Set $V := \hat{V} \times \mathbf{R}^n \subset \mathbf{R}^n \times \mathbf{R}^n \simeq \mathbf{R}^{2n}$ and define $\eta : V \to U$ by $(x,y) \mapsto (\beta(x), ((D\beta(x))^\tau)^{-1}y)$. Then η is easily checked to be a canonical transformation. This change-of-variables is described classically as the "usual extension" of the transformation $q = \beta(x)$ of the "position variables" to a canonical transformation.

Example 3.13 (Adapted from [18].) The Hénon-Heiles Hamiltonian of Example 3.5 assumes the form discussed in (a) by taking

$$V(q) = \frac{A}{2}q_1^2 + \frac{B}{2}q_2^2 + \frac{1}{3}q_1^3 + \lambda q_1 q_2^2.$$

To illustrate transformations of the type discussed in (a) assume $A = B = 1$, $\lambda > \frac{1}{2}$, set $\kappa := \sqrt{2 - 1/\lambda}$ and consider the change of (position) variables $\beta : \mathbf{R}^2 \to \mathbf{R}^2$ given by

$$q_1 = (1/\sqrt{1 + \kappa^2})(x_1 + \kappa x_2),$$
$$q_2 = (1/\sqrt{1 + \kappa^2})(\kappa x_1 - x_2).$$

Then

$$\frac{d\beta}{dx}(x) = (1/\sqrt{1 + \kappa^2})\begin{pmatrix} 1 & \kappa \\ \kappa & -1 \end{pmatrix} = \left(\left(\frac{d\beta}{dx}(x)\right)^\tau\right)^{-1},$$

and so a canonical transformation $\eta : \mathbf{R}^4 \to \mathbf{R}^4$ extending the position transformation β is provided by

$$q_1 = (1/\sqrt{1 + \kappa^2})(x_1 + \kappa x_2), \qquad p_1 = (1/\sqrt{1 + \kappa^2})(y_1 + \kappa y_2),$$
$$q_2 = (1/\sqrt{1 + \kappa^2})(\kappa x_1 - x_2), \qquad p_2 = (1/\sqrt{1 + \kappa^2})(\kappa y_1 - y_2).$$

The transformed Hamiltonian $\hat{H} := H \circ \eta$ is

$$\hat{H}(x, y) = \frac{1}{2}(y_1^2 + y_2^2) + \frac{1}{2}(x_1^2 + x_2^2) + \frac{\hat{\theta}}{3}x_1^3 + \hat{\lambda}x_1 x_2^2 + \hat{\tau}x_2^3,$$

where

$$\hat{\theta} := \frac{2\lambda}{\sqrt{3 - 1/\lambda}}, \qquad \hat{\lambda} := \frac{1 - \lambda}{\sqrt{3 - 1/\lambda}}, \qquad \hat{\tau} := \frac{\lambda + 1}{3}\sqrt{\frac{2 - 1/\lambda}{3 - 1/\lambda}}.$$

The *Poisson bracket* $\{f, g\}$ of functions $f, g : U \to \mathbf{R}$ is the function $\{f, g\} : U \to \mathbf{R}$ defined by

$$\{f, g\}(x) := \langle f_x(x), Jg_x(x)\rangle. \tag{3.14}$$

Note from the chain rule that, along any solution $x = x(t)$ of (3.8), one has

$$\frac{d}{dt}f(x) = \langle f_x(x), \dot{x}\rangle = \langle f_x(x), JH_x(x)\rangle = \{f, H\}(x).$$

In particular, f is an integral of (3.8) if and only if $\{f, H\} \equiv 0$.

Let K be a field and let A be a K-Lie algebra which is also an associative commutative K-algebra with multiplicative identity. Then A is a K-*Poisson algebra* if for each $a \in A$, the *adjoint mapping* $ad_a : b \in A \mapsto [a, b] \in A$ associated with the Lie bracket $[\,,\,] : A \times A \to A$ is a derivation on A.

Proposition 3.15 *The Poisson bracket gives the algebra $C^\infty(U, \mathbf{R})$ of C^∞-functions $f : U \to \mathbf{R}$ the structure of a real Poisson algebra.*

Proof. By straightforward calculation. ∎

Corollary 3.16 (Conservation of Energy) *The Hamiltonian is always an integral of the associated Hamiltonian system.*

The value of the Hamiltonian along a solution is the *energy* of that solution.

Proof. Proposition 3.15 implies, in particular, that $\{H, H\} \equiv 0$. ∎

Corollary 3.17 *Suppose $W \subset \mathbf{R}$ is open, $f : U \to W$ is an integral of a Hamiltonian system defined on U, and g is a function $W \to \mathbf{R}$. Then $g \circ f : U \to \mathbf{R}$ is an integral of the same system.*

Less formally: any function of an integral is an integral.

Proposition 3.18 *Suppose $U, V \subset \mathbf{R}^{2n}$ are non-empty open sets and $\eta : V \to U$ is a canonical transformation. If $f, g : U \to \mathbf{R}$ are any two functions, then $\{f \circ \eta, g \circ \eta\} = \{f, g\} \circ \eta$.*

Remark In fancier language: the mapping $\eta^* : C^\infty(U, \mathbf{R}) \to C^\infty(V, \mathbf{R})$ defined by $\eta^* : f \mapsto f \circ \eta$ is a "morphism of real Poisson algebras". The displayed identity is written in accordance with this notation as $\{\eta^* f, \eta^* g\} = \eta^* \{f, g\}$.

Proof. We have already noted that a $2n \times 2n$ matrix M satisfying (3.10) must be non-singular, and that $J^{-1} = -J$, hence $M^\tau = -JM^{-1}J$. It follows that

$$MJM^\tau = MJ(-J)M^{-1}J = MM^{-1}J = J,$$

and we conclude that M^τ is symplectic if and only if M has this property.

Now note from the chain-rule that for any $h : U \to \mathbf{R}$ we have $\frac{d(h \circ \eta)}{dy}(y) = \frac{dh}{dx}(\eta(y))\frac{d\eta}{dy}(y)$, and therefore $(h \circ \eta)_y(y) = (\frac{d\eta}{dy}(y))^\tau h_x(\eta(y))$. From this last observation and that ending the previous paragraph we conclude that

$$
\begin{aligned}
\{f \circ \eta, \, g \circ \eta\}(y) &= \langle (f \circ \eta)_y(y), J(g \circ \eta)_y(y) \rangle \\
&= \langle (\tfrac{d\eta}{dy}(y))^\tau f_x(\eta(y)), \, J(\tfrac{d\eta}{dy}(y))^\tau g_x(\eta(y)) \rangle \\
&= \langle f_x(\eta(y)), \, (\tfrac{d\eta}{dy}(y))J(\tfrac{d\eta}{dy}(y))^\tau g_x(\eta(y)) \rangle \\
&= \langle f_x(\eta(y)), \, Jg_x(\eta(y)) \rangle \\
&= (\{f, g\} \circ \eta)(y) .
\end{aligned}
$$
∎

Corollary 3.19 *Suppose* $H : U \to \mathbf{R}$ *is a Hamiltonian and* $f : U \to \mathbf{R}$ *is an integral of the associated equations. Then* $\eta^* f := f \circ \eta : V \to \mathbf{R}$ *is an integral of the Hamiltonian system associated with* $\eta^* H := H \circ \eta : V \to \mathbf{R}$, *and is functionally independent of* $\eta^* H$ *when* f *is functionally independent of* H.

Less formally: when $f : U \to \mathbf{R}$ is an integral of a Hamiltonian system on U and $\eta : V \to U$ is canonical, the transformed function $\eta^* f : V \to \mathbf{R}$ is an integral of the transformed equations on V.

Proof. By Proposition 3.18 we have $\{\eta^* f, \eta^* H\} = \eta^* \{f, H\} = \eta^* 0 \equiv 0$, and the integral statement follows. The functional independence assertion is an immediate consequence of the non-singular nature of the derivative $D\eta(y) : \mathbf{R}^{2n} \to \mathbf{R}^{2n}$ at each $y \in V$. $\qquad\square$

Integrals for Hamiltonian systems can often be associated with symmetries ("Noether's Theorem"), e.g., the vector-valued angular momentum integral for the Kepler problem introduced in Example 1.7 can be regarded as a consequence of the existence of the $SO(3)$-action on solutions (see, e.g., page 54 of [14]). Since our main concern will be with non-integrable systems this relationship will not be pursued.

When functionally independent integrals $f_1, \ldots, f_r$ of the equations (3.1) satisfy $\{f_i, f_j\} \equiv 0$, they are said to be *in involution*. The n-degree of freedom Hamiltonian system (3.1) is *integrable* when there are n functionally independent integrals; it is *completely integrable* when there are n such integrals in involution, and in that case any such collection of functions is called a *complete set of integrals in involution*. (We warn the reader that for many authors the terms "integrable" and "completely integrable" are interchangeable; both would mean what we call completely integrable. In this connection see Remark 3.33.)

Corollary 3.20 *Suppose* $H : U \to \mathbf{R}$, *the associated Hamiltonian system on* U *is (completely) integrable, and* $\eta : V \to U$ *is canonical. Then the transformed system on* V *is also (completely) integrable.*

Proposition 3.21 *In* (3.1) *assume* $n > 1$ *and consider the following assertions:*

(a) *the system is completely integrable;*

(b) *the system is integrable;*

(c) *there is at least one integral functionally independent of* H.

Then (a) $\Rightarrow$ (b) $\Rightarrow$ (c), *and for two degrees of freedom, the three statements are equivalent.*

Proof. (a) ⇒ (b) is immediate from the definitions, and (b) ⇒ (c) is immediate from the observation that when f_1 and f_2 are functionally independent at least one must be functionally independent of H. To see that (c) ⇒ (a) when $n = 2$ simply observe that when f is an integral the involutivity condition $\{f, H\} \equiv 0$ is automatic. □

Example 3.22 Any one-degree of freedom Hamiltonian system is completely integrable.

Example 3.23 The linear harmonic oscillator of Example 3.2 is completely integrable: the collection $f_j(x) = \frac{\omega_j}{2}(q_j^2 + p_j^2)$ for $j = 1, \ldots, n$ provides a complete set of integrals in involution.

Example 3.24 In Example 1.7 the angular momentum integral was used to produce three functionally independent integrals $\omega_1, \omega_2, \omega_3$ for the Kepler problem when $n = 3$, but these are not in involution, e.g., $\{\omega_1, \omega_2\} = \omega_3$. The system is, nevertheless, completely integrable, e.g., H, ω_1 and E_1 (as defined in that example) provide a complete set of integrals in involution.

Example 3.25 The Kepler problem is also completely integrable in the case $n = 2$: the angular momentum integral Ω is functionally independent of H, and Proposition 3.21 therefore applies.

Example 3.26 For arbitrary real constants A, B the Hénon-Heiles Hamiltonian

$$H(q_1, q_2, p_1, p_2) = \tfrac{1}{2}p_1^2 + \tfrac{1}{2}p_2^2 + \tfrac{A}{2}q_1^2 + \tfrac{B}{2}q_2^2 + \tfrac{1}{3}q_1^3 + \lambda q_1 q_2^2$$

is completely integrable for the values $\lambda = 0$ and $1/6$, and when $A = B$ this is also the case for $\lambda = 1$. Indeed, by Proposition 3.21 it suffices to exhibit integrals functionally independent of the Hamiltonian corresponding to these values, and one checks that such integrals are provided by

$$\lambda = 0 \quad f = p_2^2 + Bq_2^2,$$

$$\lambda = 1/6 \quad f = \tfrac{3}{2}(4B - A)(Bq_2^2 + p_2^2) + Bq_1 q_2^2 + p_2(q_2 p_1 - q_1 p_2)$$
$$+ \tfrac{1}{6}q_2^2(q_1^2 + \tfrac{1}{4}q_2^2),$$

$$\lambda = 1 \quad f = p_1 p_2 + q_1 q_2(A + q_1) + \tfrac{1}{3}q_2^3$$

(see, e.g., [11,13]). The case $A = 16B$, $\lambda = 1/16$ is also completely integrable (see [16]); an integral functionally independent of the Hamiltonian is provided by

$$f = \tfrac{1}{6}q_2^6 + q_1^2 q_2^4 + 16Bq_1 q_2^4 - 48q_1 q_2^2 p_2^2 + 16q_2^3 p_1 p_2$$
$$- 192B^2 q_2^4 - 384Bq_2^2 p_2^2 - 192p_2^4.$$

We will eventually see how these parameter values can be detected using differential Galois theory (although this was not the method originally employed).

Complete integrability is a rare (see, e.g., [28,23]) but highly desirable attribute for a Hamiltonian system, and is the first property one investigates when confronted with a new equation. Specifically, knowledge of a complete set of integrals reduces the study of a Hamiltonian system to that of straight-line flows on cylinders and/or tori (the "Liouville-Arnol'd-Jost Theorem" - see, e.g., Appendix 26, pages 210-214, of [3]). In contrast, solution behavior in non-integrable systems tends to be random ("stochastic"; "chaotic"): knowledge of the long term behavior of the solution through a particular point generally has no implications for the long term behavior of solutions through nearby points.

Example 3.27 Perhaps the easiest way to gain some appreciation for the concept of complete integrability is to see how known integrals can be used to obtain information about solutions, even when those solutions have not (or cannot) be determined explicitly. Here we illustrate this idea with the Kepler problem for the case $n = 2$, which in Example 3.24 was seen to be completely integrable. Indeed, we already noted in Example 1.7 that the angular momentum integral for $n = 3$ can be used to reduce that problem to the present context.

For ease of reference we recall a few pertinent facts: the equations

$$q' = p, \quad p' = -q/|q|^3, \tag{3.28}$$

which can also be written

$$q'' = -q/|q|^3, \tag{3.29}$$

are defined on the open subset $\hat{U} = (\mathbf{R}^2\backslash\{0\}) \times \mathbf{R}^2 \subset \mathbf{R}^4$; the Hamiltonian function is $H(q,p) = \frac{1}{2}|p|^2 - 1/|q|$; the angular momentum integral is $\Omega :$ $(q,p) \mapsto \langle q, Jp \rangle$, where J is clockwise rotation by $\pi/2$ radians; and the Runge-Lenz vector (integral) is $E : (q,p) \mapsto q/|q| - \Omega(q,p)Jp$.

For the remainder of the example, fix any (maximal) solution $(q,p) = (q(t),p(t))$ of (3.28), and set $h := H(q(t),p(t))$, $\omega := \Omega(q(t),p(t))$ and $\epsilon :=$ $|E(q(t),p(t))|$; these entities are constant (real numbers) by virtue of the fact that H, Ω and E are integrals (and the condition $q(t) \neq 0$, which would otherwise result in $h = \infty$). Note that by calculating $\langle E, E \rangle = \epsilon^2$ directly one obtains the relation

$$\epsilon^2 = 1 + 2h\omega^2. \tag{3.30}$$

Assume $\epsilon \neq 0$, introduce the basis $e_1 := E(q(t), p(t))/|E(q(t), p(t))|$, $e_2 = J^{-1}e_1$ of $\mathbf{R}^2$ (note that e_1 is independent of t), and write $q = q(t) = re^{i\theta}$, where $r = r(t) := |q(t)|$ and the angle $\theta = \theta(t)$ is that measured counterclockwise from e_1 to q.

Theorem (Kepler's Laws) *Suppose $\omega > 0$ and that the curve $q = q(t) = r(t)e^{i\theta(t)}$ is bounded. Then:*

(a) **(Kepler's First Law)** *the path of $q = q(t)$ is an ellipse with one focus at the origin and eccentricity ϵ (< 1), and q traverses this ellipse repeatedly, always in counterclockwise direction;*

(b) **(Kepler's Second Law)** *for $A(t) := \frac{1}{2}\int_0^t r^2(u)\dot\theta(u)\,du$ one has $\dot A(t) = \frac{1}{2}\omega$;*

(c) **(Kepler's Third Law)** *$q = q(t)$ makes one circuit of the ellipse in time $T = 2\pi a^{3/2}$, where $a = \omega^2/(1 - \epsilon^2)$ is the semi-major axis of this conic.*

The usual informal statement of the second law is: the vector q sweeps out equal areas in equal times.

From the following proof one can easily determine those modifications of the statement necessary for describing the case $\omega < 0$.

Proof. In the proof we suppress the t in writing $q = q(t)$, $r = r(t)$, etc.

By elementary analytical geometry we have $\epsilon r \cos\theta = |E||q|\cos\theta = \langle E, q\rangle = |q| - \omega^2 = r - \omega^2$, and therefore

$$r = \omega^2/(1 - \epsilon\cos\theta). \tag{3.31}$$

This is a standard polar-coordinate equation for a conic with eccentricity ϵ and one focus at the origin: it is a circle if $\epsilon = 0$; an ellipse with semi-major axis $a = \omega^2/(1 - \epsilon^2)$ and semi-minor axis $b = \omega^2/\sqrt{1 - \epsilon^2}$ if $0 < \epsilon < 1$; a parabola if $\epsilon = 1$; a hyperbola if $\epsilon > 1$. It is immediate from (3.31) and $\omega > 0$ that r is bounded away from 0.

Now observe that $\omega = \langle q, Jp\rangle = -Im(q\bar p)$ and from $\dot q = p$ that

$$q\bar p = re^{i\theta}(-ire^{-i\theta}\dot\theta + \dot re^{-i\theta}) = -ir^2\dot\theta + r\dot r\,;$$

hence

$$\omega = r^2\dot\theta. \tag{3.32}$$

Because $\omega > 0$ and r never vanishes we conclude that $q = re^{i\theta}$ must trace out the entire conic (or a branch thereof, in the case of a hyperbola) in a counterclockwise direction, and since $q = q(t)$ is assumed bounded this forces the conclusions of the first law.

(One can easily say more, e.g., it is evident from (3.30) and (3.31) that when $\omega > 0$ the curve $q = q(t)$ will be an ellipse if and only if $h < 0$, a parabola if and only if $h = 0$, and one branch of a hyperbola if and only if $h > 0$.)

The second law is immediate from (3.32).

For the third law note from (3.32) that $A(t) = \frac{1}{2}\omega t$, implying that the area of the ellipse must be $\frac{1}{2}\omega T$. But that area can also be written πab, hence $\pi\omega a^{3/2} = \frac{1}{2}\omega T$, and the third law follows. $\qquad\square$

Remark 3.33 Hamiltonian systems with n degrees-of-freedom admitting n functionally independent integrals "tend to be" completely integrable, i.e., one can usually find n such functions in involution (as was illustrated in Example 3.24). But there are exceptions, e.g., the problem of four point vortices in the plane (see the discussion at the top of page 124 (§2.3) of [2]).

For discussions of integrability and non-integrability extending beyond the Hamiltonian context see, e.g., [15] and [26].

4   Normal Variational Equations

Suppose

$$\dot{x} = JH_x(x) \qquad\qquad (4.1)$$

is a Hamiltonian system defined on a non-empty open set $U \subset \mathbf{R}^{2n}$ and $x = x(t)$ is a non-equilibrium solution with image Γ contained in the plane P determined by $q_j = p_j = 0$, $j = 2,\ldots,n$. Assume the vector field JH_x is tangent to P, in which case the restriction $h := H|P$ defines a one-degree of freedom Hamiltonian system on this plane. In these circumstances the variational equation

$$\dot{y} = JH_{xx}(x)\,y, \qquad x = x(t) \qquad\qquad (4.2)$$

(H_{xx} denotes the Hessian matrix) along Γ decouples: by reordering variables it can be written in a block diagonal form consisting of a 2×2 block and a $(2n - 2) \times (2n - 2)$ block, with the first block representing the variational equation of $x = x(t)$ when considered as a solution of the Hamiltonian system on P determined by h. The linear system defined by the second reflects variations of the solution $x = x(t)$ (to first order) in directions orthogonal to P, and as a consequence is known as the *normal variational equation* (NVE) along Γ.

For the two-degree-of-freedom case the situation can be described as follows. (Our formulation is borrowed from [7]; for the n degree-of-freedom case in general, see §4.1 of [8].) Let $U \subset \mathbf{R}^4 = \{(q_1, q_2, p_1, p_2)\}$ be an open set containing the plane $P := \{(q_1, 0, p_1, 0)\}$ and consider an analytic Hamiltonian $H : U \to \mathbf{R}$ of the form

$$\begin{cases} H(q_1, q_2, p_1, p_2) = h(q_1, p_1) + \frac{1}{2}k(q_1, p_1)q_2^2 + \ell(q_1, p_1)q_2 p_2 \\ \qquad + \frac{1}{2}m(q_1, p_1)p_2^2 + \mathcal{O}_3(q_2, p_2) \,. \end{cases} \tag{4.3}$$

In this case the general remarks of the previous paragraph have the following consequences: the associated Hamiltonian system

$$\dot{x} = JH_x(x) \tag{4.4}$$

restricts to the one-degree of freedom Hamiltonian system

$$\dot{q} = \frac{\partial h}{\partial p}, \qquad \dot{p} = -\frac{\partial h}{\partial q} \tag{4.5}$$

on P, where $q = q_1$, $p = p_1$, and h is identified with $h|P$; any solution $x = x(t)$ of (4.4) with image Γ contained in P may also be viewed as a solution of (4.5), and when c is the energy of the solution we have

$$\Gamma \subset \{ (q, p) \,:\, h(q, p) = c \} \tag{4.6}$$

(by conservation of energy); the variational equation along Γ decouples as (i.e., can be written)

$$(a) \begin{cases} \dot{\alpha} = h_{qp}\alpha + h_{pp}\beta \,, \\ \dot{\beta} = -h_{qq}\alpha - h_{qp}\beta \,, \end{cases}$$
$$(b) \begin{cases} \dot{\xi} = \ell\xi + m\eta \,, \\ \dot{\eta} = -k\xi - \ell\eta \,, \end{cases} \tag{4.7}$$

where (4.7a) is the variational equation along Γ when $x = x(t)$ is considered a solution of (4.5), and (4.7b) represents the NVE for $x = x(t)$ when k, ℓ and m are evaluated on Γ.

Remark 4.8 Contemporary mathematical treatments of Hamiltonian systems are inevitably in terms of vector fields on symplectic manifolds, e.g., the spinning top of Example 1.10 restricts to a Hamiltonian vector field, with Hamiltonian f_1, on the symplectic cotangent bundle T^*S^2 of the sphere. (See, e.g., [6] or [14].) In this more general context, the NVE along a non-equilibrium solution can be formulated in terms of a flat connection on the symplectic "normal bundle" of Γ ((2.3) is needed here, e.g., see [8, §1]). In particular, the existence of a plane P tangent to the vector field is not an essential ingredient in the definition of the NVE. But without this assumption, calculations with the equation can be unwieldy, and since one of our goals is to treat specific examples in detail, we have limited consideration to this particular context.

5 Differential Galois Theory and Non-Integrability

Practically all we have stated thus far has an obvious reformulation in the complex setting (time included), where we now take "function" to mean "meromorphic function". The single item requiring clarification is the meaning of the "image of a solution". From standard existence theory we know that (1.1) admits unique solutions locally, i.e., defined on open discs in the complex plane: by a solution image we simply mean a component in the topology generated by the images of such local solutions. In the case of a non-equilibrium solution, such an image is a Riemann surface with local parameter t. (For additional detail see page 7 of [8].)

We have reached the point where we can bring differential Galois theory into the picture.

First note from (3.10) and the opening paragraph of the proof of Proposition 3.18 that the collection $Sp(n, \mathbf{C})$ of $2n \times 2n$ complex symplectic matrices forms an algebraic group under the usual matrix multiplication. We also note, from the observation that for $n = 1$ condition (3.10) is equivalent to $det(M) = 1$, that

$$Sp(1, \mathbf{C}) = SL(2, \mathbf{C}). \tag{5.1}$$

We need the standard fact that the spectrum of any element $g \in Sp(n, \mathbf{C})$ must have the form $\{\lambda_1, \ldots, \lambda_n, \lambda_1^{-1}, \ldots, \lambda_n^{-1}\}$ (see, e.g., page 226 of [1]). We say that g is *resonant* if there are integers m_j, not all zero, such that $\prod_{j=1}^{n-1} \lambda_j^{m_j} = 1$; otherwise g is *non-resonant*.

Henceforth we assume the situation described in the first paragraph of the preceding section, but now in the complex setting, and we allow H to be meromorphic (hence not necessarily defined at all points of U).

Theorem 5.2 *Suppose $\Gamma \subset P$ is the image of a non-equilibrium solution of (4.1). Let G denote the differential Galois group of the corresponding NVE, let $G^0 \triangleleft G$ denote the component of the identity, and regard both as matrix groups acting on $\mathbf{C}^{2n-2}$. Then the following statements hold.*

(a) *Meromorphic integrals of (4.1) induce non-trivial G-invariant rational functions on $\mathbf{C}^{2n-2}$. More precisely, suppose $2 \leq k \leq n$ and $f_1 = H, f_2, \ldots, f_k$ are functionally independent meromorphic integrals of (4.1). Then there are $(k-1)$ non-trivial rational functions on $\mathbf{C}^{2n-2}$*

which are algebraically independent over $\mathbf{C}$ and invariant under the action of G.

(b) Suppose (4.1) is integrable and G contains a non-resonant element. Then the group G^0 is an $(n-1)$-torus. In particular, G^0 is abelian.

(c) **(Morales-Ramis)** When (4.1) is completely integrable the group G^0 is abelian.

Assertion (a) was originally established by S.L. Ziglin in terms of the monodromy group of the NVE rather than the differential Galois group [34,35] (additional detail is found in [8]); in fact the monodromy version is all we require for the examples in §7. For a discussion of the statement as given above see page 90 of [25].

The basic idea behind the proof of (a) is rather simple: if f is an analytic integral functionally independent of H, choose any $\gamma \in \Gamma$ and expand f in a power series at γ. Without loss of generality the constant term vanishes; the first non-zero homogeneous term (the "junior part" of f) can then be interpreted as a polynomial function on $\mathbf{C}^{2n-2}$, and that function will be invariant under the action of G. In the case of meromorphic integrals the idea passes to quotients, thereby producing a G-invariant rational function. Unfortunately, functionally independent integrals can induce algebraically dependent G-invariant rational functions, and overcoming this problem can be regarded as a major technical detail in Ziglin's proof ("Ziglin's Lemma").

To prove assertion (b) we first note that the differential Galois group of the NVE is contained in the symplectic group $Sp(n-1, \mathbf{C})$ (see §3.4.4 of [6]); one then combines Theorems 2.14 and 2.15 of [8].

For a proof of (c) see [27] or Chapter 4 of [25]. The idea is: complete integrability is a Lie (in fact Poisson) algebraic concept (recall Proposition 3.15, which also holds in the complex case) which will be mirrored in the Lie algebra of G; that Lie algebra must be abelian.

Theorem 5.2 can be formulated more generally in accordance with Remark 4.8; this is done in the cited references.

With Theorem 5.2(a) as motivation, define a subgroup $H \subset Gl(m, \mathbf{C})$ to be k-*Ziglin* if there are at least k rational functions on $\mathbf{C}^m$ algebraically independent over $\mathbf{C}$ and invariant under the action of H. A subgroup $H \subset Gl(m, \mathbf{C})$ is *not Ziglin* if if is not k-Ziglin for all integers $k \geq 1$.

Corollary 5.3 (a) When G is not k-Ziglin there are at most $k-1$ integrals f_j of (4.1) such that the collection $H, f_1, \ldots, f_{k-1}$ is functionally independent.

(b) In particular, when G is not Ziglin there is no integral of (4.1) functionally independent of H.

For the applications we have in mind we only need to understand the 1-Ziglin subgroups of $SL(2, \mathbf{C})$, which for simplicity we refer to as *Ziglin groups*. Recall from (5.1) that the elements of this group are symplectic. Moreover, note that an element is resonant if and only if the eigenvalues are roots of unity.

6 Preliminaries to the Applications

Theorem 5.2 can be difficult to apply directly: one needs a non-equilibrium solution $(q, p) = (q(t), p(t))$ of the corresponding system (3.1) in explicit form to be able to calculate the NVE along the corresponding image Γ. But in practice it is often possible to ignore $(q, p) = (q(t), p(t))$ completely and replace the NVE with a far simpler equation. The idea will be illustrated in this section and will be used in all our examples; additional techniques can be found in [25].

In the following statement P denotes the $p_1 q_1$-plane, i.e., the plane defined by $p_2 = q_2 = 0$, and to ease notation we occasionally write (q_1, p_1) as (q, p).

Proposition 6.1 *Let U be a non-empty open subset of $\mathbf{C}^4$ and let H be a meromorphic function on U of the form*

$$H(q_1, q_2, p_1, p_2) = \tfrac{1}{2} p_1^2 + \tfrac{m}{2} p_2^2 - r(q_1^n) + \tfrac{n^2}{2} q_1^{2n-2} s(q_1^n) q_2^2 + \mathcal{O}_3(q_2, p_2). \quad (6.2)$$

That is, assume the complex (meromorphic) analogue of (4.3) and impose the following additional hypotheses on the functions h, k, ℓ and m appearing in that formula:

(1) *$h(q_1, p_1)$ has the form $\tfrac{1}{2} p_1^2 - r(q_1^n)$, where $r(z)$ is a non-constant polynomial and $n \geq 1$ is an integer;*

(2) *$k = k(q_1, p_1)$ has the form $k(q_1) = n^2 q_1^{2n-2} s(q_1^n)$ with n as in (a) and $s(z)$ rational;*

(3) *$\ell = \ell(q_1, p_1)$ vanishes identically and $m = m(q_1, p_1)$ is constant.*

Then the following assertions hold:

(a) *for any complex number c there is a non-equilibrium solution of (4.5) of energy c having as image $\Gamma \subset P$ any prescribed non-singular component of the algebraic curve defined by*

$$p^2 = 2(r(q^n) + c); \quad (6.3)$$

(b) *the monodromy group of the NVE (4.7b) along* Γ *based at an arbitrary point* $(q_0, p_0) \in \Gamma$ *embeds into that of the second order equation*

$$y'' + \left(\frac{r'(z)}{2 \cdot (r(z) + c)} + \frac{n-1}{nz} \right) y' + \frac{m \cdot s(z)}{2 \cdot (r(z) + c)} y = 0, \quad ' = \frac{d}{dz}, \qquad (6.4)$$

on the Riemann sphere $\mathbf{P}^1$ *based at* q_0^n, *and the identity component of the Zariski closure* $G_R \subset GL(2, \mathbf{C})$ *of the latter group is a subgroup of* $SL(2, \mathbf{C})$;

(c) *when the Hamiltonian system associated with (6.2) is (meromorphically) integrable, the group* $G_R \subset GL(2, \mathbf{C})$ *defined in (b) is Ziglin.*

Equation (6.4) is often called the *algebraic form* of the NVE (4.7b). Specifically, the NVE is the pull-back of (6.4) under the mapping $\Gamma \to \mathbf{P}^1$, where $(q_1, p_1) \mapsto z = q_1^n$, with this mapping commonly interpreted as a change of the time variable.

For the differential Galois group analogue of (b), see [25, Theorem 2.5].

Proof. (a) : First recall that a solution $y = y(t)$ to any second order ordinary differential equation

$$\ddot{y} = f(y) \qquad (6.5)$$

with f meromorphic on an open subset of $\mathbf{C}$ can always be constructed as follows: choose F locally such that $F'(z) = f(z)$; then define $y = y(t)$, in a small neighborhood of some distinguished point where F does not vanish, by

$$t = \int^{y(t)} \frac{1}{\sqrt{2F(u)}} \, du. \qquad (6.6)$$

Differentiating this last equation w.r.t. t and squaring yields

$$(\dot{y}(t))^2 = 2F(y(t)),$$

whereupon from a subsequent differentiation we see that $y = y(t)$ satisfies (6.5). One then extends the domain of $y = y(t)$ by analytic continuation.

Now note from hypothesis (1) that the one-degree of freedom system (4.5) assumes the form

$$\begin{aligned} \dot{q} &= p, \\ \dot{p} &= nr'(q^n)q^{n-1}, \end{aligned} \qquad (6.7)$$

which is more conveniently written

$$\ddot{q} = nr'(q^n)q^{n-1}.$$

This is as in (6.5), and using the construction of the previous paragraph with the choice $F(z) = r(z^n) + c$ we can manufacture a non-equilibrium local solution $(q, p) = (q(t), p(t)) := (y(t), \dot{y}(t))$ of (6.7) satisfying $(\dot{y}(t))^2 = 2(r(z^n) + c)$. Assertion (a) then follows by analytic continuation.

(b) and (c) : The pull-back assertion immediately following the proposition statement can be verified by direct calculation (for greater generality see Theorem 4.4 of [7]); the embedding property is an easy corollary (e.g., see page 42 of [8]). One then needs the following two facts: the components of the identity of the Zariski closures of the two monodromy groups are identified by this embedding (Proposition 4.7 of [8]); an algebraic subgroup of $SL(2, \mathbf{C})$ is Ziglin if and only if the component of the identity has this property (e.g., see Proposition 2.9 of [8]). □

One can see from (6.6) of the preceding proof why elliptic functions are often encountered in applications of the methods we are describing. Indeed, virtually all successes were restricted to such contexts, and were accomplished by means of a direct analysis of the NVE, until H. Yoshida initiated the use of the algebraic form of that linear equation in the spirit of Proposition 6.1 (e.g., see [33]). In the best circumstances judicious choices for c render that algebraic form amenable to study, e.g., the equation is Fuchsian. Since this last property is realized in all our examples it proves useful to recall a few facts about such entities.

The most general second-order Fuchsian equation on the Riemann sphere $\mathbf{P}^1$ has the form

$$y'' + \left(\sum_{j=1}^{m} \frac{A_j}{z - a_j} \right) y' + \left(\sum_{j=1}^{m} \frac{B_j}{(z - a_j)^2} + \sum_{j=1}^{m} \frac{C_j}{z - a_j} \right) y = 0, \qquad (6.8)$$

where the (finite) singularities a_j are distinct and the only relation on the complex constants A_j, B_j and C_j is $\sum_{j=1}^{m} C_j = 0$. The *normal form* of (6.8) is

$$y'' + \left(\sum_{j=1}^{m} \frac{\hat{B}_j}{(z - a_j)^2} + \sum_{j=1}^{m} \frac{\hat{C}_j}{z - a_j} \right) y = 0, \qquad (6.9)$$

where

$$\begin{cases} \text{(a)} \ \hat{B}_j = \frac{1}{4}(1 + 4B_j - (1 - A_j)^2), \\[2mm] \text{(b)} \ \hat{C}_j = C_j - \frac{1}{2}A_j \left[\sum_{i \neq j} \frac{A_i}{a_j - a_i} \right]. \end{cases} \qquad (6.10)$$

More generally, the *normal form* of any second order equation

$$y'' + c_1(z)y' + c_2(z)y = 0 \qquad (6.11)$$

on $\mathbf{P}^1$ is

$$y'' + \left(c_2(z) - \left(\frac{c_1(z)}{2} \right)^2 - \frac{c_1'(z)}{2} \right) y = 0. \qquad (6.12)$$

Note from (6.10) that the normal form of a Fuchsian equation is again Fuchsian. The converse, however, is false, e.g., the normal form of the non-Fuchsian equation $y'' + \frac{1}{z^2}y' + \frac{1-4z}{4z^4}y = 0$ is the Fuchsian equation $y'' = 0$.

Proposition 6.13 *The differential Galois group G of (6.9) can be identified with the Zariski closure of the monodromy group based at any non-singular point of $\mathbf{P}^1$. Moreover, when $G^0 \subset SL(2, \mathbf{C})$ the differential Galois group of the normal form (6.9) is Ziglin if and only if this is the case for G.*

For a proof see Theorem 3.16 and Proposition 4.25 of [8].

The work of the section thus far is summarized by the following result, wherein $P \subset \mathbf{C}^4 \simeq \{(q_1, q_2, p_1, p_2)\}$ is the $q_1 p_1$-plane, i.e., the plane determined by the vanishing of q_2 and p_2 (see the first paragraph of §4).

Theorem 6.14 *Let $U \subset \mathbf{C}^4$ be an open neighborhood of P and let $H : U \to \mathbf{C}$ be a meromorphic Hamiltonian of the form*

$$H(q_1, q_2, p_1, p_2) = \tfrac{1}{2}p_1^2 + \tfrac{m}{2}p_2^2 - r(q_1^n) + \tfrac{n^2}{2}q_1^{2n-2}s(q_1^n)q_2^2 + \mathcal{O}_3(q_2, p_2), \quad \text{(i)}$$

where $n \geq 1$ is an integer, $m \in \mathbf{C}$ is a constant, $r(z)$ is a polynomial, and $s(z)$ is a rational function. Suppose $c \in \mathbf{C}$ is such that the equation

$$y'' + \left(\frac{r'(z)}{2 \cdot (r(z) + c)} + \frac{n-1}{nz} \right) y' + \frac{m \cdot s(z)}{2 \cdot (r(z) + c)} y = 0, \qquad ' = \frac{d}{dz}, \quad \text{(ii)}$$

on $\mathbf{P}^1$ is Fuchsian. Then the differential Galois group of the normal form of (ii) must be Ziglin when the Hamiltonian system associated with (i) is (meromorphically) integrable.

While reading this result one should recall Proposition 3.21.

The advantage in dealing with the normal form of (ii) is that the differential Galois group is always a subgroup of $SL(2, \mathbf{C})$ (see, e.g., page 41 of [20]). This will be crucial in our examples.

Proposition 6.15 *A subgroup of $SL(2, \mathbf{C})$ is either:*

(a) *reducible, i.e., conjugate to a subgroup in lower triangular form;*

242

(b) *imprimitive, i.e., conjugate to a subgroup of*

$$\left\{\begin{pmatrix} \lambda & 0 \\ 0 & \lambda^{-1} \end{pmatrix}\right\}_{\lambda \in \mathbf{C}\setminus\{0\}} \bigcup \left\{\begin{pmatrix} 0 & \eta \\ -\eta^{-1} & 0 \end{pmatrix}\right\}_{\eta \in \mathbf{C}\setminus\{0\}};$$

(c) *projectively tetrahedral, octahedral or icosahedral, i.e., projectively iso-morphic to A_4, S_4 or A_5; or*

(d) *has Zariski closure $SL(2, \mathbf{C})$.*

Moreover, the group is Ziglin if and only if it is reducible with all elements resonant, or is imprimitive, or is one of the three groups in (c). If it contains a non-resonant element it is Ziglin if and only if it is imprimitive.

For a proof of the classification in (w.l.o.g.) the case of an algebraic sub-group of $SL(2, \mathbf{C})$, see page 7 of [21]; for the Ziglin assertions, see page 25 of [8].

In the notation of (6.8) define

$$\begin{cases} \text{(a) } A_{m+1} = A_\infty := 2 - \sum_{j=1}^m A_j, \\ \text{(b) } B_{m+1} = B_\infty := \sum_{j=1}^m (B_j + C_j a_j). \end{cases} \qquad (6.16)$$

Moreover, let $a_{m+1} := \infty$ and include this point within the singularities of (6.8) and (6.9).

Proposition 6.17 *The trace of the monodromy generator of the normal form (6.9) at the singularity a_j is given by*

$$t_j = -2\cos\left(\pi\sqrt{(A_j - 1)^2 - 4B_j}\right), \quad j = 1, \dots, m+1.$$

In particular, when $(A_j - 1)^2 - 4B_j$ is a strictly negative real number, we have

$$t_j = -2\cosh\left(\pi\sqrt{4B_j - (A_j - 1)^2}\right),$$

in which case $|t_j| > 2$.

The result is standard. (For references see §2 of [12].)

When the Fuchsian equation (6.8) has precisely three singular points (in-cluding ∞, i.e., $m = 2$), the nature of the differential Galois group of the normal form (6.9) (in the sense of Proposition 6.15) can be determined com-pletely from the traces t_j given in Proposition 6.17. Specifically, the main theorem of [12] has the following consequence, wherein

$$\sigma := t_1^2 + t_2^2 + t_3^2 - t_1 t_2 t_3. \qquad (6.18)$$

Theorem 6.19 *Let* $G \subset SL(2, \mathbf{C})$ *denote the differential Galois group of* (6.9) *in the case* $m = 2$. *Then:*

(a) G *is reducible if and only if* $\sigma = 4$;

(b) *if* G *is not reducible then* G *is imprimitive if and only if at least two of* t_1, t_2, t_3 *are zero, and is (isomorphic to) the eight element quaternion group if all three traces vanish;*

(c)   (i) G *is projectively tetrahedral if and only if* $\sigma = 2$ *and* $t_j \in \{0, \pm 1\}$ *for* $j = 1, 2, 3$,

     (ii) G *is projectively octahedral if and only if* $\sigma = 3$ *and one has* $t_j \in \{0, \pm 1, \pm \sqrt{2}\}$ *for* $j = 1, 2, 3$, *and*

     (iii) G *is projectively icosahedral if and only if* $\sigma \in \{2 - \mu_2, 3, 2 + \mu_1\}$ *and* $t_j \in \{0, \pm \mu_2, \pm 1, \pm \mu_1\}$ *for* $j = 1, 2, 3$, *where* $\mu_1 := \frac{1}{2}(1 + \sqrt{5})$ *and* $\mu_2 = \mu_1^{-1} = -\frac{1}{2}(1 - \sqrt{5})$; *and*

(d) *when* (a)–(c) *fail,* $G = SL(2, \mathbf{C})$.

Corollary 6.20 *When at least two of* t_1, t_2, t_3 *are non-zero and*

$$\sigma \notin \{2 - \mu_2 \simeq 1.381966, 2, 3, 2 + \mu_1 \simeq 3.618034, 4\}$$

we have $G = SL(2, \mathbf{C})$. *In particular, this is the case when* σ *is real and greater than* 4.

Note that $\sigma \notin \{2 - \mu_2, 2, 3, 2 + \mu_1, 4\}$ is equivalent to σ not being a zero of the polynomial $(z - 2)(z - 3)(z - 4)(z^2 - 5z + 5)$.

Corollary 6.21 *Suppose equation* (ii) *of Theorem* 6.14 *is Fuchsian on* $\mathbf{P}^1$ *with precisely three singular points and normal form* (6.9) *(with* $m = 2$*), and let* t_1, t_2, t_3 *and* σ *be defined as in* Proposition 6.17 *and* (6.18) *respectively. If the Hamiltonian system associated with* (i) *of Theorem* 6.14 *is completely integrable (in terms of meromorphic functions), then one of the following conditions must hold:*

(a) $\sigma = 4$;

(b) *at least two of* t_1, t_2 *and* t_3 *are zero;*

(c)   (i) $\sigma = 2$ *and* $t_j \in \{0, \pm 1\}$ *for* $j = 1, 2, 3$,

     (ii) $\sigma = 3$ *and* $t_j \in \{0, \pm 1, \pm \sqrt{2}\}$ *for* $j = 1, 2, 3$, *or*

     (iii) $\sigma \in \{2 - \mu_2, 3, 2 + \mu_1\}$ *and* $t_j \in \{0, \pm \mu_2, \pm 1, \pm \mu_1\}$ *for* $j = 1, 2, 3$, *where* $\mu_1 := \frac{1}{2}(1 + \sqrt{5})$ *and* $\mu_2 = \mu_1^{-1} = -\frac{1}{2}(1 - \sqrt{5})$.

In particular, the system has no (meromorphic) integral functionally independent of H *when at least two of* t_1, t_2, t_3 *are non-zero and* σ *is not a zero of* the polynomial $(z-2)(z-3)(z-4)(z^2-5z+5)$, e.g., when at least two of t_1, t_2, t_3 are non-zero and σ is real and greater than 4.

Proof. Combine Proposition 3.21, Corollary 5.3, Proposition 6.15 and Theorem 6.19. $\qquad\square$

7 Applications

In our applications the Hamiltonian will always have the form considered in Theorem 6.14, i.e.,

$$H(q_1, q_2, p_1, p_2) = \tfrac{1}{2}p_1^2 + \tfrac{m}{2}p_2^2 - r(q_1^n) + \tfrac{n^2}{2}q_1^{2n-2}s(q_1^n)q_2^2 + \mathcal{O}_3(q_2, p_2), \quad (7.1)$$

where $n \geq 1$ is an integer, $m \in \mathbf{C}$ is a constant, $r(z)$ is a polynomial, and $s(z)$ is a rational function. Recall (from Proposition 6.1) that this assumes the form (4.3) by taking

$$\begin{aligned} h(q_1, p_1) &= \tfrac{1}{2}p_1^2 - r(q_1^n)\,, \\ \ell(q_1, p_1) &\equiv 0\,, \\ m(q_1, q_2) &\equiv m\,, \\ k(q_1, p_1) &= n^2 q_1^{2n-2} s(q_1^n)\,. \end{aligned} \qquad (7.2)$$

In the following examples we drop the $\mathcal{O}_3(q_2, p_2)$ terms from the Hamiltonians to ease the notation; their inclusion would have no effect on calculations or results. Moreover, in Examples (7.3) and (7.9) we remark on the geometry of the relevant images $\Gamma \subset P$, so as to illustrate (4.6), but we note that this information is of no consequence in applications of Theorem 6.14.

Example 7.3 (Γ = Elliptic Curve) Choose distinct complex numbers a_1, a_2, a_3 satisfying $a_1a_2 + a_1a_3 + a_2a_3 = 0$; note that none can be 0, and that $a_i \neq -a_j$ for $i \neq j$. Equivalently, choose complex numbers a_1, a_2 satisfying $a_2 \neq -a_1$ and set $a_3 := -a_1a_2/(a_1 + a_2)$. Then for the choices

$$\begin{aligned} m &= 1, \\ n &= 1, \\ r(z) &= -\tfrac{1}{3}\left((z-a_1)(z-a_2)(z-a_3) + a_1a_2a_3\right), \\ s(z) &= \mu + 2\lambda z, \end{aligned}$$

where $\mu, \lambda \in \mathbf{R}$ are also arbitrary, the Hamiltonian (7.1) becomes

$$H(x, x_2, y, y_2) = \tfrac{1}{2}(p_1^2 + p_2^2) - \tfrac{a_1^2 + a_1a_2 + a_2^2}{3(a_1+a_2)}\,q_1^2 + \tfrac{\mu}{2}q_2^2 + \tfrac{1}{3}q_1^3 + \lambda q_1 q_2^2\,. \quad (7.4)$$

Note that by defining

$$A := -\tfrac{2(a_1^2 + a_1a_2 + a_2^2)}{3(a_1+a_2)}, \quad B := \mu, \qquad (7.5)$$

we can write (7.4) in the Hénon-Heiles form

$$H(x, x_2, y, y_2) = \tfrac{1}{2}y^2 + \tfrac{1}{2}y_2^2 + \tfrac{A}{2}x^2 + \tfrac{B}{2}x_2^2 + \tfrac{1}{3}x^3 + \lambda x x_2^2 \qquad (7.6)$$

(recall Examples 3.6 and 3.26). Nevertheless, Hamiltonian (7.4) is preferable for calculations.

We choose a non-equilibrium solution of energy $c = (1/3)a_1 a_2 a_3$ (with image in P). This gives $r(z) + c = -\tfrac{1}{3}(z - a_1)(z - a_2)(z - a_3)$, and the equation appearing in (4.6) becomes $p^2 = -\tfrac{2}{3}(q - a_1)(q - a_2)(q - a_3)$; by means of analytic continuation we conclude that the solution image Γ is an affine elliptic curve. This is satisfying geometrically, but, as it turns out, not computationally. Indeed, here the second order equation (ii) of Theorem 6.14 is the Fuchsian equation

$$y'' + \left(\frac{1}{2(z-a_1)} + \frac{1}{2(z-a_2)} + \frac{1}{2(z-a_3)} \right) y'$$
$$+ \left(\frac{3(2\lambda a_1 + \mu)}{(a_1 - z)(a_1 - a_2)(a_1 - a_3)} + \frac{3(2\lambda a_2 + \mu)}{(a_2 - a_1)(a_2 - z)(a_2 - a_3)} \right. \qquad (7.7)$$
$$\left. + \frac{3(2\lambda a_3 + \mu)}{(a_3 - a_1)(a_3 - a_2)(a_3 - z)} \right) y = 0 \,,$$

and the presence of four regular singular points precludes a direct application of Corollary 6.21 for investigating non-integrability.

In fact it is possible to determine the Ziglin nature of the differential Galois group of the corresponding normal form

$$y'' + \frac{3}{16} \left(\frac{1}{(z-a_1)^2} + \frac{1}{(z-a_2)^2} + \frac{1}{(z-a_3)^2} \right.$$
$$- \frac{(2/3)(a_2 + a_3 - 2a_1) - 16\lambda a_1 - 8\mu}{(a_1 - z)(a_1 - a_2)(a_1 - a_3)} - \frac{(2/3)(a_1 + a_3 - 2a_2) - 16\lambda a_2 - 8\mu}{(a_2 - a_1)(a_2 - z)(a_2 - a_3)}$$
$$\left. - \frac{(2/3)(a_1 + a_2 - 2a_3) - 16\lambda a_3 - 8\mu}{(a_3 - a_1)(a_3 - a_2)(a_3 - z)} \right) y = 0$$

of (7.7) by applying the Kovacic algorithm (e.g., as in [8, §4]), but we will not do so. (The calculations are tedious.) Instead we settle for a less ambitious result: we indicate how parameter values which might correspond to integrable cases of (7.4), i.e., which are worthy of closer study in this regard, can be detected from the associated traces t_j.

To this end note from (7.7) that in the notation of (6.8) and (6.16) we have, at the singular points a_1, a_2, a_3 and $a_4 := \infty$,

$$A_1 = 1/2,$$
$$A_2 = 1/2,$$
$$A_3 = 1/2,$$
$$A_4 = 1/2,$$

$$B_1 = 0,$$
$$B_2 = 0,$$
$$B_3 = 0,$$
$$B_4 = -3\lambda,$$
$$C_1 = \frac{-3(2\lambda a_1 + \mu)}{2(a_1 - a_2)(a_1 - a_3)},$$
$$C_2 = \frac{-3(2\lambda a_2 + \mu)}{2(a_2 - a_1)(a_2 - a_3)},$$
$$C_3 = \frac{-3(2\lambda a_3 + \mu)}{2(a_3 - a_1)(a_3 - a_2)},$$

and therefore

$$t_1 = t_2 = t_3 = 0, \quad t_4 = -2\cos^2\left(\tfrac{\pi}{2}\sqrt{1 + 48\lambda}\right). \tag{7.8}$$

The equality (and simplicity) of the first three traces suggests (but certainly does not prove!) that there might be something special about those values of λ for which $t_4 = 0$. That condition is easily seen to be the case if and only if $1 + 48\lambda$ is the square of an odd integer, i.e., is of the form $(2n+1)^2$ for some integer n, w.l.o.g. non-negative, which in turn is equivalent to $0 \le \lambda = \frac{1}{12}n(n+1) = \frac{1}{6}(1 + 2 + \cdots + n)$ (where this last quantity is defined to be 0 when $n = 0$). Those λ corresponding to

$$n = 0, 1, 2, 3, 4, 5, 6, 7, 8, 9, 10, 11, \ldots$$

are

$$\lambda = 0, \tfrac{1}{6}, \tfrac{1}{2}, 1, \tfrac{5}{3}, \tfrac{5}{2}, \tfrac{7}{2}, \tfrac{14}{3}, 6, \tfrac{15}{2}, \tfrac{55}{6}, 11, \ldots,$$

and so we might expect these values to correspond to integrable cases of (7.4) and (7.6). Indeed, recall from Example 3.26 that $\lambda = 0, 1/6$ and 1 do have this property.

Such an investigation is suggestive, but hardly satisfactory. We obtain far better results for (7.6) in the next example (with a more judicious choice for c).

Example 7.9 (Γ = Singular Curve) Choose any nonzero complex numbers a_2, θ and let $\mu, \lambda \in \mathbf{C}$ be arbitrary. Then for the choices

$$n = 1,$$
$$r(z) = -\tfrac{\theta}{3}z^2(z - \tfrac{3}{2}a_2),$$
$$s(z) = \mu + 2\lambda z,$$

the Hamiltonian (7.1) becomes

$$H(q_1, q_2, p_1, p_2) = \tfrac{1}{2}(p_1^2 + p_2^2) - \tfrac{\theta a_2}{2}q_1^2 + \tfrac{\mu}{2}q_2^2 + \tfrac{\theta}{3}q_1^3 + \lambda q_1 q_2^2. \tag{7.10}$$

The case $\theta = 1$ with a_2 and μ real once again leads to the Hénon-Heiles Hamiltonian

$$H(q_1, q_2, p_1, p_2) = \tfrac{1}{2}p_1^2 + \tfrac{1}{2}p_2^2 + \tfrac{A}{2}q_1^2 + \tfrac{B}{2}q_2^2 + \tfrac{1}{3}q_1^3 + \lambda q_1 q_2^2 \qquad (7.11)$$

of Example 3.5, where we now have $A = -a_2$ and $B = \mu$.

Here we choose a non-equilibrium solution of energy $c = 0$ (with image contained in P). This gives $r(z) + c = -\tfrac{\theta}{3}z^2(z - \tfrac{3}{2}a_2)$, and it follows from (4.6) and analytic continuation that Γ must be a component of the collection of smooth points of the singular curve defined by $p^2 = -\tfrac{2\theta}{3}q^2(q - \tfrac{3}{2}a_2)$. The second order equation (ii) of Theorem 6.14 is the Fuchsian equation

$$y'' + \left(\tfrac{1}{z} + \tfrac{1}{2(z - \frac{3a_2}{2})} \right) y' + \left(\tfrac{\mu}{a_2\theta z^2} + \tfrac{2(3\lambda a_2 + \mu)}{3a_2^2\theta z} - \tfrac{2(3\lambda a_2 + \mu)}{3a_2^2\theta(z - \frac{3a_2}{2})} \right) y = 0 \,,$$

the normal form is

$$y'' + \tfrac{1}{4} \left(\tfrac{a_2\theta + 4\mu}{a_2\theta z^2} + \tfrac{3}{4(z - \frac{3a_2}{2})} + \tfrac{2(a_2\theta + 12\lambda a_2 + 4\mu)}{a_2^2\theta z} - \tfrac{2(a_2\theta + 12\lambda a_2 + 4\mu)}{a_2^2\theta(z - \frac{3a_2}{2})} \right) y = 0 \,,$$

and the respective traces at the singularities $0, 3a_2/2$ and ∞ are

$$t_1 = -2\cos\left(2\pi\sqrt{\tfrac{-\mu}{a_2\theta}} \right), \quad t_2 = 0, \quad t_3 = -2\cos\left(\tfrac{\pi}{2}\sqrt{1 + \tfrac{48\lambda}{\theta}} \right). \qquad (7.12)$$

Writing a_2 as $-A$ and μ as B, this becomes

$$t_1 = -2\cos\left(2\pi\sqrt{\tfrac{B}{A\theta}} \right), \quad t_2 = 0, \quad t_3 = -2\cos\left(\tfrac{\pi}{2}\sqrt{1 + \tfrac{48\lambda}{\theta}} \right), \qquad (7.13)$$

and the quantity σ of (6.18) is therefore given by

$$\sigma = 4\cos^2\left(2\pi\sqrt{\tfrac{B}{A\theta}} \right) + 4\cos^2\left(\tfrac{\pi}{2}\sqrt{1 + \tfrac{48\lambda}{\theta}} \right). \qquad (7.14)$$

Now make the additional restrictions $A = B$ and $\theta = 1$, in which case t_1 becomes -2 and the quantity σ simplifies to

$$\sigma = 4 + 4\cos^2\left(\tfrac{\pi}{2}\sqrt{1 + 48\lambda} \right).$$

For λ real it is then immediate from Corollary 6.21 and the calculations ending Example 7.3 that when $A = B \neq 0$ the only possible integrable cases of the Hénon-Heiles system associated with (7.11) occur when λ has the form $\tfrac{1}{12}n(n+1)$ with $0 \leq n \in \mathbf{Z}$.

Under the assumptions $A = B = 1$ and $\lambda \in \mathbf{R}$, we can eliminate all $\lambda > 1$ on this list by examining the alternate form of the Hamiltonian described in Example 3.13, i.e.,

$$\hat{H}(x, y) = \tfrac{1}{2}(y_1^2 + y_2^2) + \tfrac{1}{2}(x_1^2 + x_2^2) + \tfrac{\hat{\theta}}{3}x_1^3 + \hat{\lambda}x_1 x_2^2 + \hat{\tau}x_2^3, \qquad (7.15)$$

248

where

$$\lambda > 1/2, \qquad \hat{\theta} := \frac{2\lambda}{\sqrt{3-1/\lambda}}, \qquad \hat{\lambda} := \frac{1-\lambda}{\sqrt{3-1/\lambda}}, \qquad \hat{\tau} := \frac{\lambda+1}{3}\sqrt{\frac{2-1/\lambda}{3-1/\lambda}}.$$

Note this also has the form considered in (7.10). (The $\hat{\tau}$ term is $\mathcal{O}_3(x_2, y_2)$ and can therefore be ignored.) Here one sees from (7.12) that the relevant traces are

$$t_1 = -2\cos\left(2\pi\sqrt{\frac{\sqrt{3-1/\lambda}}{2\lambda}}\right), \qquad t_2 = 0, \qquad t_3 = -2\cos\left(\frac{\pi}{2}\sqrt{\frac{24}{\lambda}-23}\right).$$

We first examine the trace t_3. Since $\frac{24}{\lambda} - 23 < 0 \Leftrightarrow \lambda > \frac{24}{23}$, and since $\frac{5}{3} > \frac{24}{23} > 1$, Proposition 6.17 implies that for any $\lambda > 1$ on the list we have $|t_3| > 2$, hence $\sigma > 4$, and from Corollary 6.21 we deduce that the system associated with (vi) can be integrable only if $t_1 = 0$. But one checks easily that $t_1 \neq 0$ for $\lambda = 5/3$ and all $\lambda \geq 2$, and as a result for all $\lambda > 1$ on the list. We thereby establish the non-integrability of $\hat{H}$ for all $\lambda > 1$, and by Corollary 3.19 the same property holds for the system associated with (7.11). From the earlier work in this example we can now conclude that $0, \frac{1}{6}, \frac{1}{2}$ and 1 are the only possible values for the real parameter λ in (7.11) (assuming $A = B = 1$) which can correspond to completely integrable systems. This result was originally established by H. Ito [18] using arguments, based on Ziglin's work, involving the analysis of elliptic functions. (Also see [29].)

We have already verified that the cases $\lambda = 0, \frac{1}{6}$ and 1 are completely integrable (recall Example 3.26); only $\lambda = \frac{1}{2}$ remains for discussion. That case has long evaded a rigorous analysis. Numerical studies have suggested non-integrability, but a proof of that fact, using methods extending those described here, has only recently been announced by J.J. Morales-Ruiz and co-workers.

We encountered another completely integrable case of (7.11) in Example 3.26, i.e., $A = 16B$ and $\lambda = 1/16$. Note this gives $t_2 = 0, t_3 = -2$ in (7.13), hence $\sigma = 4$. Even without the knowledge of a second integral Corollary 6.21 would suggest that this case be examined for the existence of such a function.

Example 7.16 The two-degree of freedom Hamiltonian system associated with

$$H(q_1, q_2, p_1, p_2) = \tfrac{1}{2}(p_1^2 + p_2^2) + \tfrac{1}{2}(q_1^2 - q_2^2) + \epsilon q_1 q_2^2 + \tfrac{1}{4}q_1^4 \qquad (7.17)$$

is shown in [24] to be non-integrable for all complex ϵ in some punctured open neighborhood of $\epsilon = 0$. This Hamiltonian assumes the form considered in (7.1) by choosing

$$
\begin{aligned}
m &= 1, \\
n &= 1, \\
r(z) &= -\tfrac{1}{2}z^2 - \tfrac{1}{4}z^4, \\
s(z) &= -1 + 2\epsilon z,
\end{aligned}
$$

and with $c = -1/4$ one computes the second-order equation (ii) of Theorem 6.14 to be

$$
y'' + \left(\tfrac{1}{z-i} + \tfrac{1}{z+i} \right) y' + \left(\tfrac{2i\epsilon-1}{2(z-i)^2} - \tfrac{2i\epsilon+1}{2(z+i)^2} - \tfrac{i}{2(z-i)} + \tfrac{i}{2(z+i)} \right) y = 0,
$$

which we note is Fuchsian. The relevant traces at the respective singularities $a_1 = i$, $a_2 = -i$ and $a_3 = \infty$ are

$$
t_1 = -2\cos(\pi\sqrt{2 - 4i\epsilon}), \quad t_2 = -2\cos(\pi\sqrt{2 + 4i\epsilon}), \quad t_3 = 2,
$$

and

$$
\sigma = 4 + 4 \left(\cos(\pi\sqrt{2 + 4i\epsilon}) - \cos(\pi\sqrt{2 - 4i\epsilon}) \right)^2.
$$

This last expression is easily seen to represent a non-constant (complex) analytic function of ϵ, at the very least for ϵ near 0, and since $\sigma(0) = 4$ the result from [24] now follows from Corollary 6.21. Note that when $\epsilon = 0$ the system is integrable, e.g., a functionally independent integral is provided by $f = q_2^2 - p_2^2$.

Example 7.18 $(m = 1, \ n > 1$ arbitrary) The choices $r(z) = -(1/n)z$ and $s(z) = 2\lambda/(n^2 z)$ in (7.1) result in the Hamiltonian

$$
H(q_1, q_2, p_1, p_2) = \tfrac{1}{2}(p_1^2 + p_2^2) + \tfrac{1}{n}q_1^n + \lambda q_1^{n-2} q_2^2,
$$

and we take $c = 1/n$. The resulting NVE is the pull-back of the Fuchsian equation

$$
y'' + \left(\tfrac{n-1}{nz} + \tfrac{1}{2(z-1)} \right) y' + \left(\tfrac{\lambda}{nz} - \tfrac{\lambda}{n(z-1)} \right) y = 0
$$

with corresponding normal form

$$
y'' + \tfrac{1}{4} \left(\tfrac{n^2-1}{n^2 z^2} + \tfrac{3}{4(z-1)^2} + \tfrac{n-1+4\lambda}{nz} - \tfrac{n-1+4\lambda}{n(z-1)} \right) y = 0.
$$

The relevant traces at $a_1 = 0$, $a_2 = 1$ and $a_3 = \infty$ are given by

$$
t_1 = -2\cos(\tfrac{\pi}{n}), \quad t_2 = 0, \quad t_3 = -2\cos\left(\tfrac{\pi}{2}\sqrt{(\tfrac{n-2}{n})^2 + \tfrac{16\lambda}{n}} \right)
$$

respectively, and the quantity σ of (6.18) is therefore

$$\sigma = 4 \left(\cos^2\left(\frac{\pi}{n}\right) + \cos^2\left(\frac{\pi}{2}\sqrt{\left(\frac{n-2}{n}\right)^2 + \frac{16\lambda}{n}}\right) \right).$$

In the remaining examples we investigate specific values of n.

Examples 7.19 (a) ($n = 2$) The Hamiltonian is

$$H(q_1, q_2, p_1, p_2) = \tfrac{1}{2}(p_1^2 + p_2^2) + \tfrac{1}{2}q_1^2 + \lambda q_2^2,$$

and the equations admit the independent integral $q_1^2 + p_1^2$ no matter what value one assumes for λ. The possible existence of such integrals is suggested by the relevant traces $t_1 = t_2 = 0$ and $t_3 = -2\cos(\pi\sqrt{2\lambda})$, which are associated only with the reducible and imprimitive cases.

Integrability for this example would be no surprise to experts: all Hamiltonian systems with quadratic Hamiltonians are completely integrable (e.g., see Example 2, page 60, of [4]).

(b) ($n = 3$) The Hamiltonian is

$$H(q_1, q_2, p_1, p_2) = \tfrac{1}{2}(p_2^2 + p_2^2) + \tfrac{1}{3}q_1^3 + \lambda q_1 q_2^2,$$

the relevant traces become $t_1 = -1$, $t_2 = 0$ and $t_3 = -2\cos\left(\frac{\pi}{6}\sqrt{1 + 48\lambda}\right)$, and $\sigma = 1 + 4\cos^2\left(\frac{\pi}{6}\sqrt{1 + 48\lambda}\right)$. Here one encounters all the cases for σ which occur in Corollary 6.21. Indeed, using Theorem 6.19 one can check that the group is

- reducible for $\lambda = 0, \frac{1}{2}, \ldots$,
- imprimitive but not reducible for $\lambda = \frac{1}{6}, \ldots$,
- projectively tetrahedral for $\lambda = \frac{1}{16}, \frac{5}{16}, \ldots$,
- projectively octahedral for $\lambda = \frac{6}{192}, \frac{77}{192}, \ldots$,
- projectively icosahedral for $\lambda = \frac{11}{1200}, \ldots$, and is
- generically Zariski dense in $SL(2, \mathbf{C})$.

Note that $\sigma > 4$ for all $\lambda < 0$; for these values the group is $SL(2, \mathbf{C})$ (see Corollary 6.20). This is true, in particular, for $\lambda = -1$ (the "Monkey Saddle").

(c) ($n = 4$) The Hamiltonian is

$$H(q_1, q_2, p_1, p_2) = \tfrac{1}{2}(p_1^2 + p_2^2) + \tfrac{1}{4}q_1^4 + \lambda q_1^2 q_2^2,$$

the relevant traces are $t_1 = -\sqrt{2}$, $t_2 = 0$ and $t_3 = -2\cos\left(\frac{\pi}{4}\sqrt{1 + 16\lambda}\right)$, and $\sigma = 2 + 4\cos^2\left(\frac{\pi}{4}\sqrt{1 + 16\lambda}\right)$. The differential Galois group is

- reducible for $\lambda = 0, \frac{1}{2}, \ldots$,
- imprimitive but not reducible for $\lambda = \frac{3}{16}, \ldots$,
- never projectively tetrahedral,
- projectively octahedral for $\lambda = \frac{7}{144}, \frac{55}{144}, \ldots$, and

$$\vdots$$

(There are icosahedral cases, but the expressions are somewhat complicated.)

(d) ($n = 5$) The Hamiltonian is

$$H(q_1, q_2, p_1, p_2) = \tfrac{1}{2}(p_1^2 + p_2^2) + \tfrac{1}{5}q_1^5 + \lambda q_1^3 q_2^2,$$

the relevant traces are

$$t_1 = -\tfrac{1}{2}(1 + \sqrt{5}),$$
$$t_2 = 0,$$
$$t_3 = -2\cos\left(\tfrac{\pi}{10}\sqrt{9 + 80\lambda}\right),$$

and $\sigma = \frac{3}{2} + \frac{\sqrt{5}}{2} + 4\cos^2\left(\frac{\pi}{10}\sqrt{9 + 80\lambda}\right) \simeq 2.618033989 + 4\cos^2\left(\frac{\pi}{10}\sqrt{9 + 80\lambda}\right)$.
The group is reducible for $\lambda = 0$ and for other values with rather complicated expressions; the group is imprimitive for $\lambda = 1/5$; it is never projectively tetrahedral and never projectively octahedral; it is projectively icosahedral for $\lambda = 7/80$; it is generically $SL(2, \mathbf{C})$.

(e) ($n = 6$) One uncovers reducibility at $\lambda = 0, \frac{1}{2}, \ldots$, and an imprimitive group at $\lambda = \frac{5}{24}$. We omit further analysis.

(f) ($n = 7$) One obtains a reducible group when $\lambda = 0, \ldots$, and an imprimitive group when $\lambda = \frac{3}{14}, \ldots$.

(g) ($n = 8$) $\lambda = 0, \frac{1}{2}, \ldots$ are reducible cases; $\lambda = \frac{7}{32}, \ldots$ is imprimitive; etc.

(h) ($n = 9$) $\lambda = 0, \ldots$ is reducible; $\lambda = 2/9, \ldots$ is imprimitive; etc.

(i) ($n = 10$) $\lambda = 0, \ldots$ is reducible; $\lambda = \frac{9}{40}, \ldots$ is imprimitive; etc.

Remark 7.20 Section 7 was designed to exhibit a large number of easily verifiable examples, and with the exception of Example 7.3 this was achieved (if at all) by dealing only with second-order Fuchsian equations on $\mathbf{P}^1$ having three singular points (in essence, hypergeometric equations). In the general second-order case one must compute the relevant group using other methods, e.g., the Kovacic algorithm or the Stokes multiplier techniques of J.-P. Ramis. Both approaches are described and illustrated in [25]. For systems with more than two-degrees of freedom the calculation of the group is generally far more difficult, but there are exceptions (see, e.g., §5.1.2 of [25]). Fortunately, computer algebraists are currently hard at work devising algorithms for computing the differential Galois groups of linear equations of order greater than two (as one can see from other contributions to this volume).

Remark 7.21 Work by this author and co-workers emphasizes Ziglin groups, as in this paper; work by J.J. Morales-Ruiz and his co-authors treats integrability in terms of the "solvability" of the NVE, which (by Liouville's Theorem) is essentially equivalent. (In fact Morales and co-workers replace "solvable" with the the more suggestive term "integrable".)

The two viewpoints are illustrated with the Hénon-Heiles Hamiltonian

$$H(q_1, q_2, p_1, p_2) = \tfrac{1}{2}p_1^2 + \tfrac{1}{2}p_2^2 + \tfrac{1}{2}q_1^2 + \tfrac{1}{2}q_2^2 + \tfrac{1}{3}q_1^3 + \lambda q_1 q_2^2. \tag{7.22}$$

In our treatment of this problem (see Examples 7.3 and 7.9) we bypassed the NVE (4.7b) in favor of the algebraic form of that equation (i.e., equation (ii) of Theorem 6.14) and isolated λ-values corresponding to Ziglin differential Galois groups. In contrast, Morales works directly with the NVE, which for most energy values is a Lamé equation involving λ as a parameter. He shows this equation to be (Liouville) solvable only for $\lambda = 0, 1/6, 1/2$ and 1 (see page 124 of [25]), and thereby provides one more independent derivation of the non-integrability results of Ito, relating to (7.22), described near the end of Example 7.9.

Readers need to be aware that some caution must be exercised when switching between the Ziglin group and solvability viewpoints: the reducible group $\left\{ \begin{pmatrix} \lambda & 0 \\ \rho & \lambda^{-1} \end{pmatrix} : \lambda \in \mathbf{C} \backslash \{0\}, \rho \in \mathbf{C} \right\} \subset SL(2, \mathbf{C})$ is solvable but not Ziglin.

Remark 7.23 The book [25] contains many references to applications of the methods outlined in this paper, but, as one would expect, that list is quickly becoming outdated. For more recent work see the review [26]. To round out our discussion of the complete integrability of the two-body problem in §3 (e.g., recall Example 3.27) I would like to point out recent work of Boucher, Julliard Tosel and Tsygvintsev [9], [10], [19], [31] and [32], who use these methods to study the non-integrability of the three-body problem.

Those wishing to pursue the ideas discussed in this paper should consult [25], [6], and [26].

References

1. Arnol'd, V. I. *Mathematical Methods of Classical Mechanics*, Springer-Verlag, New York, 1978.
2. Arnol'd, V. I. (ed.) *Dynamical Systems* III, Encyclopaedia of Mathematical Sciences, Vol. **3**, Springer-Verlag, Berlin, 1988.
3. Arnol'd, V. I., Avez, A. *Ergodic Problems of Classical Mechanics*, W.A. Benjamin, New York, 1968.
4. Arnol'd, V. I., Novikov, S. P. (eds.) *Dynamical Systems* IV, Encyclopaedia of Mathematical Sciences, Vol. **4**, Springer-Verlag, Berlin, 1990.
5. Audin, M. *Spinning Tops, A Course on Integrable Systems*, Cambridge University Press, Cambridge, 1996.
6. Audin, M. *Les systèmes hamiltoniens et leur intégrabilité*, Cours Spécialisés, 8, Société Mathématique de France & EDP Sciences, 2001.
7. Baider, A., Churchill, R. C., Rod, D. L. *Monodromy and non-integrability in complex Hamiltonian systems*, J. Dynamics Differential Equations **2**, (1990), 451–481.
8. Baider, A., Churchill, R. C., Rod, D. L., Singer, M. F. *On the infinitesimal geometry of integrable systems*, Fields Institute Communications **7**, American Mathematical Society, Providence, RI, 1996.
9. Boucher, D. *Sur les équations différentielles linéaires paramétrées, une application aux systèmes hamiltoniens*, thesis, Univ.de Limoges, 2000.
10. Boucher, D. *Sur la non-intégrabilité du problème plan des trois corps de masses égales*, C.R. Acad. Sci. Paris, t. **331**, Série I, (2000), 391–394.
11. Bountis, T, Segur, H., Vivaldi, F. *Integrable Hamiltonian systems and the Painlevé property*, Phys. Rev. A (3) **25** (1982), 1257–1264.
12. Churchill, R. C. *Two generator subgroups of $SL(2, \mathbf{C})$ and the hypergeometric, Riemann and Lamé equations*, J. Symbolic Comput. **28** (1999), 521-546.
13. Churchill, R. C., Falk, G. T. *Lax pairs in the Hénon-Heiles and related families, Hamiltonian Dynamical Systems; History, Theory, and Applications*, H.S. Dumas, K.R. Meyer and D.S. Schmidt, eds., IMA Volumes in Mathematics and its Applications **63**, Springer-Verlag, New York, 1995, 89–98.
14. Cushman, R. H., Bates, L. M. *Global Aspects of Classical Integrable Systems*, Birkhäuser Verlag, Basel, 1997.
15. Goriely, A. *Integrability, partial integrability, and nonintegrability for systems of ordinary differential equations*, J. Math. Phys. **37** (1996), 1871–1893.

16. Hall, L. S. *A theory of exact and approximate configurational invariants,* Phys. D **8** (1983), 90–116.

17. Hénon, M., Heiles, C. *The applicability of the third integral of motion; some numerical experiments,* Astronom. J. **69** (1964), 73–79.

18. Ito, H. *Non-integrability of Hénon-Heiles system and a theorem of Ziglin,* Kodai Math. J. **8** (1985), 120–138.

19. Julliard Tosel, E. *Meromorphic Parametric Non-Integrability; the Inverse Square Potential,* Arch. Rational Mech. Anal. **152** (2000), 187–205.

20. Kaplansky, I. *An Introduction to Differential Algebra,* Second Edition. Actualités Sci. Ind., No. 1251, Publications de l'Inst. de Mathématique de l'Université de Nancago, No. V., Hermann, Paris, 1976.

21. Kovacic, J. *An algorithm for solving second order linear homogeneous differential equations,* J. Symbolic Comput. **2** (1986), 3–43.

22. Lorenz, E. N. *Deterministic Nonperiodic Flow,* J. Atmospheric Sci. **20** (1963), 130–141.

23. Markus, L., Meyer, K. R. *Generic Hamiltonian dynamical systems are neither integrable nor ergodic,* Memoirs of the AMS **144**, Amer. Math. Soc., Providence, RI, 1974.

24. Meletlidou, E. *A non-integrability test for perturbed separable planar Hamiltonians,* Phys. Letters A **270** (2000), 47–54.

25. Morales-Ruiz, J. J. *Differential Galois Theory and Non-Integrability of Hamiltonian Systems,* Birkhäuser Verlag, Basel, 1999.

26. Morales-Ruiz, J. J. *Kovalevskaya, Liapounov, Painlevé, Ziglin and the Differential Galois Theory,* Regul. Chaotic Dyn. **5**(3) (2000), 251–272.

27. Morales-Ruiz, J.J., Ramis, J.-P. *Galoisian obstructions to integrability of Hamiltonian systems,* Methods Appl. Anal., 8, issue 1, 2001.

28. Poincaré, H. *Les Méthodes Nouvelles de la Méchanique Céleste,* Vol. I., Gauthiers-Villars, Paris, 1892.

29. Rod, D. L. *On a theorem of Ziglin in Hamiltonian dynamics,* Proceedings of a Conference on Hamiltonian Systems (K. Meyer and D. Saari, eds.), Contemp. Math. **81**, American Mathematical Society, Providence, RI, 1988, 259–270.

30. Tabor, M. *Chaos and Integrability in Nonlinear Dynamics,* John Wiley & Sons, New York, 1989.

31. Tsygvintsev, A. *La non-intégrabilité méromorphe du problème plan des trois corps,* C.R. Acad. Sci. Paris, t. **331**, Série I (2000), 241–244.

32. Tsygvintsev, A. *Sur l'absence d'une intégrale première méromorphe supplémentaire dans le problème plan des trois corps,* C.R. Acad. Sci. Paris, t. **333**, Série I (2001), 125–128.

33. Yoshida, H. *Ziglin analysis for proving non-integrability of Hamiltonian systems, Proceedings of the Workshop on Finite Dimensional Integrable Non-linear Dynamical Systems* (P.G.L. Leach and W.H. Steeb, eds.), World Scientific, Singapore, 1988, 74–93.
34. Ziglin, S. L. *Branching of solutions and non-existence of first integrals in Hamiltonian mechanics I,* Funct. Anal. Appl. **16** (1982), 181–189.
35. Ziglin, S. L. *Branching of solutions and non-existence of first integrals in Hamiltonian mechanics II,* Funct. Anal. Appl. **17** (1983), 6–17.

Differential Algebra and Related Topics, pp. 257–279
Proceedings of the International Workshop
Eds. L. Guo, P. J. Cassidy, W. F. Keigher & W. Y. Sit
© 2002 World Scientific Publishing Company

MOVING FRAMES AND DIFFERENTIAL ALGEBRA

ELIZABETH L. MANSFIELD

Institute of Mathematics and Statistics,
University of Kent,
Canterbury CT2 7NF, UK
E-mail: E.L.Mansfield@ukc.ac.uk

The purpose of this article is to explore the features of differential algebras comprised of differential invariants together with invariant differential operators. A tutorial on moving frames, used to construct complete sets of invariants and invariant differential operators, is followed by a discussion of how these sets can differ in structure from standard differential rings. We conclude with a brief description of a conjectured characteristic set algorithm for invariant over-determined differential systems.

Introduction

Sophus Lie introduced and applied continuous group actions to differential systems; this was his life's work and is now the subject of modern texts [19,24]. Tresse used the invariants of such actions to study the equivalence problem for ordinary differential equations [27], a tour de force which has assumed new importance in the search for on-line symbolic integrators of nonlinear ordinary differential equations [1,4,5,6,25]. Cartan introduced the notion of a moving frame [3], to solve equivalence problems in differential geometry, relativity, and so on. Moving frames were further developed and applied in a substantial body of work, in particular to differential geometry and (exterior) differential systems, see for example papers by Bryant [2], Green [10] and Griffiths [11]. From the point of view of symbolic computation, a breakthrough in the understanding of Cartan's methods came in a series of papers by Fels and Olver [8,9], Olver [21,22,23] and Kogan and Olver [15], which provide a coherent, rigorous and constructive moving frame method free from any particular application, and hence applicable to a huge range of examples, from classical invariant theory to numerical schemes.

The purpose of this article is to explore the features and consequences of moving frames for (non-exterior) differential systems, which are of interest to symbolic analysts and differential algebraists. There is, as yet, no formulation of the differential algebraic structures enjoyed by collections of invariants and the invariant differential operators which act on them, which gives insight into what kinds of theorems might be true. The usual assumption in differential algebra, that the derivative operators commute, does not hold. The huge range

258

of applicability of moving frames, and the need to have a theoretical underpinning of their implementation in symbolic computer algebra environments, make the algebraic understanding of moving frames an important task.

The first section is essentially a tutorial on the moving frame method, and shows how the method yields complete sets of invariants and invariant differential operators. Some applications are described. This section is, of necessity, in the language of undergraduate calculus. We then discuss the differences between a typical differential algebra and one formed by invariants. Finally we discuss how two central processes in symbolic differential algebra, namely cross-differentiation and reduction, are affected, and briefly describe a conjectured characteristic set algorithm for invariant differential systems.

We refer the reader to the original articles by Fels and Olver [8,9] for details and proofs of theorems concerning moving frames not given here, and for an exposition of the moving frame method for exterior differential systems and for pseudo-group actions.

1 Moving frames, a tutorial

Throughout this section, we will look in detail at two different examples. We then discuss some applications at the end.

1.1 Group Actions and differential invariants

We begin with a specified smooth left group action

$$G \times M \to M, \qquad (g,z) \mapsto g \cdot z, \qquad (h, g \cdot z) \mapsto (hg) \cdot z$$

on the manifold M by a Lie group G. Our interest here is in the case when M is a jet bundle over $\mathcal{X} \times \mathcal{U}$, where $\mathcal{X}$ is the space of independent variables and $\mathcal{U}$ the dependent variables. The group actions which are important in applications are often defined only locally, or defined only for a neighbourhood of the identity of the group, and indeed most of the calculations we perform in the sequel are valid only locally. In the remainder of this article, the qualification "local" is assumed to hold where necessary.

Definition 1.1 Given a group action on $\mathcal{X} \times \mathcal{U}$, there is an induced action on the jet bundle $J^n(\mathcal{X} \times \mathcal{U})$ called the *prolongation* of the group action, which is calculated using the chain rule of differentiation.

Example 1.2 (Lie group actions for ODE systems) Consider the action of $SL(2)$, the group of 2×2 real matrices of determinant one, on $\mathcal{X} \times \mathcal{U} = \mathbb{R} \times \mathbb{R} = M_1$ as

$$x^* = ax + bu(x),$$
$$u^*(x^*) = cx + du(x) \tag{1.3}$$

with $ad - bc = 1$. With u_x denoting $\mathrm{d}u/\mathrm{d}x$ etc, the prolonged group action is obtained using the chain rule as

$$u_{x^*}^* = (\mathrm{d}u^*/\mathrm{d}x) / (\mathrm{d}x^*/\mathrm{d}x) = (c + du_x)/(a + bu_x),$$
$$u_{x^*x^*}^* = (\mathrm{d}u_{x^*}^*/\mathrm{d}x) / (\mathrm{d}x^*/\mathrm{d}x) = u_{xx}/(a + bu_x)^3, \tag{1.4}$$

and so forth.

Example 1.5 (Lie group actions for PDE systems) This is the simplest case of the actions studied by Gonzalez Lopez *et al.* [12]. Consider the action of $SL(2)$, the set of 2×2 matrices of determinant one, on $\mathcal{X} \times \mathcal{U} = \mathbb{R}^2 \times \mathrm{P}^1\mathbb{R} = M_2$

$$x^* = x,$$
$$t^* = t,$$
$$u^*(x^*, t^*) = \frac{au(x,t) + b}{cu(x,t) + d}$$

with $ad - bc = 1$. The prolonged group action is, in this case, easy to derive. We have $\partial u^*/\partial x^* = u_{x^*}^* = u_x^*$ and so forth, so for example,

$$u_{x^*}^* = \frac{u_x}{(cu + d)^2},$$
$$u_{t^*}^* = \frac{u_t}{(cu + d)^2},$$
$$u_{x^*x^*}^* = \frac{u_{xx}(cu + d) - 2cu_x^2}{(cu + d)^3}.$$

If x and t were not invariant under the group action, we would have

$$u_{x^*}^* = \frac{u_x^* t_t^* - u_t^* t_x^*}{x_x^* t_t^* - x_t^* t_x^*},$$

where we have used the fact that the inverse of the jacobian of a transformation is the jacobian of the inverse transformation. In general, for a system with p independent variables x_i, $i = 1, \ldots, p$, we have

$$\frac{\partial F}{\partial x_j^*} = \frac{D(x_1^*, \ldots, x_{j-1}^*, F, x_{j+1}^*, \ldots, x_p^*)}{D(x_1^*, \ldots, x_p^*)}, \tag{1.6}$$

where $D(\zeta_1, \ldots, \zeta_p)$ is the determinant of the matrix whose (k,i)-th element is $D_k \zeta_i$ and D_k is the (total) derivative with respect to x_k. By the chain rule, we need only to specify D_k to be the (total) derivative with respect to the k-th component of an arbitrary "base" coordinate system in the quotient in (1.6), and the equality of the two sides of (1.6) is proved by taking both the starred and non-starred x coordinates as the base coordinates.

Definition 1.7 An invariant of a group action on a manifold M with local coordinates $\mathbf{z}$ is a function f on M such that $f(\mathbf{z}) = f(g \cdot \mathbf{z})$ for all $g \in G$.

In our examples, local coordinates on the jet bundle are given by $x, u,$ $u_x, u_{xx}, \ldots$ and $x, t, u, u_x, u_t, \ldots$ respectively, and $x^* = g \cdot x$, $u_{x^*}^* = g \cdot u_x$, etc.

Definition 1.8 An invariant of a prolonged group action is called a *differential invariant*.

Example 1.9 For the action given in Example 1.5, the Schwarzian of u with respect to x,

$$\{u, x\} = \frac{u_{xxx}}{u_x} - \frac{3}{2}\frac{u_{xx}^2}{u_x^2}$$

is a differential invariant. Indeed it is straightforward to check that

$$\frac{u_{x^*x^*x^*}^*}{u_{x^*}^*} - \frac{3}{2}\left(\frac{u_{x^*x^*}^*}{u_{x^*}^*}\right)^2 = \frac{u_{xxx}}{u_x} - \frac{3}{2}\left(\frac{u_{xx}}{u_x}\right)^2 .$$

Not all group actions are of interest! We now define some properties of group actions we will be using.

Definiton 1.10 (Isotropy groups) For $S \subset M$, the *isotropy group* of S is defined to be $G_S = \{g \in G \,|\, g \cdot S \subset S\}$ and the *global isotropy group* of S is defined to be $G_S^* = \bigcap_{z \in S} G_z$.

Definition 1.11 (Free and effective actions) A group action on M is said to be:

$$
\begin{array}{rll}
\textit{free} & \text{if} & G_z = \{e\}, \text{ for all } z \in M; \\
\textit{locally free} & \text{if} & G_z \text{ is a discrete subgroup of } G, \text{ for all } z \in M; \\
\textit{effective} & \text{if} & G_M^* = \{e\}; \\
\begin{array}{r}\textit{locally effective} \\ \textit{on subsets}\end{array} & \text{if} & \begin{array}{l}G_U^* \text{ is a discrete subgroup of G,} \\ \text{for every open subset } U \subset M.\end{array}
\end{array}
$$

Both actions of $SL(2)$ in Examples 1.2 and 1.5 are effective on their respective $\mathcal{X} \times \mathcal{U}$ but neither is free. The prolongation of the first action is free on $J^n(M_1)$ if $n \geq 1$ while the second action becomes free on $J^n(M_2)$ if $n \geq 2$. A theorem of Ovsiannikov [24], later corrected by Olver [20], guarantees that the prolongation of actions, which are locally effective on subsets of $\mathcal{X} \times \mathcal{U}$, will be free on an open dense subset of $J^n(\mathcal{X} \times \mathcal{U})$ for sufficiently large n.

1.2 Constructing moving frames

Definition 1.12 (Moving frame) If $G \times M \to M$ is a smooth action of G on M, then a right moving frame on M is a right G-equivariant map $\rho : M \to G$, that is, $\rho(g \cdot z) = \rho(z)g^{-1}$.

Moving frames exist if and only if the action is free and regular. The technical details concerning regularity are given fully in the second of the Fels and Olver papers [9]. Here we describe, and illustrate in Figure 1, the meaning of "free and regular" and the construction of a moving frame.

In practice, frames are defined, and used, only locally, so we restrict our attentions to an open neighbourhood $U \subset M$. If the action is free and regular, then the group orbits all have the dimension of the group and foliate U. Further, there will be a surface $\mathcal{K} \subset U$ which crosses the group orbits transversally, and for which the intersection of a given group orbit with $\mathcal{K}$ is a single point. Finally, a group element taking an element $k \in \mathcal{K}$ to another point in U on the same orbit as k is unique.

For $z \in U$, let k be the intersection of $\mathcal{K}$ with the orbit through z, and suppose $h \in G$ is the unique group element such that $h \cdot z = k$. Define $\rho : U \to G$ by $\rho(z) = h$. Then ρ is a moving frame on U.

Figure 1: construction of a moving frame.

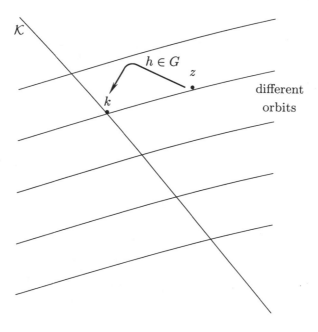

In practice, a right moving frame is obtained by solving a set of equations of the form $\Phi(g \cdot z) = 0$ for g as a function of $z \in M$; one requires Φ to satisfy the conditions of the implicit function theorem so that the solution is unique. The system Φ is known as the set of *normalisation equations*. In terms of the construction of the moving frame given above, the surface $\mathcal{K}$ is the surface defined by $\Phi(z) = 0$, while the frame ρ satisfies $\Phi(\rho(z) \cdot z) = 0$. Since $\Phi(\rho(z) \cdot z) = 0 = \Phi(\rho(g \cdot z) \cdot (g \cdot z)) = \Phi((\rho(g \cdot z) \cdot g) \cdot z)$ and since $h = \rho(z)$ is the unique element satisfying $0 = \Phi(h \cdot z)$, by assumption, it must be that $\rho(g \cdot z) \cdot g = \rho(z)$ or $\rho(g \cdot z) = \rho(z) \cdot g^{-1}$ as required.

Example 1.2 (Part 2) The prolonged group action is free and regular on a dense open subset of $J^n(M_1)$ if $n \geq 1$. If we take the normalisation equations to be

$$\Phi: \qquad x^* = 0, \qquad u^* - 1 = 0, \qquad u_{x^*}^* = 0,$$

we obtain from (1.3) and (1.4),

$$a = u, \qquad b = -x, \qquad c = \frac{u_x}{xu_x - u}, \qquad d = -\frac{1}{xu_x - u}. \qquad (1.13)$$

To show that the map $\rho: J^n \to SL(2)$, $n \geq 1$, given by

$$\rho(x, u, u_x, u_{xx}, \ldots) = \begin{pmatrix} u & -x \\ \dfrac{u_x}{xu_x - u} & -\dfrac{1}{xu_x - u} \end{pmatrix} \qquad (1.14)$$

is a moving frame, observe that

$$x^* u_{x^*}^* - u^* = \frac{xu_x - u}{a + bu_x}$$

so that

$$\rho(g \cdot (x, u, u_x, u_{xx}, \ldots)) = \rho(x^*, u^*, u_{x^*}^*, \ldots)$$

$$= \begin{pmatrix} u^* & -x^* \\ \dfrac{u_{x^*}^*}{x^* u_{x^*}^* - u^*} & -\dfrac{1}{x^* u_{x^*}^* - u^*} \end{pmatrix}$$

$$= \begin{pmatrix} cx + du & -ax - bu \\ \dfrac{c + du_x}{xu_x - u} & -\dfrac{a + bu_x}{xu_x - u} \end{pmatrix}$$

$$= \begin{pmatrix} u & -x \\ \dfrac{u_x}{xu_x - u} & -\dfrac{1}{xu_x - u} \end{pmatrix} \begin{pmatrix} d & -b \\ -c & a \end{pmatrix}$$

$$= \rho(x, u, u_x, \dots) \cdot g^{-1}$$

Example 1.5 (Part 2) We can uniquely solve normalisation equations on $J^n(M_2)$, if $n \geq 2$. If we take the normalisation equations to be

$$\Phi: \qquad u^* = 0, \qquad u^*_{x^*} - 1 = 0, \qquad u^*_{x^* x^*} = 0,$$

then the frame is

$$\rho(x, t, u, u_x, u_t, u_{xx}, u_{xt}, u_{tt}, \dots) = \begin{pmatrix} \dfrac{1}{\sqrt{u_x}} & -\dfrac{u}{\sqrt{u_x}} \\ \dfrac{u_{xx}}{2(u_x)^{3/2}} & \dfrac{2u_x^2 - uu_{xx}}{2(u_x)^{3/2}} \end{pmatrix}.$$

This is equivariant by a calculation similar to that for Example 1.2 above.

The existence of non-trivial denominators in the expressions for the frames shows why the frames are essentially local.

In these relatively simple examples, we can explicitly solve the normalisation equations for the group parameters. In general, the normalisation equations can be highly nonlinear, and so a variety of methods for particular group actions have been developed, for example for groups which factor [14]. In a symbolic computing environment, if the normalisation equations are polynomial in their arguments, then powerful algorithms exist that allow the equations to be used implicitly.

1.3 Construction of differential invariants

Given a moving frame $\rho : M \to G$, the following theorem, coupled with Theorem 1.19, allows a complete set of differential invariants to be readily constructed.

Theorem 1.15 The components of $I(z) = \rho(z) \cdot z$ are invariants.

Proof. $I(g \cdot z) = \rho(g \cdot z)g \cdot z = \rho(z)g^{-1}g \cdot z = \rho(z) \cdot z = I(z)$. $\square$

Thus, once the frame is known, invariants are straightforward to obtain; one simply evaluates the components of $g \cdot z$ on the frame.

Example 1.2 (Part 3) Local coordinates on the jet bundle over M_1 are $(x, u, u_x, u_{xx}, u_{xxx}, \dots)$. The components of $\rho(z) \cdot z$ for a typical point are obtained by evaluating the expression for $u^*_{x^*,\ldots,x^*}$ on the frame (1.13). Hence we have

$$u^*_{x^*x^*}\big|_{\text{frame}} = \frac{u_{xx}}{(u - xu_x)^3}\,,$$

$$u^*_{x^*x^*x^*}\big|_{\text{frame}} = \frac{u_{xxx}}{(u - xu_x)^4} + \frac{3xu^2_{xx}}{(u - xu_x)^5} \qquad (1.16)$$

and so forth. It is straightforward to check that these are indeed invariants.

Example 1.5 (Part 3) Local coordinates on the jet bundle over M_2 are $(x, t, u, u_x, u_t, u_{xx}, u_{xt}, u_{tt}, u_{xxx}, \ldots)$. We already have that x and t are invariant and we have normalised $g \cdot u$, $g \cdot u_x$ and $g \cdot u_{xx}$. The invariants we obtain by evaluating the remaining $g \cdot u_{x\ldots xt\ldots t}$ on the frame are,

$$u^*_{t^*}\big|_{\text{frame}} = \frac{u_t}{u_x}\,,$$

$$u^*_{x^*t^*}\big|_{\text{frame}} = \frac{u_{xt}u_x - u_t u_{xx}}{u^2_x}\,,$$

$$u^*_{t^*t^*}\big|_{\text{frame}} = \frac{u_{tt}u^2_x - u^2_t u_{xx}}{u^3_x}\,, \qquad (1.17)$$

$$u^*_{x^*x^*x^*}\big|_{\text{frame}} = \frac{u_{xxx}}{u_x} - \frac{3}{2}\left(\frac{u_{xx}}{u_x}\right)^2$$

and so forth. It is straightforward to check that these are indeed invariants.

Let us now introduce some notation. Suppose we have a set of dependent variables u^α and a set of independent variables x_j, $j = 1, \ldots, p$. Let the general derivative term be denoted as

$$\frac{\partial^{|K|}u^\alpha}{\partial x_1^{K_1}\ldots\partial x_p^{K_p}} = u^\alpha_K.$$

Definition 1.18 We define J_i and I^α_K to be

$$J_i = (g \cdot x_i)\big|_{\text{frame}}\,, \qquad I^\alpha_K = (g \cdot u^\alpha_K)\big|_{\text{frame}}\,.$$

If there is only one dependent variable, we drop the α suffix. Note that while we have used a multi-index of differentiation here which is a p-tuple of integers, the original Fels and Olver papers use a different form of multi-index which is a string of integers between 1 and p. The correspondence is

$$(K_1, \ldots, K_p) \equiv \underbrace{1\ldots1}_{K_1 \text{ terms}} \quad \ldots \quad \underbrace{p\ldots p}_{K_p \text{ terms}}$$

and we shall use either notation interchangeably. The need for notation which can remember the order of differentiation for some quantities is important, as we shall demonstrate below that the invariant differential operators may not commute.

By Theorem 1.15, the moving frame method yields a differential invariant

J_i for each independent variable x_i, $i = 1, \ldots, p$,

I^α for each dependent variable u^α, and

I_K^α for each derivative term u_K^α.

A simple replacement rule enables any invariant function to be written in terms of these invariants. Indeed, the following theorem yields the fact that the invariants listed above are a complete set of invariants.

Theorem 1.19 (Fels-Olver-Thomas Replacement Theorem) *Let $F(z)$ be an invariant function, that is, $F(g \cdot z) = F(z)$ for all $g \in G$. If local coordinates of a point in $z \in K$ are $(y_1, \ldots, y_m)$, and $I(z) = \rho(z) \cdot z = (Y_1, \cdots, Y_m)$, then $F(y_1, \ldots, y_m) = F(Y_1, \ldots, Y_m)$.*

In other words, to find the expression of F in terms of the invariants, simply replace each y_i with the corresponding component of I.

Proof. If $F(z) = F(g \cdot z)$ for all $g \in G$, then we have $F(y_1, \ldots, y_m) = F(z) = F(\rho(z) \cdot z) = F(Y_1, \ldots, Y_m)$. □

Finally, the method yields p independent invariant differential operators, $\mathcal{D}_i$, $i = 1, \ldots, p$. These will not commute in general.

Theorem 1.20 *The set*

$$\left\{ \mathcal{D}_i = \frac{\partial}{\partial x_i^*}\Big|_{frame} \Big| i = 1, \ldots, p \right\}$$

is a maximal set of independent invariant differential operators.

It should be noted that when these operators are applied to differential expressions, they act as total derivative operators, that is, the usual chain rule holds.

Example 1.2 (Part 4) We have that

$$\mathcal{D} = \frac{d}{dx^*}\Big|_{frame} = \left(\frac{dx^*}{dx}\right)^{-1}\Big|_{frame}\frac{d}{dx} = \frac{1}{u - xu_x}\frac{d}{dx}$$

is an invariant operator. Indeed, it is simple to check that

$$\frac{1}{u^* - x^* u_{x^*}^*} \frac{\mathrm{d}}{\mathrm{d}x^*} = \frac{1}{u - x u_x} \frac{\mathrm{d}}{\mathrm{d}x}.$$

Example 1.5 (Part 4) Since x and t are invariants, the invariant differential operators are $\mathcal{D}_1 = \frac{\partial}{\partial x}$ and $\mathcal{D}_2 = \frac{\partial}{\partial t}$.

There are two ways in which one can generate arbitrary n-th order differential invariants. One is to evaluate $g \cdot u_K$ where $|K| = n$ on the frame, and another is to start with a lower order invariant and to operate on it with invariant operators.

Example 1.2 (Part 5) Denote $(g \cdot \mathrm{d}^n u / \mathrm{d}x^n)\big|_{\text{frame}}$ by I_n. The normalisation equations give $I_0 = 0$ and $I_1 = 0$; while I_2 and I_3 are calculated above in (1.16). The following identities hold:

$$\begin{aligned}
\mathcal{D}I_2 &= I_3, \\
\mathcal{D}I_3 &= I_4 + 3I_2^2, \\
\mathcal{D}I_4 &= I_5 + 10I_2 I_3.
\end{aligned} \tag{1.21}$$

In general, invariant differentiation requires "correction terms", that is,

$$\mathcal{D}_j I_K^\alpha = I_{Kj}^\alpha + M_{Kj}^\alpha. \tag{1.22}$$

The correction terms M_{Kj}^α are given in terms of the induced prolonged action of the Lie algebra of the group and the normalisation equations in remarkable, explicit recurrence formulae, by Fels and Olver [9]. The formulae for general normalisation equations appear in Mansfield [17], and have been implemented in Maple [18]. (Their derivation and exposition, while not difficult, requires knowledge of the prolonged, induced action of the associated Lie algebra, so we do not reproduce them here.) The existence of the correction terms, M_{Kj}^α, has profound consequences for the study of systems of differential invariants, not only in applications but when calculations with invariants are done in a symbolic manipulation environment. These will be discussed in subsequent sections. Here we simply describe some of the more apparent features that can be discerned from simple examples and the formulae themselves.

An examination of the formulae for the M_{Kj}^α shows that even when $|K|$ is less than the order of the normalisation equations, M_{Kj}^α may contain terms whose order is up to 1 more than the order of the normalisation equations. The M_{Kj}^α will cancel the I_{Kj}^α term if I_K^α is a normalised invariant, that is, if $g \cdot u_K^\alpha = c$ is a normalisation equation. Further, depending on the group action, M_{Kj}^α may not be of polynomial type. In particular, they may contain denominators. If the normalisation equations are of polynomial type and the induced action of the Lie algebra is of rational type, then the correction terms will be rational. Further, the multiplicative set generated by factors of the possible denominators can be determined in advance.

It should be noted that the invariants I_K^α are **not** necessarily the result of differentiating some other quantity. Further, the invariant operators may not commute, so that $M_{jK}^\alpha \neq M_{Kj}^\alpha$. The correction terms are one of the quantities for which the string index rather than a multi-index is necessary. In the next example, we illustrate these remarks.

Example 1.23 If the group is $SO(2) \ltimes \mathbb{R}^2$ acting as

$$\begin{pmatrix} x^* \\ y^* \end{pmatrix} = \begin{pmatrix} \cos\theta & \sin\theta \\ -\sin\theta & \cos\theta \end{pmatrix} \begin{pmatrix} x - a \\ y - b \end{pmatrix}$$

with $u^* = u$ an invariant, then a moving frame

$$\rho(x, y, u, u_x, u_y, \ldots) = (\arctan(-u_x/u_y), (x, y))$$

is obtained from the normalisation equations $x^* = 0$, $y^* = 0$ and $u_{x^*}^* = 0$. Thus

$$J_1 = 0, \qquad J_2 = 0, \qquad I = u, \qquad I_1 = 0,$$

while the invariants corresponding to u_y and u_{xx} are

$$I_2 = \sqrt{u_x^2 + u_y^2},$$

$$I_{11} = \frac{u_y^2 u_{xx} - 2u_x u_y u_{xy} + u_x^2 u_{yy}}{u_x^2 + u_y^2}.$$

The two invariant differential operators obtained by evaluating $\partial/\partial x^*$ and $\partial/\partial y^*$ on the frame are

$$\mathcal{D}_1 = \frac{1}{\sqrt{u_x^2 + u_y^2}} (u_y \partial_x - u_x \partial_y),$$

$$\mathcal{D}_2 = \frac{1}{\sqrt{u_x^2 + u_y^2}} (u_x \partial_x + u_y \partial_y).$$

These operators do not commute. In fact,

$$[\mathcal{D}_1, \mathcal{D}_2] = I_{11}\mathcal{D}_1 + I_{12}\mathcal{D}_2.$$

Despite the fact that $I_1 = 0$ and thus $\mathcal{D}_1 I_1 = 0$, the invariant I_{11} is **not** zero. As an example of the correction terms arising with invariant differentiation, we have

$$\mathcal{D}_1 I_{11} = I_{111} - 2\frac{I_{11}I_{12}}{I_2},$$

which is not of polynomial type.

1.4 Applications

One major application of moving frames is in deciding if two differential equations are equivalent to each other under the action of a specified Lie group. This is the basis of a new breed of on-line integrators for nonlinear differential equations, although these typically use pseudo-groups (group actions involving arbitrary functions) rather than the finite dimensional Lie groups discussed here. Another classical application is the *reduction* of an ordinary differential equation (ODE) which is invariant under a Lie group action; this provides a "divide and conquer" mechanism for the integration of such ODEs.

Example 1.2 (Part 6) Here we show how the solution to a specific ODE invariant under the action of Example 1.2 can be reduced to several smaller integration problems by the use of the invariants and the formulae for their invariant differentiation provided by the frame. Suppose for the sake of argument that the ODE we are studying is of fifth order and is given in terms of the invariants $I_n = \mathrm{d}^n u^* / \mathrm{d}x^{*n}|_{\mathrm{frame}}$ by

$$I_5 - I_4^2 = 0; \tag{1.24}$$

an expression for an invariant ODE in terms of the I_n can always be obtained by the straightforward application of the Fels-Olver-Thomas replacement rule. (In terms of the derivatives of u, the numerator of this equation has 113 summands and so we do not reproduce it here.)

Expressing I_5 as a differential expression of I_2 and its invariant derivatives is not actually that helpful. Instead, let

$$I_3 = f(\zeta), \qquad \zeta = I_2, \tag{1.25}$$

where f is a function to be determined. Then, from the formulae (1.21) we have $I_4 + 3\zeta^2 = \mathcal{D}I_3 = f'(\zeta)\mathcal{D}(I_2) = f'f$ so that

$$I_4 = f'f - 3\zeta^2.$$

Taking $\mathcal{D}$ of both sides of this equation, using the formula for $\mathcal{D}I_4$ in (1.21), and inserting everything into (1.24), we obtain

$$f'' f^2 + (f')^2 f - 16\zeta f - (f'f - 3\zeta^2)^2 = 0, \tag{1.26}$$

which is a second order ODE for f as a function of $\zeta = I_2$ and is called the "reduction" of the equation (1.24). Thus, the fifth order ODE is reduced to the solution of a third order ODE (1.25) and a second order ODE (1.26) which is decoupled from (1.24). This system is in principle simpler than the original fifth order ODE. One can further subdivide (1.25) by considering invariants of a subgroup of $SL(2)$. For particular group actions, this and related methods can lead to the complete solution of the original ODE by quadratures (for example if the group is solvable [19,24]) or in terms of well-known lower order equations [7].

We will consider an application of the frame for Example 1.5 in the next section.

2 Comparison of $\{u_K^\alpha \,||K| \geq 0\}$ with $\{I_K^\alpha \,||K| \geq 0\}$

The "differential structure" of the invariants considered together with the invariant differential operators has several unusual features. In this section we explore some of these.

Consider a typical situation in which we consider a dependent variable u^α with two arguments, x and y. Every derivative term u_K^α is generated by applying the commuting derivative operators $\partial/\partial x$ and $\partial/\partial y$ the requisite number of times. All the terms u_K^α are independent coordinates of the jet bundle; there are no a priori relations between them. This situation is represented in Figure 2.

Figure 2: the "differential structure" of $\{u_K^\alpha \,||K| \geq 0\}$

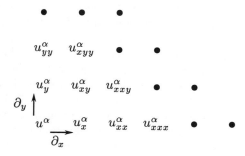

By contrast, the set $\{I_K^\alpha \,||K| \geq 0\}$ may not be generated by I^α.

Example 1.23 (Part 2) In Figure 3, we show the "differential structure" of the set of invariants for $SO(2) \ltimes \mathbb{R}^2$ acting as in Example 1.23. In this example there are **two** generating invariants, namely I and I_{11}. It is not possible to obtain I_{11} from I using invariant differentiation.

If a normalisation equation takes the form $\phi(g \cdot x, g \cdot u, g \cdot u_x, \dots) = 0$, then by construction, $\phi(J, I, I_1, \dots) = 0$. Thus, if the dimension of G is r, we will have r normalisation equations which solve for the r parameters. So, there will be $r = \dim(G)$ relations between the invariants J_i, I^α and I^α_K. In many applications, these take the form of r of the invariants being constants.

Figure 3: the "differential structure" of $\{I^\alpha_K \,|\, |K| \geq 0\}$ for Ex. 1.23

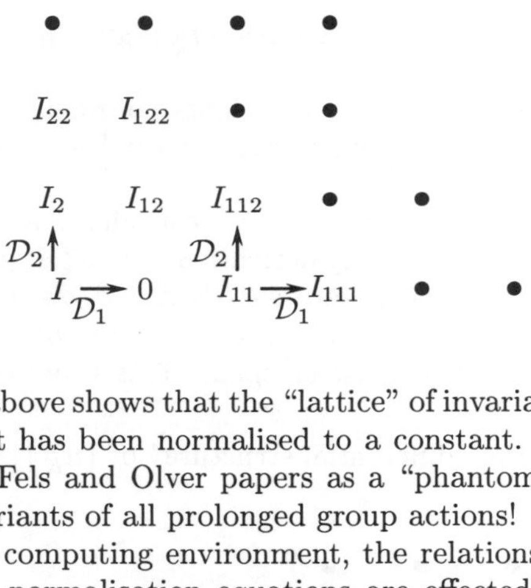

$$
\begin{array}{cccc}
\bullet & \bullet & \bullet & \bullet \\[4pt]
I_{22} & I_{122} & \bullet & \bullet \\[4pt]
I_2 & I_{12} & I_{112} & \bullet \quad \bullet \\[2pt]
\mathcal{D}_2\uparrow & & \mathcal{D}_2\uparrow & \\[2pt]
I \xrightarrow{\;\mathcal{D}_1\;} 0 & & I_{11} \xrightarrow{\;\mathcal{D}_1\;} I_{111} & \bullet \quad \bullet
\end{array}
$$

The example above shows that the "lattice" of invariants can have "holes", where an invariant has been normalised to a constant. Such an "invariant" is denoted in the Fels and Olver papers as a "phantom" invariant. Indeed, constants are invariants of all prolonged group actions!

In a symbolic computing environment, the relations between the invariants given by the normalisation equations are effected by simplifying with respect to the equations. We will assume that the normalisation equations are polynomials in their arguments, and then, without further loss of generality, that the normalisation equations are an algebraic Gröbner basis for the algebraic ideal they generate. It should be borne in mind, however, that not all Gröbner bases are suitable to define a frame. The solution space of the equations $\Phi = 0$ needs to be a unique surface of dimension equal to that of the ambient space less the dimension of the group. Further considerations apply when the equations are used in symbolic calculations; normalisation equations such as $(I_1)^2 = 0$ will lead to undetected zero coefficients of leading terms, and even zero denominators. So, the normalisation equations need to form a prime ideal.

Definition 2.1 For a given term ordering on the set $\{I_K^\alpha \,|\, |K| \geq 0\}$, a normalisation equation $\phi(J_i, I^\alpha, I_K^\alpha) = 0$ (containing terms other than J_j, $j = 1, \ldots, p$) will have a leading invariant term. This term is called a "highest normalised invariant". Denote the set of such highest normalised invariants by $\mathcal{HNI}$.

The problems noted above are eliminated if we assume that the $\mathcal{HNI}$ occur linearly in the normalisation equations. In this case, standard simplification procedures have the effect of eliminating the highest normalised invariants from all results of all calculations. Simply put, the highest normalised invariants never appear, and thus can be considered to be "holes" in the lattice of invariants.

Similarly, we may take an ordering on the terms J_j, $j = 1, \ldots, p$, and obtain, from those normalisation equations not containing any of the I_K^α, a set $\mathcal{HJ}$ of highest normalised invariants deriving from the independent variables.

It should not be assumed that invariants J_i, which correspond to the independent variables, play the role of the independent variables in the standard differential rings. The following example serves as a warning.

Example 2.2 Suppose we consider the rotation group $SO(2)$ acting as

$$x^* = \cos(\theta)x - \sin(\theta)u(x),$$
$$u^*(x^*) = \sin(\theta)x + \cos(\theta)u(x),$$

and suppose we take the normalisation equation

$$u_{x^*}^* = 0.$$

Then the invariants corresponding to x, u and u_{xx} are

$$J = \frac{uu_x + x}{\sqrt{1 + u_x^2}}, \qquad I_0 = \frac{xu_x - u}{\sqrt{1 + u_x^2}}, \qquad I_2 = \frac{u_{xx}}{(1 + u_x^2)^{3/2}}.$$

The invariant operator is

$$\mathcal{D} = \frac{1}{\sqrt{1 + u_x^2}}\frac{\mathrm{d}}{\mathrm{d}x},$$

and we have

$$\mathcal{D}J = I_0 I_2 + 1, \qquad \mathcal{D}I_0 = -J I_2.$$

In other words, the invariant derivative of the invariant corresponding to the independent variable is in terms of the second order invariant $\{I_2\}$! The zero-th order invariant, in terms of these invariants, is given by $x^2 + u^2 = J^2 + I_0^2$, which can be obtained by a simple application of the Fels-Olver-Thomas replacement rule, or verified directly.

Theorem 2.3 *The set*

$$\{J_k, I^\alpha, I^\alpha_{Kj} \,|\, j = 1, \ldots, p, J_k \notin \mathcal{HJ}, I^\alpha \notin \mathcal{HNI}, I^\alpha_K \in \mathcal{HNI}\}$$

generates the set of all invariants, in the sense that all invariants are functions of these and their invariant derivatives.

Nontrivial correction terms in the invariant differentiation formulae yield the existence of nontrivial "differential syzygies".

Example 1.23 (Part 3) From Figure 3 one can see there are at least two ways of obtaining I_{112}, from $\mathcal{D}_2 I_{11}$ and from $\mathcal{D}_1^2 \mathcal{D}_2 I$. The differential syzygy is given by

$$\mathcal{D}_2 I_{11} - \mathcal{D}_1^2 \mathcal{D}_2 I = \frac{I_{11} I_{22} - I_{11}^2 - 2I_{12}^2}{I_2}. \tag{2.4}$$

"Differential syzygies" can be written explicitly in terms of the generating invariants. They are

(a) $\mathcal{D}_j \phi(J_i, I^\alpha, I^\alpha_k, \ldots) = 0$ for all j and for all $\phi \in \Phi$, and

(b) $\mathcal{D}^{Kj} I^\alpha_{Jk} - \mathcal{D}^{Jk} I^\alpha_{Kj} = M^\alpha_{JkKj} - M^\alpha_{KjJk}$ for all multi-indices Kj and Jk such that $I^\alpha_K, I^\alpha_J \in \mathcal{HNI}$.

Example 1.5 (Part 5) A striking example of the application of differential syzygies is provided by Example 1.5. The generating invariants are x, y, I_2 and I_{111} given in (1.17). The differential syzygy is obtained by calculating $\mathcal{D}_1^3 I_2 - \mathcal{D}_2 I_{111}$ using the relevant recurrence formulae. One obtains

$$\mathcal{D}_1^3 I_2 - \mathcal{D}_2 I_{111} = -2I_{111} I_{12} - I_2 I_{1111}.$$

Noting $\mathcal{D}_1 = \partial/\partial x$, $\mathcal{D}_2 = \partial/\partial t$, $I_{12} = \mathcal{D}_1 I_2$ and $\mathcal{D}_1 I_{111} = I_{1111}$, and setting $v = I_{111}$, the syzygy takes the form

$$\frac{\partial v}{\partial t} = \left(\frac{\partial^3}{\partial x^3} + v_x + 2v \frac{\partial}{\partial x} \right) I_2. \tag{2.5}$$

The operator acting on I_2 is the Hamiltonian operator for the Korteweg-de Vries equation [19] (after the variables have been suitably scaled)! Further, if u is a solution of the PDE $I_{111} = I_2$, then $v = I_{111} = \{u, x\}$, the Schwarzian of u with respect to x, satisfies the Korteweg-de Vries equation. The meaning of this remarkable fact is fully explored in Gonzalez Lopez *et al.* [12].

Solution spaces of differential equations which are invariant under a group action have a *structure* which can be exploited in applications; this is well known. This idea of evaluating relations between invariants on the system of interest, to obtain further information, has been noted and used but not as clearly understood or as amenable to routine symbolic calculations. We refer the reader to the excellent article by Svinolupov and Sokolov [26] for further examples (but not in the language of differential syzygies). Another classical construction which seems allied with the concept of differential syzygies is that of the *resolvent system* [24].

In summary, there are several features enjoyed by collections of invariants which differ from those pertaining to standard differential rings.

A. More than one generator may be needed to obtain the full set of the I_K^α under invariant differentiation.

B. There are functional relations between the invariants, given by the normalisation equations.

C. Nontrivial "differential syzygies" may exist.

D. The invariant differential operators may not commute.

E. Invariant derivatives of the I_K^α may not be of polynomial type. If the normalisation equations are polynomial and the induced action of the Lie algebra is rational, then the correction terms are rational. Their denominators consist of products of a finite number of factors which can be determined in advance.

3   Calculations with invariants

One major application of differential algebra is to the proofs of correctness and termination of algorithms used in symbolic computation [13], in particular in the calculation of characteristic sets of over-determined differential systems. An example of an application of this is given in Mansfield, Reid and Clarkson [16]. A lengthy discussion of how such an algorithm might work for invariantised systems is given in Mansfield [17]. Here we briefly look at the main points.

The idea of writing an invariant system in terms of the invariants generated by a moving frame is an attractive one:

(a) The structure of the solution space, as evidenced by the group action, is more easily discerned.

(b) The normalisation equations act as constraints, making the system (more) over-determined and therefore (more) amenable to characteristic set calculations.

(c) The invariantized system may be less complex, especially if large expressions can be written as a single symbolic invariant.

(d) It is *not* necessary to solve for the group parameters, that is, to know the frame in explicit detail, in order to calculate with an invariant system symbolically. Indeed, it is not necessary to know the invariants in terms of the original variables! This is because the correction terms can be calculated from the normalisation equations and the induced action of the Lie algebra alone.

(e) The choice of normalisation equations can be adapted to the application at hand.

Unfortunately, mitigating against these good features are some bad ones, at least for calculations involving cross-differentiation and differential reduction, two calculations which are central to characteristic set algorithms.

Suppose we consider a term ordering, $<$, on the symbolic derivatives which is compatible with differentiation, that is,

$$u_K^\alpha < u_J^\beta \implies u_{KL}^\alpha < u_{JL}^\beta.$$

Then it is not true that the induced ordering on the symbolic invariants, given by

$$u_K^\alpha < u_J^\beta \iff I_K^\alpha < I_J^\beta,$$

will be compatible with invariant differentiation! This means that differential simplification procedures may not terminate if carried out on expressions whose order is less than or equal to the order of the normalisation equations. What we can say is the following.

Lemma 3.1 *If the order of the normalisation equations is N, and if $<$ is a compatible ordering on the derivatives $\{u_k^\alpha \mid \alpha, K\}$, then provided $|K|$ and $|J|$ are greater than N,*

$$I_K^\alpha < I_J^\beta \implies \mathcal{D}^L I_K^\alpha < \mathcal{D}^L I_J^\beta.$$

As remarked in the last section, highest normalised invariants can act as "holes" in the lattice of invariants. Consider the standard calculation of a "diffSpolynomial" or "cross-derivative" of two differential expressions, f_1 and f_2. Suppose the leading derivative term of f_i is u_{K^i}, and that the least common derivative of these is u_L. We take for simplicity the simplest case, where the leading derivative terms occur to degree 1 in the f_i. Then the cross-derivative is obtained by differentiating f_i by D^{L-K^i}, cross-multiplying by the leading coefficients, and subtracting. In other words, two expressions for u_L are obtained, which is then eliminated.

Now suppose our expressions are written in terms of the invariants. If a cross-differentiation of two equations f_1 and f_2 involves a path of differentiation which passes through a highest normalised invariant, so that derivatives of the highest normalised invariant cannot appear (as they are simplified out using the normalisation equations), two expressions for the least common derivative of the highest derivative terms in the f_i will not be obtained. Thus, the cross-derivative of the two equations may not exist.

Figure 4: cross-derivatives may not exist

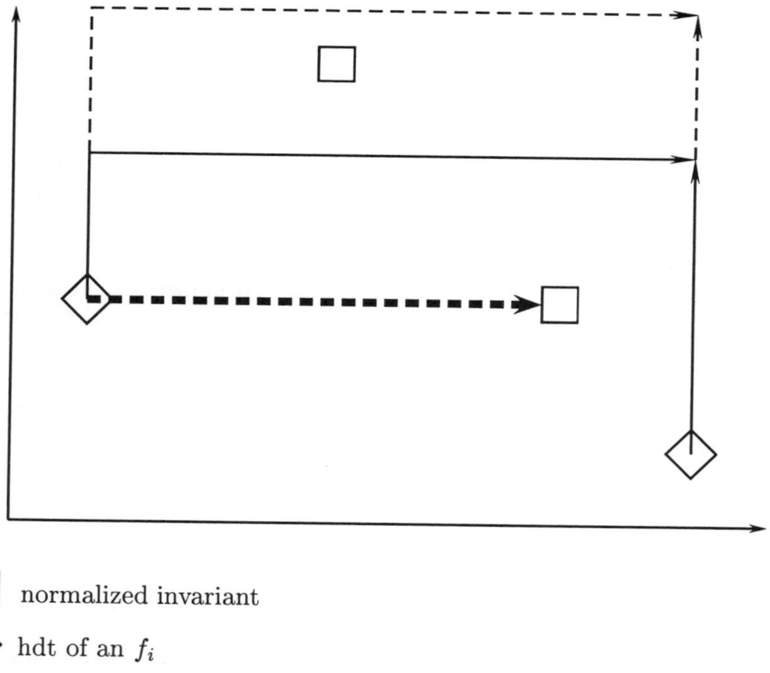

☐ normalized invariant

◇ hdt of an f_i

However, one can calculate cross-differentiations for derivatives of the f_i. Further, different cross-differentiations for different derivatives of the f_i may be fundamentally different, if one cannot be obtained from the other without a path of differentiation passing though a highest normalised invariant. These remarks are illustrated in Figure 4 for equations involving two independent variables. The invariant diffSpolynomial between two expressions whose highest terms (marked by diamonds) cannot be calculated, as the path of differentiation of one (marked by a fat dashed arrow) passes through a normalised invariant (marked by squares). However, two diffSpolynomials of derivatives of these can be calculated (marked by thin unbroken and broken arrows) and these cannot be related to each other, as the paths of differentiation cannot be deformed to each other without passing through a normalised invariant.

For three or more variables, it is possible that more than one cross-derivative may exist, depending on whether the paths by which one differentiates the f_i to obtain the least common derivative term pass one way round a highest normalised invariant or the other.

The problems occur only for equations whose order is less than or equal to $N + 1$, where N is the order of the normalisation equations. Similar remarks apply to differential reduction, for the same reasons.

Now suppose we wish to calculate a "completion algorithm" for a system given in terms of invariants, by which is meant, all possible cross-derivatives and simplifications obtainable from the given set. A typical scenario is depicted in Figure 5, where leading or highest derivative terms (hdt's) are depicted by diamonds at the position in the lattice given by the multi-index which defines them. Highest normalised invariants are depicted by squares, which of necessity lie beneath the dotted line depicting those derivative terms of order $N + 1$, where N is the order of the normalised invariants. In this "inner region" of the lattice, differential cross-derivatives and reductions are problematic.

Thus, one can divide a calculation of a (conjectured) characteristic set into two sections. For equations of order less than $N+1$, since cross-differentiations and differential reductions may not be defined, or well-defined, or may not terminate, we differentiate all equations up to and including order $N + 1$. Then, a Gröbner basis of the result can be calculated, with all the I_K^α regarded as separate indeterminates. In this way, all feasible cross-derivatives of equations in the inner region are calculated and simplified. In the outer region, where the correction terms M_{Kj} are of order strictly less than $|Kj|$, we can calculate the usual cross-derivatives and their reductions.

Now, the calculations in the outer region can lead to new integrability conditions in the inner region, and calculations in the inner region can lead to integrability conditions of order $N + 1$ which therefore need to be included in the calculations for the outer region. So, a looping mechanism is required to ensure all feasible cross-derivatives and their reductions are calculated. Full details of this algorithm, together with a significant application, appear elsewhere [17], and we do not repeat the details here.

One problem with the above completion algorithm is that multiple large Gröbner basis calculations are required mainly because we don't know the relationship between cross derivatives and reductions calculated via different paths of differentiation around a highest normalised invariant. To conclude this section, we put forward the conjecture that an algorithm which calculates cross-derivatives and reductions for one (arbitrary) path of differentiation, but which includes the evaluation of the differential syzygies of the frame on the input system, will yield information equivalent to what the "all path" algorithm described above yields.

Figure 5: Sketch of a conjectured "characteristic set algorithm" for invariantized differential systems

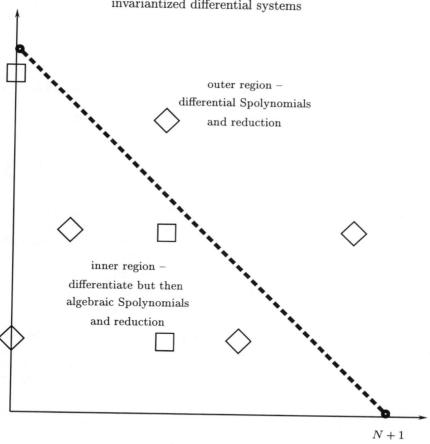

outer region –
differential Spolynomials
and reduction

inner region –
differentiate but then
algebraic Spolynomials
and reduction

$N+1$

normalized invariant

hdt of an f_i

278

Acknowledgments

This article is based on my talk given at the International Workshop on Differential Algebra and Related Topics, Rutgers, Newark, USA, in November, 2000. I wish to thank Phyllis Cassidy, Li Guo, William Keigher and William Sit for the invitation to speak.

I am grateful to Peter Olver who explained the moving frame method to me, answering endless questions. I would like to acknowledge conversations with Evelyne Hubert, Andrew Martin, and Agnes Szanto, which have helped to clarify various issues for me. I also thank the referees for their comments and suggestions.

References

1. Berth, M. *Invariants of ordinary differential equations*, PhD Thesis, Ernst-Mortiz-Arndt Universität, Greifswald, 1999.
2. Bryant, R.L. *Submanifolds and special structures on the octonians*, J. Diff. Geom. **17** (1982), 185–232.
3. Cartan, E. *Oeuvres complètes*, Gauthier-Villiars, 1952-55.
4. Cheb-Terrab, E.S., Duarte, L.G.S., daMota, L.A.C.P. *Computer algebra solving of first order ODEs using symmetry methods*, Comput. Phys. Comm. **101** (1997), 254–268.
5. Cheb-Terrab, E.S., Roche, A.D. *Abel ODEs: Equivalence and Integrable Classes*, Comput. Phys. Comm. **130** (2000), 204–231.
6. Cheb-Terrab, E.S., Roche, A.D. *Integrating Factors for Second Order ODEs*, J. Symbolic Comput. **27** (1999), 501–519.
7. Clarkson P.A., Olver, P.J. *Symmetry and the Chazy Equation*, J. Differential Equations **124** (1996), 225–246.
8. Fels M., Olver, P.J. *Moving Coframes I*, Acta Appl. Math. **51** (1998), 161–213.
9. Fels, M., Olver, P.J. *Moving Coframes II*, Acta Appl. Math. **55** (1999), 127–208.
10. Green, M.L. *The moving frame, differential invariants and rigidity theorems for curves in homogeneous spaces*, Duke Math. J. **45** (1978), 735–779.
11. Griffiths, P. *On Cartan's methods of Lie groups and moving frames as applied to uniqueness and existence questions in differential geometry*, Duke Math. J. **41** (1974), 775–814.
12. Gonzalez Lopez, A., Heredero, R.H., Mari Beffa, G. *Invariant differential equations and the Adler-Gel'fand-Dikii bracket*, J. Math. Phys. **38** (1997), 5720–5738.

13. Hubert, E. *Factorization-free decomposition algorithms in differential algebra*, J. Symbolic Comput., **29** (2000), 641–662.
14. Kogan, I. *Inductive approach to Cartan's moving frame method with applications to classical invariant theory*, Ph.D. thesis, University of Minnesota, 2000.
15. Kogan, I.,Olver, P.J. *Invariant Euler-Lagrange equations and the invariant variational bicomplex*, preprint, University of Minnesota, 2001.
16. Mansfield, E.L., Reid, G.J., Clarkson, P.A. *Nonclassical reductions of a 3+1 cubic nonlinear Schrödinger system*, Comp. Phys. Comm. **115** (1998), 460–488.
17. Mansfield, E.L. *Algorithms for Symmetric Differential Systems*, preprint, University of Kent, **UKC/IMS/99/34**, 1999.
18. Mansfield, E.L. `Indiff`: *a* MAPLE *package for over-determined differential systems with Lie symmetry*, preprint, University of Kent, **UKC/IMS/99/37**, 1999.
19. Olver, P.J. *Applications of Lie groups to differential equations*, Second edition, Springer-Verlag, New York, 1992.
20. Olver, P.J. *Moving frames and singularities of prolonged group actions*, Selecta Math. **6** (2000), 41-77.
21. Olver, P.J. *Joint invariant signatures*, Found. Comput. Math. **1** (2001), 3-67.
22. Olver, P.J. *Moving frames – in geometry, algebra, computer vision, and numerical analysis*, preprint, University of Minnesota, 2000.
23. Olver, P.J. *Classical Invariant Theory*, LMS Student Texts **44**, Cambridge University Press, Cambridge, 1999.
24. Ovsiannikov, L.V. *Group Analysis of Differential Equations*, (trans. Ames, W.F.), Academic Press, New York, 1982.
25. Schwarz, F. *Solving third order differential equations with maximal symmetry group*, Computing **65** (2000), 155–167.
26. Svinolupov, S.I., Sokolov, V.V. *Factorization of evolution equations*, Russian Math Surveys **47** (1992), 127–162.
27. Tresse, A.M. *Détermination des Invariants Ponctuels de l'Équation Différentielle Ordinaire du Second Ordre* $y'' = \omega(x, y, y')$, Gekrönte Preissschrift, S. Hirzel, Leipzig, 1896.

Differential Algebra and Related Topics, pp. 281–305
Proceedings of the International Workshop
Eds. L. Guo, P. J. Cassidy, W. F. Keigher & W. Y. Sit
© 2002 World Scientific Publishing Company

BAXTER ALGEBRAS AND DIFFERENTIAL ALGEBRAS

LI GUO

Department of Mathematics and Computer Science,
Rutgers University at Newark,
Newark, NJ 07102, USA
E-mail: liguo@newark.rutgers.edu

In memory of Professor Chuan-Yan Hsiong

A Baxter algebra is a commutative algebra A that carries a generalized integral operator. In the first part of this paper we review past work of Baxter, Miller, Rota and Cartier in this area and explain more recent work on explicit constructions of free Baxter algebras that extended the constructions of Rota and Cartier. In the second part of the paper we will use these explicit constructions to relate Baxter algebras to Hopf algebras and give applications of Baxter algebras to the umbral calculus in combinatorics.

0 Introduction

This is a survey article on Baxter algebras, with emphasis on free Baxter algebras and their applications in probability theory, Hopf algebra and umbral calculus. This article can be read in conjunction with the excellent introductory article of Rota [28]. See also [31,29].

A Baxter algebra is a commutative algebra R with a linear operator P that satisfies the Baxter identity

$$P(x)P(y) = P(xP(y)) + P(yP(x)) + \lambda P(xy), \forall x, y \in R, \qquad (0.1)$$

where λ is a pre-assigned constant, called the weight, from the base ring of R.

0.1 Relation with differential algebra

The theory of Baxter algebras is related to differential algebra just as the integral analysis is related to the differential analysis.

Differential algebra originated from differential equations, while the study of Baxter algebras originated from the algebraic study by Baxter[5] on integral equations arising from fluctuation theory in probability theory. Differential algebra has provided the motivation for some of the recent studies on Baxter algebras. A Baxter algebra of weight zero is an integration algebra, i.e., an algebra with an operator that satisfies the integration by parts formula (see Example 1 below). The motivation of the recent work in [13,14,15] is to extend the beautiful theory of differential algebra [19] to integration algebras. On the other hand, the Baxter operators, regarded as a twisted family of integration operators, motivated the study of a twisted family of differential operators, generalizing the differential operator (when the weight is 0) and the difference operator (when the weight is 1).

0.2 Some history

In the 1950's and early 1960's, several spectacular results were obtained in the fluctuation theory of sums of independent random variables by Anderson [1], Baxter [4], Foata [9] and Spitzer [32]. The most important result is Spitzer's identity (see Proposition 2.16) which was applied to show that certain functionals of sums of independent random variables, such as the maximum and the number of positive partial sum, were independent of the particular distribution. In an important paper [5], Baxter deduced Spitzer's identity and several other identities from identity (0.1). This identity was further studied by Wendel [34], Kingman [18] and Atkinson [3].

Rota [27] realized the algebraic and combinatorial significance of this identity and started a systemic study of the algebraic structure of Baxter algebras. Free Baxter algebras were constructed by him [27,31] and Cartier [6]. Baxter algebras were also applied to the study of Schur functions [33,22,35], hypergeometric functions and symmetric functions, and are closely related to several areas in algebra and geometry, such as quantum groups and iterated integrals, as well as differential algebra. The two articles by Rota [28,29] include surveys in this area and further references. Rota's articles helped to revive the study of Baxter algebras in recent years: Baxter sequences in [35], free Baxter algebras in [14,15,10,11] and applications in [2,12]

Despite the close analogy between differential algebras and Baxter algebras, in particular integration algebras, relatively little is known about Baxter algebras in comparison with differential algebras. It is our hope that this article will further promote the study of Baxter algebras and related algebraic structures.

0.3 Outline

After introducing notations and examples in Section 1, we will focus on the construction of free Baxter algebras in Section 2. We will give three constructions of free Baxter algebras. We first explain Cartier's construction using brackets, followed by a similar construction using a generalization of shuffle products. These two constructions are "external" in the sense that each is a free Baxter algebra obtained without reference to any other Baxter algebra.

We then explain Rota's standard Baxter algebra which chronologically came first. Rota's construction is an "internal" construction, obtained as a Baxter subalgebra inside a naturally defined Baxter algebra whose construction traces back to Baxter [5].

In Section 3, we give two applications of Baxter algebras. We use free Baxter algebras to construct a new class of Hopf algebras, generalizing the classical divided power Hopf algebra. We then use Baxter algebras to give an interpretation and generalization of the umbral calculus. Other applications of free Baxter algebras can be found in Section 1, relating Baxter operators to integration and summation, and in Section 2, proving the famous formula of Spitzer.

We are not able to include some other work on Baxter algebras, for example on Baxter sequences and the Young tableau [33,35], and on zero divisors and chain conditions in free Baxter algebras [10,11]. We refer the interested readers to the original literature.

1 Definitions, examples and basic properties

1.1 Definitions and examples

We will only consider rings and algebras with identity in this paper. If R is the ring or algebra, the identity will be denoted by $\mathbf{1}_R$, or by 1 if there is no danger of confusion.

Definition 1.1 Let C be a commutative ring. Fix a λ in C. A *Baxter C-algebra* (*of weight λ*) is a commutative C-algebra R together with a *Baxter operator* (*of weight λ*) on R, that is, a C-linear operator $P : R \to R$ such that

$$P(x)P(y) = P(xP(y)) + P(yP(x)) + \lambda P(xy),$$

for any $x, y \in R$.

Let $\mathbf{Bax}_C = \mathbf{Bax}_{C,\lambda}$ denote the category of Baxter C-algebras of weight λ in which the morphisms are algebra homomorphisms that commute with the Baxter operators.

There are many examples of Baxter algebras.

Example 1.2 (Integration) Let R be $\mathrm{Cont}(\mathbb{R})$, the ring of continuous functions on $\mathbb{R}$. For f in $\mathrm{Cont}(\mathbb{R})$, define $P(f) \in \mathrm{Cont}(\mathbb{R})$ by

$$P(f)(x) = \int_0^x f(t)dt, \ x \in \mathbb{R}.$$

Then $(\mathrm{Cont}(\mathbb{R}), P)$ is a Baxter algebra of weight zero.

Example 1.3 (Divided power algebra) This is the algebra

$$R = \bigoplus_{n \geq 0} Ce_n$$

on which the multiplication is defined by

$$e_m e_n = \binom{m+n}{m} e_{m+n}, \quad m, \ n \geq 0.$$

The operator $P : R \rightarrow R$, where $P(e_n) = e_{n+1}$, $n \geq 0$, is a Baxter operator of weight zero.

Example 1.4 (Hurwitz series) Let R be

$$HC := \{(a_n) | a_n \in C, n \in \mathbb{N}\},$$

the ring of Hurwitz series [17]. The addition is defined componentwise, and the multiplication is given by $(a_n)(b_n) = (c_n)$, where $c_n = \sum_{k=0}^{n} \binom{n}{k} a_k b_{n-k}$. Define

$$P : HC \rightarrow HC, \quad P((a_n)) = (a_{n-1}), \quad \text{where } a_{-1} = 0.$$

Then HC is a Baxter algebra of weight 0 which is the completion of the divided power algebra.

We will return to Example 1.3 and 1.4 in Section 2.2.

Example 1.5 (Scalar multiplication) Let R be any C-algebra. For a given $\lambda \in C$, define

$$P_\lambda : R \rightarrow R, x \mapsto -\lambda x, \forall x \in R.$$

Then P_λ is a Baxter operator of weight λ on R.

Example 1.6 (Partial sums) This is one of the first examples of a Baxter algebra, introduced by Baxter [5]. Let A be any C-algebra. Let

$$R = \prod_{n \in \mathbb{N}_+} A = \{(a_1, a_2, \ldots) | a_n \in A, \ n \in \mathbb{N}_+\}.$$

with addition, multiplication and scalar product defined entry by entry. Define $P : R \rightarrow R$ to be the "partial sum" operator:

$$P(a_1, a_2, \ldots) = \lambda(0, a_1, a_1 + a_2, a_1 + a_2 + a_3, \ldots).$$

Then P is a Baxter operator of weight λ on R.

We will return to this example in Section 2.3

Example 1.7 (Distributions [5]) Let R be the Banach algebra of functions

$$\varphi(t) = \int_{-\infty}^{\infty} e^{itx} dF(x)$$

where F is a function such that $||\varphi|| := \int_{-\infty}^{\infty} |dF(x)| < \infty$ and such that $F(-\infty) := \lim_{x \to -\infty} F(x)$ exists. The addition and multiplication are defined pointwise. Let $P(\varphi)(t) = \int_0^{\infty} e^{itx} dF(x) + F(0) - F(-\infty)$. Then (R, P) is a Baxter algebra of weight -1.

We will come back to this example in Proposition 2.16.

Example 1.8 This is an important example in combinatorics [31]. Let R be the ring of functions $f : \mathbb{R} \to \mathbb{R}$ with finite support in which the product is the convolution:

$$(fg)(x) := \sum_{y \in \mathbb{R}} f(y)g(x - y), \; x \in \mathbb{R}.$$

Define $P : R \to R$ by

$$P(f)(x) = \sum_{y \in \mathbb{R}, \max(0,y)=x} f(y), \; x \in \mathbb{R}.$$

Then (R, P) is a Baxter algebra of weight -1.

1.2 Integrations and summations

Define a system of polynomials $\Phi_n(x) \in \mathbb{Q}[x]$ by the generating function

$$\frac{t(e^{xt} - 1)}{e^t - 1} = \sum_{n=0}^{\infty} \Phi_n(x) \frac{t^n}{n!}.$$

Then we have

$$\Phi_n(x) = B_n(x) - B_n,$$

where $B_n(x)$ is the n-th Bernoulli polynomial, defined by the generating function

$$\frac{te^{xt}}{e^t - 1} = \sum_{n=0}^{\infty} B_n(x) \frac{t^n}{n!},$$

and $B_n = B_n(0)$ is the n-th Bernoulli number. It is well-known (see [16, Section 15.1]) that

$$\Phi_{n+1}(k + 1) = (n + 1) \sum_{r=1}^{k} r^n \tag{1.9}$$

for any integer $k \geq 1$. The following property of Baxter algebras is due to Miller [21].

Proposition 1.10 *Let C be a $\mathbb{Q}$-algebra and let $R = C[t]$. Let P be a C-linear operator on R such that $P(1) = t$.*

(a) *The operator P is a Baxter operator of weight 0 if and only if $P(t^n) = \frac{1}{n+1}t^{n+1}$.*

(b) *The operator P is a Baxter operator of weight -1 if and only if $P(t^n) = \frac{1}{n+1}\Phi_{n+1}(t+1)$.*

(c) *The operator P is a Baxter operator of weight -1 if and only if $P(t^n)(k) = \sum_{r=1}^{k} r^n$ for every integer $k \geq 1$.*

Proof. (a): The only if part can be easily proved by induction. For details see the proof of Theorem 3 in [21]. The if part is obvious.

(b) is Theorem 4 in [21].

(c): Because of (b), we only need to show

$$P(t^n) = \frac{1}{n+1}\Phi_{n+1}(t+1) \Leftrightarrow P(t^n)(k) = \sum_{r=1}^{k} r^n, \forall k \geq 1.$$

($\Rightarrow$) follows from (1.9). ($\Leftarrow$) can be seen easily, say from [21, Lemma 5]. $\square$

Using this proposition and free Baxter algebras that we will construct in the next section, we will prove the following property of Baxter algebras in Section 2.2.

Proposition 1.11 *Let C be a $\mathbb{Q}$-algebra and let (R, P) be a Baxter C-algebra of weight λ. Let t be $P(1)$.*

(a) *If $\lambda = 0$, then $P(t^n) = \frac{1}{n+1}t^{n+1}$ for all $n \geq 1$.*

(b) *If $\lambda = -1$, then $P(t^n) = \frac{1}{n+1}\Phi_{n+1}(t+1)$.*

Because of Proposition 1.10 and 1.11, a Baxter operator of weight zero is also called an anti-derivation or integration, and a Baxter operator of weight -1 is also called a summation operator [21].

2 Free Baxter algebras

Free objects are usually defined to be generated by sets. We give the following more general definition.

Definition 2.1 Let A be a C-algebra. A Baxter C-algebra $(F_C(A), P_A)$, together with a C-algebra homomorphism $j_A : A \to F_C(A)$, is called a *free Baxter C-algebra on A (of weight λ)*, if, for any Baxter C-algebra (R, P) of weight λ and any C-algebra homomorphism $\varphi : A \to R$, there exists a unique Baxter C-algebra homomorphism $\tilde{\varphi} : (F_C(A), P_A) \to (R, P)$ such that the diagram

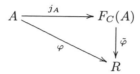

commutes.

Let X be a set. One can define a free Baxter C-algebra $(F_C(X), P_X)$ on X in a similar way. This Baxter algebra is naturally isomorphic to $(F_C(C[X]), P_{C[X]})$, where $C[X]$ is the polynomial algebra over C generated by the set X.

Free Baxter algebras play a central role in the study of Baxter algebras. Even though the existence of free Baxter algebras follows from the general theory of universal algebras [20], in order to get a good understanding of free Baxter algebras, it is desirable to find concrete constructions. This is in analogy to the important role played by the ring of polynomials in the study of commutative algebra.

Free Baxter algebras on sets were first constructed by Rota [27] and Cartier [6] in the category of Baxter algebras with no identities (with some restrictions on the weight and the base ring). Two more general constructions have been obtained recently [14,15]. The first construction is in terms of *mixable shuffle products* which generalize the well-known *shuffle products* of path integrals developed by Chen [7] and Ree [23]. The second construction is modified after the construction of Rota.

The shuffle product construction of a free Baxter algebra has the advantage that its module structure and Baxter operator can be easily described. The construction of a free Baxter algebra as a standard Baxter algebra has the advantage that its multiplication is very simple. There is a canonical isomorphism between the shuffle Baxter algebra and the standard Baxter algebra. This isomorphism enables us to make use of properties of both constructions.

2.1 Free Baxter algebras of Cartier

We first recall the construction of free Baxter algebras by Cartier[6,14]. The original construction of Cartier is for free objects on a set, in the category of algebras with no identity and with weight -1. We will extend his construction to free objects in the category of algebras with identity and with arbitrary weight. We will still consider free objects on sets. In order to consider free objects on a C-algebra A, it will be more convenient to use the tensor product notation to be introduced in Section 2.2.

Let X be a set. Let λ be a fixed element in C. Let M be the free commutative semigroup with identity on X. Let $\widetilde{X}$ denote the set of symbols of the form

$$u_0 \cdot [\ ], \quad u_0 \in M,$$

and

$$u_0 \cdot [u_1, \ldots, u_m], m \geq 1, \ u_0, u_1, \ldots, u_m \in M.$$

Let $\mathfrak{B}(X)$ be the free C-module on $\widetilde{X}$. Cartier gave a C-bilinear multiplication $\diamond_c$ on $\mathfrak{B}(X)$ by defining

$$(u_0 \cdot [\ ]) \diamond_c (v_0 \cdot [\ ]) = u_0 v_0 \cdot [\ ],$$
$$(u_0 \cdot [\ ]) \diamond_c (v_0 \cdot [v_1, \ldots, v_n]) = (v_0 \cdot [v_1, \ldots, v_n]) \diamond_c (u_0 \cdot [\ ])$$
$$= u_0 v_0 \cdot [v_1, \ldots, v_n],$$

and

$$(u_0 \cdot [u_1, \ldots, u_m]) \diamond_c (v_0 \cdot [v_1, \ldots, v_n])$$
$$= \sum_{(k,P,Q) \in \overline{S}_c(m,n)} \lambda^{m+n-k} u_0 v_0 \cdot \Phi_{k,P,Q}([u_1, \ldots, u_m], [v_1 \ldots, v_n]).$$

Here $\overline{S}_c(m,n)$ is the set of triples (k, P, Q) in which k is an integer between 1 and $m+n$, P and Q are ordered subsets of $\{1, \ldots, k\}$ with the natural ordering such that $P \cup Q = \{1, \ldots, k\}$, $|P| = m$ and $|Q| = n$. For each $(k, P, Q) \in \overline{S}_c(m,n)$, $\Phi_{k,P,Q}([u_1, \ldots, u_m], [v_1, \ldots, v_n])$ is the element $[w_1, \ldots, w_k]$ in $\widetilde{X}$ defined by

$$w_j = \begin{cases} u_\alpha, & \text{if } j \text{ is the } \alpha\text{-th element in } P \text{ and } j \notin Q; \\ v_\beta, & \text{if } j \text{ is the } \beta\text{-th element in } Q \text{ and } j \notin P; \\ u_\alpha v_\beta, & \text{if } j \text{ is the } \alpha\text{-th element in } P \text{ and the } \beta\text{-th element in } Q. \end{cases}$$

Define a C-linear operator P_X^c on $\mathfrak{B}(X)$ by

$$P_X^c(u_0 \cdot [\,]) = 1 \cdot [u_0],$$
$$P_X^c(u_0 \cdot [u_1, \ldots, u_m]) = 1 \cdot [u_0, u_1, \ldots, u_m].$$

The following theorem is a modification of Theorem 1 in [6] and can be proved in the same way.

Theorem 2.2 *The pair* $(\mathfrak{B}(X), P_X^c)$ *is a free Baxter algebra on* X *of weight* λ *in the category* **Bax$_C$**.

2.2 Mixable shuffle Baxter algebras

We now describe the mixable shuffle Baxter algebras. It gives another construction of free Baxter algebras. An advantage of this construction is that it is related to the well-known shuffle products. It also enables us to consider the free object on a C-algebra A.

Intuitively, to form the shuffle product, one starts with two decks of cards and puts together all possible shuffles of the two decks. Similarly, to form the mixable shuffle product, one starts with two decks of charged cards, one deck positively charged and the other negatively charged. When a shuffle of the two decks is taken, some of the adjacent pairs of cards with opposite charges are allowed to be merged into one card. When one puts all such "mixable shuffles" together, with a proper measuring of the number of pairs that have been merged, one gets the mixable shuffle product. We now give a precise description of the construction.

Mixable shuffles. For $m, n \in \mathbb{N}_+$, define the set of (m,n)-*shuffles* by

$$S(m,n) = \left\{ \sigma \in S_{m+n} \left| \begin{array}{l} \sigma^{-1}(1) < \sigma^{-1}(2) < \ldots < \sigma^{-1}(m), \\ \sigma^{-1}(m+1) < \sigma^{-1}(m+2) < \ldots < \sigma^{-1}(m+n) \end{array} \right. \right\}.$$

Here S_{m+n} is the symmetric group on $m+n$ letters. Given an (m,n)-shuffle $\sigma \in S(m,n)$, a pair of indices $(k, k+1)$, $1 \leq k < m+n$, is called an *admissible pair* for σ if $\sigma(k) \leq m < \sigma(k+1)$. Denote T^σ for the set of admissible pairs for σ. For a subset T of T^σ, call the pair (σ, T) a *mixable* (m,n)-*shuffle*. Let $|T|$ be the cardinality of T. Identify (σ, T) with σ if T is the empty set. Denote

$$\overline{S}(m,n) = \{(\sigma, T) \mid \sigma \in S(m,n), \ T \subset T^\sigma\}$$

for the set of *mixable* (m,n)-*shuffles*.

Example 2.3 There are three $(2,1)$ shuffles:

$$\sigma_1 = \begin{pmatrix} 1\,2\,3 \\ 1\,2\,3 \end{pmatrix}, \ \sigma_2 = \begin{pmatrix} 1\,2\,3 \\ 1\,3\,2 \end{pmatrix}, \ \sigma_3 = \begin{pmatrix} 1\,2\,3 \\ 3\,1\,2 \end{pmatrix}.$$

The pair $(2,3)$ is an admissible pair for σ_1. The pair $(1,2)$ is an admissible pair for σ_2. There are no admissible pairs for σ_3.

For $A \in \mathbf{Alg}_C$ and $n \geq 0$, let $A^{\otimes n}$ be the n-th tensor power of A over C with the convention $A^{\otimes 0} = C$. For $x = x_1 \otimes \ldots \otimes x_m \in A^{\otimes m}$, $y = y_1 \otimes \ldots \otimes y_n \in A^{\otimes n}$ and $(\sigma, T) \in \overline{S}(m,n)$, the element

$$\sigma(x \otimes y) = u_{\sigma(1)} \otimes u_{\sigma(2)} \otimes \ldots \otimes u_{\sigma(m+n)} \in A^{\otimes(m+n)},$$

where

$$u_k = \begin{cases} x_k, & 1 \leq k \leq m, \\ y_{k-m}, & m+1 \leq k \leq m+n, \end{cases}$$

is called a *shuffle* of x and y; the element

$$\sigma(x \otimes y; T) = u_{\sigma(1)} \widehat{\otimes} u_{\sigma(2)} \widehat{\otimes} \ldots \widehat{\otimes} u_{\sigma(m+n)} \in A^{\otimes(m+n-|T|)},$$

where for each pair $(k, k+1)$, $1 \leq k < m+n$,

$$u_{\sigma(k)} \widehat{\otimes} u_{\sigma(k+1)} = \begin{cases} u_{\sigma(k)} u_{\sigma(k+1)}, & (k, k+1) \in T \\ u_{\sigma(k)} \otimes u_{\sigma(k+1)}, & (k, k+1) \notin T, \end{cases}$$

is called a *mixable shuffle* of x and y.

Example 2.4 For $x = x_1 \otimes x_2 \in A^{\otimes 2}$ and $y = y_1 \in A$, there are three shuffles of x and y:

$$x_1 \otimes x_2 \otimes y_1, \ x_1 \otimes y_1 \otimes x_2, \ y_1 \otimes x_1 \otimes x_2.$$

For the mixable shuffles of x and y, we have in addition,

$$x_1 \otimes x_2 y_1, \ x_1 y_1 \otimes x_2.$$

Now fix $\lambda \in C$. Define, for x and y above,

$$x \diamond^+ y = \sum_{(\sigma, T) \in \overline{S}(m,n)} \lambda^{|T|} \sigma(x \otimes y; T) \in \bigoplus_{k \leq m+n} A^{\otimes k}. \tag{2.5}$$

Example 2.6 For x, y in our previous example,

$$x \diamond^+ y = x_1 \otimes x_2 \otimes y_1 + x_1 \otimes y_1 \otimes x_2 + y_1 \otimes x_1 \otimes x_2 + \lambda(x_1 \otimes x_2 y_1 + x_1 y_1 \otimes x_2).$$

Shuffle Baxter algebras. The operation $\diamond^+$ extends to a map

$$\diamond^+ : A^{\otimes m} \times A^{\otimes n} \to \bigoplus_{k \leq m+n} A^{\otimes k}, \; m, \, n \in \mathbb{N}$$

by C-linearity. Let

$$\text{III}_C^+(A) = \text{III}_C^+(A, \lambda) = \bigoplus_{k \in \mathbb{N}} A^{\otimes k} = C \oplus A \oplus A^{\otimes 2} \oplus \dots.$$

Extending by additivity, the binary operation $\diamond^+$ gives a C-bilinear map

$$\diamond^+ : \text{III}_C^+(A) \times \text{III}_C^+(A) \to \text{III}_C^+(A)$$

with the convention that

$$C \times A^{\otimes m} \to A^{\otimes m}$$

is the scalar multiplication.

Theorem 2.7 [14] *The mixable shuffle product* $\diamond^+$ *defines an associative, commutative binary operation on* $\text{III}_C^+(A) = \bigoplus_{k \in \mathbb{N}} A^{\otimes k}$, *making it into a C-algebra with the identity* $\mathbf{1}_C \in C = A^{\otimes 0}$.

Define $\text{III}_C(A) = \text{III}_C(A, \lambda) = A \otimes_C \text{III}_C^+(A)$ to be the tensor product algebra. Define a C-linear endomorphism P_A on $\text{III}_C(A)$ by assigning

$$P_A(x_0 \otimes x_1 \otimes \dots \otimes x_n) = \mathbf{1}_A \otimes x_0 \otimes x_1 \otimes \dots \otimes x_n,$$

for all $x_0 \otimes x_1 \otimes \dots \otimes x_n \in A^{\otimes(n+1)}$ and extending by additivity. Let $j_A : A \to \text{III}_C(A)$ be the canonical inclusion map. Call $(\text{III}_C(A), P_A)$ the *(mixable) shuffle Baxter C-algebra on A of weight λ*.

For a given set X, we also let $(\text{III}_C(X), P_X)$ denote the shuffle Baxter C-algebra $(\text{III}_C(C[X]), P_{C[X]})$, called the *(mixable) shuffle Baxter C-algebra on X (of weight λ)*. Let $j_X : X \to \text{III}_C(X)$ be the canonical inclusion map.

Theorem 2.8 [14] *The shuffle Baxter algebra $(\text{III}_C(A), P_A)$, together with the natural embedding j_A, is a free Baxter C-algebra on A of weight λ. Similarly, $(\text{III}_C(X), P_X)$, together with the natural embedding j_X, is a free Baxter C-algebra on X of weight λ.*

Relation with Cartier's construction. The mixable shuffle product construction of free Baxter algebras is canonically isomorphic to Cartier's construction. Using the notations from Section 2.1, we define a map $f : \widetilde{X} \to \text{III}_C(X)$ by

$$f(u_0 \cdot [\,]) = u_0;$$
$$f(u_0 \cdot [u_1, \dots, u_m]) = u_0 \otimes u_1 \otimes \dots \otimes u_m,$$

and extend it by C-linearity to a C-linear map

$$f : \mathfrak{B}(X) \to \amalg_C(X).$$

The same argument as for Proposition 5.1 in [14, Prop. 5.1] can be used to prove the following

Proposition 2.9 f *is an isomorphism in* **Bax**$_C$.

Let A be a C-algebra. Using the tensor product notation in the construction of $\amalg_C(A)$, one can extend Cartier's construction (Theorem 2.2) and Proposition 2.9 for free Baxter algebras on A.

Special cases. We next consider some special cases of the shuffle Baxter algebras.

Case 1: $\lambda = 0$. In this case, $\amalg_C(A)$ is the usual shuffle algebra generated by the C-module A. It played a central role in the work of K.T. Chen [7] on path integrals and is related to many areas of pure and applied mathematics.

Case 2: $X = \phi$. Taking $A = C$, we get

$$\amalg_C(C) = \bigoplus_{n=0}^{\infty} C^{\otimes(n+1)} = \bigoplus_{n=0}^{\infty} C\mathbf{1}^{\otimes(n+1)},$$

where $\mathbf{1}^{\otimes(n+1)} = \underbrace{\mathbf{1}_C \otimes \ldots \otimes \mathbf{1}_C}_{(n+1)-\text{factors}}$. In this case the mixable shuffle product formula (2.5) gives

Proposition 2.10 For any $m, n \in \mathbb{N}$,

$$\mathbf{1}^{\otimes(m+1)} \mathbf{1}^{\otimes(n+1)} = \sum_{k=0}^{m} \binom{m+n-k}{n} \binom{n}{k} \lambda^k \mathbf{1}^{\otimes(m+n+1-k)}.$$

We are now ready to prove Proposition 1.11: By Proposition 4.2 in [12], the map

$$\amalg_C(C) \to C[x],$$
$$\mathbf{1}^{\otimes(n+1)} \mapsto \frac{x(x - \lambda) \cdots (x - \lambda(n - 1))}{n!}, \quad n \geq 0,$$

is an isomorphism of C-algebras. Then P_C enables us to define a Baxter operator Q on $C[x]$ through this isomorphism and we have $Q(1) = x$. Then by Proposition 1.10, we have

$$Q(x^n) = \begin{cases} \frac{1}{n+1}x^{n+1}, & \text{if } \lambda = 0, \\ \frac{1}{n+1}\Phi_{n+1}(x+1), & \text{if } \lambda = -1. \end{cases}$$

Now let (R, P) be any Baxter C-algebra. By the universal property of $(C[x], Q)$ ($\cong (\text{III}_C(C), P_C)$) stated in Theorem 2.8, there is a unique homomorphism $\widetilde{\varphi} : (C[x], Q) \to (R, P)$ of Baxter algebras such that $\widetilde{\varphi}(x) = P(1)$. Let $t = P(1) = \widetilde{\varphi}(Q(1)) = \widetilde{\varphi}(x)$ and since $\Phi_{n+1}(x+1)$ has coefficients in $\mathbb{Q} \subset C$, we have

$$P(t^n) = \widetilde{\varphi}(Q(x^n)) = \widetilde{\varphi}\left(\frac{1}{n+1}x^{n+1}\right) = \frac{1}{n+1}t^{n+1}$$

when $\lambda = 0$ and

$$P(t^n) = \widetilde{\varphi}(Q(x^n)) = \widetilde{\varphi}\left(\frac{1}{n+1}\Phi_{n+1}(x+1)\right) = \frac{1}{n+1}\Phi_{n+1}(t+1)$$

when $\lambda = -1$. This proves Proposition 1.11.

Case 3: $\lambda = 0$ and $X = \phi$. Taking the "pull-back" of Cases 1 and 2, we get

$$\begin{array}{ccc} \{\text{III}_C(\phi, 0)\} & \xrightarrow{\subset} & \{\text{III}_C(X, 0) | X \in \mathbf{Sets}\} \\ \cap \downarrow & & \downarrow \cap \\ \{\text{III}_C(\phi, \lambda) | \lambda \in C\} & \xrightarrow{\subset} & \{\text{III}_C(X, \lambda) | X \in \mathbf{Sets}, \lambda \in C\} \end{array} \qquad (2.11)$$

Thus $\text{III}_C(\phi, 0)$ is the *divided power algebra*

$$\text{III}_C(\phi, 0) = \bigoplus_{k \in \mathbb{N}} C e_k, \ e_n e_m = \binom{m+n}{m} e_{m+n}$$

in Example 1.3.

Variation: Complete shuffle Baxter algebras. We now consider the completion of $\text{III}_C(A)$.

Given $k \in \mathbb{N}$, $\text{Fil}^k \text{III}_C(A) := \bigoplus_{n \geq k} A^{\otimes(n+1)}$ is a Baxter ideal of $\text{III}_C(A)$. Consider the infinite product of C-modules $\widehat{\text{III}}_C(A) = \prod_{k \in \mathbb{N}} A^{\otimes(k+1)}$. It contains $\text{III}_C(A)$ as a dense subset with respect to the topology defined by the filtration $\{\text{Fil}^k \text{III}_C(A)\}$. All operations of the Baxter C-algebra $\text{III}_C(A)$ are continuous with respect to this topology. Hence they extend uniquely to operations on $\widehat{\text{III}}_C(A)$, making $\widehat{\text{III}}_C(A)$ a Baxter algebra of weight λ, with the Baxter operator denoted by $\widehat{P}$. It is called the *complete shuffle Baxter*

294

algebra on A. It naturally contains $\mathrm{III}_C(A)$ as a Baxter subalgebra and is a free object in the category of Baxter algebras that are complete with respect to a canonical filtration defined by the Baxter operator [15].

When $A = C$ and $\lambda = 0$, we have

$$\widehat{\mathrm{III}}_C(C,\lambda) = \prod_{k \in \mathbb{N}} Ce_k \cong HC,$$

the ring of Hurwitz series in Example 1.4.

2.3 Standard Baxter algebras

The standard Baxter algebra constructed by Rota in [27] is a free object in the category $\mathbf{Bax}_C^0$ of Baxter algebras not necessarily having an identity. It is described as a Baxter subalgebra of another Baxter algebra whose construction goes back to Baxter [5]. In Rota's construction, there are further restrictions that C be a field of characteristic zero, the free Baxter algebra obtained be on a finite set X, and the weight λ be 1. By making use of shuffle Baxter algebras, we will show that Rota's description can be modified to yield a free Baxter algebra on an algebra in the category $\mathbf{Bax}_C$ of Baxter algebras with an identity, with a mild restriction on the weight λ. We can also provide a similar construction for algebras not necessarily having an identity, and for complete Baxter algebras, but we will not explain it here. See [15].

We will first present Rota's construction, modified to give free objects in the category of Baxter algebras with identity. We then give the general construction.

The standard Baxter algebra of Rota. For details, see [27,31].

As before, let C be a commutative ring with an identity, and fix a λ in C. Let X be a given set. For each $x \in X$, let $t^{(x)}$ be a sequence $(t_1^{(x)}, \ldots, t_n^{(x)}, \ldots)$ of distinct symbols $t_n^{(x)}$. We also require that the sets $\{t_n^{(x_1)}\}_n$ and $\{t_n^{(x_2)}\}_n$ be disjoint for $x_1 \neq x_2$ in X. Denote

$$\overline{X} = \bigcup_{x \in X} \{t_n^{(x)} \mid n \in \mathbb{N}_+\}$$

and denote by $\mathfrak{A}(X)$ the ring of sequences with entries in $C[\overline{X}]$, the C-algebra of polynomials with variables in $\overline{X}$. Thus the addition, multiplication and scalar multiplication by $C[\overline{X}]$ in $\mathfrak{A}(X)$ are defined componentwise. Alternatively, for $k \in \mathbb{N}_+$, denote γ_k for the sequence $(\delta_{n,k})_n$, where $\delta_{n,k}$ is the

Kronecker delta. Then we can identify a sequence $(a_n)_n$ in $\mathfrak{A}(X)$ with a series

$$\sum_{n=1}^{\infty} a_n \gamma_n = a_1 \gamma_1 + a_2 \gamma_2 + \ldots.$$

Then the addition, multiplication and scalar multiplication by $C[\overline{X}]$ are given termwise.

Define

$$P_X^r = P_{X,\lambda}^r : \mathfrak{A}(X) \to \mathfrak{A}(X)$$

by

$$P_X^r(a_1, a_2, a_3, \ldots) = \lambda(0, a_1, a_1 + a_2, a_1 + a_2 + a_3, \ldots).$$

In other words, each entry of $P_X^r(a)$, $a = (a_1, a_2, \ldots)$, is λ times the sum of the previous entries of a. If elements in $\mathfrak{A}(X)$ are described by series $\sum_{n=1}^{\infty} a_n \gamma_n$ given above, then we simply have

$$P_X^r \left(\sum_{n=1}^{\infty} a_n \gamma_n \right) = \lambda \sum_{n=1}^{\infty} \left(\sum_{i=1}^{n-1} a_i \right) \gamma_n.$$

It is well-known [5,27] that, for $\lambda = 1$, P_X^r defines a Baxter operator of weight 1 on $\mathfrak{A}(X)$. It follows that, for any $\lambda \in C$, P_X^r defines a Baxter operator of weight λ on $\mathfrak{A}(X)$. Hence $(\mathfrak{A}(X), P_X^r)$ is in **Bax**$_C$.

Definition 2.12 The *standard Baxter algebra* on X is the Baxter subalgebra $\mathfrak{S}(X)$ of $\mathfrak{A}(X)$ generated by the sequences $t^{(x)} = (t_1^{(x)}, \ldots, x_n^{(x)}, \ldots)$, $x \in X$.

An important result of Rota [27,31] is

Theorem 2.13 $(\mathfrak{S}(X), P_X^r)$ *is a free Baxter algebra on* X *in the category* **Bax**$_C$.

The standard Baxter algebra in general. Given $A \in \mathbf{Alg}_C$, we now give an alternative construction of a free Baxter algebra on A in the category **Bax**$_C$.

For each $n \in \mathbb{N}_+$, denote by $A^{\otimes n}$ the n-th tensor power algebra where the tensor product is taken over C. Note that the multiplication on $A^{\otimes n}$ here is different from the multiplication on $A^{\otimes n}$ when it is regarded as a C-submodule of $\mathrm{III}_C(A)$.

Consider the direct limit algebra

$$\overline{A} = \varinjlim A^{\otimes n}$$

where the transition map is given by

$$A^{\otimes n} \longrightarrow A^{\otimes(n+1)}, \quad x \mapsto x \otimes 1_A.$$

Let $\mathfrak{A}(A)$ be the set of sequences with entries in $\overline{A}$. Thus we have

$$\mathfrak{A}(A) = \prod_{n=1}^{\infty} \overline{A}\gamma_n = \left\{ \sum_{n=1}^{\infty} a_n \gamma_n, \ a_n \in \overline{A} \right\}.$$

Define addition, multiplication and scalar multiplication on $\mathfrak{A}(A)$ componentwise, making $\mathfrak{A}(A)$ into a $\overline{A}$-algebra, with the sequence $(1, 1, \ldots)$ as the identity. Define

$$P_A^r = P_{A,\lambda}^r : \mathfrak{A}(A) \to \mathfrak{A}(A)$$

by

$$P_A^r(a_1, a_2, a_3, \ldots) = \lambda(0, a_1, a_1 + a_2, a_1 + a_2 + a_3, \ldots).$$

Then $(\mathfrak{A}(A), P_A^r)$ is in $\mathbf{Bax}_C$. For each $a \in A$, define $t^{(a)} = (t_k^{(a)})_k$ in $\mathfrak{A}(A)$ by

$$t_k^{(a)} = \otimes_{i=1}^k a_i (= \otimes_{i=1}^{\infty} a_i), \ a_i = \begin{cases} a, \ i = k, \\ 1, \ i \neq k. \end{cases}$$

Definition 2.14 The *standard Baxter algebra* on A is the Baxter subalgebra $\mathfrak{S}(A)$ of $\mathfrak{A}(A)$ generated by the sequences $t^{(a)} = (t_1^{(a)}, \ldots, t_n^{(a)}, \ldots)$, $a \in A$.

Since $\amalg_C(A)$ is a free Baxter algebra on A, the C-algebra morphism

$$A \to \mathfrak{A}(A), a \mapsto t^{(a)}$$

extends uniquely to a morphism in $\mathbf{Bax}_C$

$$\Phi : \amalg_C(A) \to \mathfrak{A}(A).$$

Theorem 2.15 [15] *Assume that $\lambda \in C$ is not a zero divisor in $\overline{A}$. The morphism in $\mathbf{Bax}_C$*

$$\Phi : \amalg_C(A) \to \mathfrak{S}(A)$$

induced by sending $a \in A$ to $t^{(a)} = (t_1^{(a)}, \ldots, t_n^{(a)}, \ldots)$ is an isomorphism.

Consequently, when λ is not a zero divisor in $\overline{A}$, $(\mathfrak{S}(A), P_A^r)$ is a free Baxter algebra on A in the category $\mathbf{Bax}_C$.

Spitzer's identity. As an application of the standard Baxter algebra, we recall the proof of Spitzer's identity by Rota [27,31]. Spitzer's identity is regarded as a remarkable stepping stone in the theory of sums of independent random variables and motivates of Baxter's identity. For other proofs of Spitzer's identity, see [5,18,34,3,6]. We first present an algebraic formulation.

Proposition 2.16 [31] *Let C be a $\mathbb{Q}$-algebra. Let (R, P) be a Baxter C-algebra of weight 1. Then for $b \in R$, we have*

$$\exp\left(-P(\log(1 + tb)^{-1})\right) = \sum_{n=0}^{\infty} t^n (Pb)^{[n]} \tag{2.17}$$

in the ring of power series $R[[t]]$. Here

$$(Pb)^{[n]} = \underbrace{P(b(P(b \ldots (Pb) \ldots)))}_{n\text{-iteration}}$$

with the convention that $(Pb)^{[1]} = P(b)$ and $(Pb)^{[0]} = 1$.

Proof. First let $x = (x_1, x_2, \ldots)$ where x_i, $i \geq 1$, are symbols and let $X = \{x\}$. Consider the standard Baxter algebra $\mathfrak{S}(X)$. It is easy to verify that

$$(Px)^{[n]} = (0, e_n(x_1), e_n(x_1, x_2), e_n(x_1, x_2, x_3), \ldots)$$

where $e_n(x_1, \ldots, x_m)$ is the elementary symmetric function of degree n in the variables $x_1, \ldots, x_m$. By definition,

$$P(x^k) = (0, x_1^k, x_1^k + x_2^k, x_1^k + x_2^k + x_3^k, \ldots, p_k(x_1, \ldots, x_m), \ldots),$$

where $p_k(x_1, \ldots, x_m) = x_1^k + x_2^k + \ldots, x_m^k$ is the power sum symmetric function of degree k in the variables $x_1, \ldots, x_m$. These two classes of symmetric functions are related by the well-known Waring's formula [31]

$$\exp\left(-\sum_{k=1}^{\infty}(-1)^k t^k p_k(x_1, \ldots, x_m)/k\right) = \sum_{n=0}^{\infty} e_n(x_1, \ldots, x_m)t^n, \ \forall \ m \geq 1.$$

This proves

$$\exp\left(-P_X^r(\log(1 + tx)^{-1})\right) = \sum_{n=0}^{\infty} t^n (P_X^r x)^{[n]}. \tag{2.18}$$

Next let (R, P) be any Baxter C-algebra and let b be any element in R. By the universal property of the free Baxter algebra $(\mathfrak{S}, P_X^r)$, there is a unique Baxter algebra homomorphism $\tilde{\varphi} : \mathfrak{S} \to R$ such that $\tilde{\varphi}(x) = b$. Since all the coefficients in the expansion of $\log(1 + u)$ and $\exp(u)$ are rational, and $\tilde{\varphi} \circ P_X^r = P \circ \tilde{\varphi}$, applying $\tilde{\varphi}$ to (2.18) gives the desired equation. $\square$

We can now specialize to the original identity of Spitzer, following Baxter [5] and Rota [31]. Consider the Baxter algebra (R, P) in Example 1.7. Let $\{X_k\}$ be a sequence of independent random variables with identical distribution function $F(x)$ and characteristic function

$$\psi(s) = \int_{-\infty}^{\infty} e^{isx} dF(x).$$

Let $S_n = X_1 + \ldots + X_n$ and let $M_n = \max(0, S_1, S_2, \ldots, S_n)$. Let $F_n(x) = \mathrm{Prob}(M_n < x)$ (Prob for probability) be the distribution function of M_n. We note that, if $f(s)$ is the characteristic function of the random variable of X, then $P(f)(s)$ is the characteristic function of the random variable $\max(0, X)$. Applying Proposition 2.16 to $b = \psi(s)$, we obtain the identity first obtained by Spitzer:

$$\sum_{n=0}^{\infty} \int_0^{\infty} e^{isx} dF_n(x) = \exp\left(\sum_{k=1}^{\infty}\left(\int_0^{\infty} e^{isx} dF(x) + F(0)\right)\right).$$

We refer the reader to [31] for the application of the standard Baxter algebra to the proof of some other identities, such as the Bohnenblust-Spitzer formula.

3 Further applications of free Baxter algebras

3.1 Overview

Recall that the free Baxter algebra $\amalg_C(A, \lambda)$ in the special case when $A = C$ and $\lambda = 0$ is the divided power algebra. The divided power algebra and its completion are known to be related to

- crystalline cohomology and rings of p-adic periods in number theory,

- shuffle products in differential geometry and topology,

- Hopf algebra in commutative algebra,

- Hurwitz series in differential algebra,

- umbral calculus in combinatorics, and

- incidence algebra in graph theory.

By the "pull-back" diagram (2.11), free (complete) Baxter algebras give a vast generalization of the (complete) divided power algebra, and so suggest a framework in which these connections and applications of the divided power algebra can be extended. We give two such connections and applications in the next two sections, one to Hopf algebras (Section 3.2) and one to the umbral calculus in combinatorics (Section 3.3).

3.2 Hopf algebra

Definition of Hopf algebra. We recall some basic definitions and facts. Recall that a *cocommutative C-coalgebra* is a triple (A, Δ, ε) where A is a C-module, and $\Delta : A \to A \otimes A$ and $\varepsilon : A \to C$ are C-linear maps that make the following diagrams commute.

$$
\begin{array}{ccc}
A & \xrightarrow{\Delta} & A \otimes A \\
\downarrow{\scriptstyle \Delta} & & \downarrow{\scriptstyle \mathrm{id} \otimes \Delta} \\
A \otimes A & \xrightarrow{\Delta \otimes \mathrm{id}} & A \otimes A \otimes A
\end{array}
\tag{3.1}
$$

$$
\begin{array}{ccccc}
C \otimes A & \xleftarrow{\varepsilon \otimes \mathrm{id}} & A \otimes A & \xrightarrow{\mathrm{id} \otimes \varepsilon} & A \otimes C \\
& {\scriptstyle \cong}\nwarrow & \uparrow{\scriptstyle \Delta} & \nearrow{\scriptstyle \cong} & \\
& & A & &
\end{array}
\tag{3.2}
$$

$$
\begin{array}{ccc}
& A & \\
{\scriptstyle \Delta}\swarrow & & \searrow{\scriptstyle \Delta} \\
A \otimes A & \xrightarrow{\tau_{A,A}} & A \otimes A
\end{array}
\tag{3.3}
$$

where $\tau_{A,A} : A \otimes A \to A \otimes A$ is defined by $\tau_{A,A}(x \otimes y) = y \otimes x$.

Recall that a *C-bialgebra* is a quintuple $(A, \mu, \eta, \Delta, \varepsilon)$ where (A, μ, η) is a C-algebra and (A, Δ, ε) is a C-coalgebra such that μ and η are morphisms of coalgebras.

Let $(A, \mu, \eta, \Delta, \varepsilon)$ be a C-bialgebra. For C-linear maps $f, g : A \to A$, the convolution $f \star g$ of f and g is the composition of the maps

$$
A \xrightarrow{\Delta} A \otimes A \xrightarrow{f \otimes g} A \otimes A \xrightarrow{\mu} A.
$$

A C-linear endomorphism S of A is called an *antipode* for A if

$$
S \star \mathrm{id}_A = \mathrm{id}_A \star S = \eta \circ \varepsilon.
\tag{3.4}
$$

A *Hopf algebra* is a bialgebra A with an antipode S.

The main theorem. On the Baxter algebra $\mathrm{III}_C(C, \lambda)$, let μ be the canonical multiplication and let $\eta : C \hookrightarrow \mathrm{III}_C(C, \lambda)$ be the unit map. Define a comultiplication Δ, a counit ε and an antipode S by

$$\Delta = \Delta_\lambda : \text{III}_C(C, \lambda) \to \text{III}_C(C, \lambda) \otimes \text{III}_C(C, \lambda),$$

$$a_n \mapsto \sum_{k=0}^{n} \sum_{i=0}^{n-k} (-\lambda)^k a_i \otimes a_{n-k-i},$$

$$\varepsilon = \varepsilon_\lambda : \text{III}_C(C, \lambda) \to C, \ a_n \mapsto \begin{cases} 1, & n = 0, \\ \lambda 1, n = 1, \\ 0, & n \geq 2, \end{cases}$$

$$S = S_\lambda : \text{III}_C(C, \lambda) \to \text{III}_C(C, \lambda), \quad a_n \mapsto (-1)^n \sum_{v=0}^{n} \binom{n-3}{v-3} \lambda^{n-v} a_v.$$

The following result is proved in [2].

Theorem 3.5 *The sextuple* $(\text{III}_C(C, \lambda), \mu, \eta, \Delta, \varepsilon, S)$ *is a Hopf C-algebra.*

3.3 The umbral calculus

Definition and examples. For simplicity, we assume that C is a $\mathbb{Q}$-algebra for the rest of the paper.

The *umbral calculus* is the study and application of *polynomial sequences of binomial type*, i.e., polynomial sequences $\{p_n(x) \mid n \in \mathbb{N}\}$ in $C[x]$ such that

$$p_n(x + y) = \sum_{k=0}^{n} \binom{n}{k} p_k(x) p_{n-k}(y)$$

in $C[x, y]$ for all n. Such a sequence behaves as if its terms are powers of x and has found applications in several areas of pure and applied mathematics, including number theory and combinatorics, since the 19th century. There are many well-known sequences of binomial types.

Examples 3.6 (a) Monomials. x^n.

(b) Lower factorial polynomials. $(x)_n = x(x - 1) \cdots (x - n + 1)$.

(c) Exponential polynomials. $\phi_n(x) = \sum_{k=0}^{n} S(n, k) x^k$, where $S(n, k)$ with $n, k \geq 0$ are the Stirling numbers of the second kind.

(d) Abel polynomials. Fix $a \neq 0$. $A_n(x) = x(x - an)^{n-1}$.

(e) Mittag-Leffler polynomials. $M_n(x) = \sum_{k=0}^{n} \binom{n}{k} \binom{n-1}{n-k} 2^k (x)_k$.

(f) Bessel polynomials. $y_n(x) = \sum_{k=0}^{n} \frac{(n+k)!}{(n-k)!k!}(\frac{x}{2})^k$ (a solution to the Bessel equation $x^2 y'' + (2x+2)y' + n(n+1)y = 0$).

There are also Bell polynomials, Hermite polynomials, Bernoulli polynomials, Euler polynomials, ...

As useful as umbral calculus is in many areas of mathematics, the foundations of umbral calculus were not firmly established for over a hundred years. Vaguely speaking, the difficulty in the study is that such sequences do not observe the *algebra* rules of $C[x]$. Rota embarked on laying down the foundation of umbral calculus during the same period of time as when he started the algebraic study of Baxter algebras. Rota's discovery is that these sequences do observe the algebra rules of the dual algebra (the umbral algebra), or in a fancier language, the coalgebra rules of $C[x]$. Rota's pioneer work [26] was completed over the next decade by Rota and his collaborators [30,25,24]. Since then, there have been a number of generalizations of the umbral calculus.

We will give a characterization of umbral calculus in terms of free Baxter algebras by showing that the umbral algebra is the free Baxter algebra of weight zero on the empty set. We also characterize the polynomial sequences studied in umbral calculus in terms of operations in free Baxter algebras.

We will then use the free Baxter algebra formulation of the umbral calculus to give a generalization of the umbral calculus, called the λ-*umbral calculus* for each constant λ. The umbral calculus of Rota is the special case when $\lambda = 0$.

Rota's umbral algebra. In order to describe the binomial sequences, Rota and his collaborators identify $C[x]$ as the dual of the algebra $C[[t]]$, called the *umbral algebra.* (algebra plus the duality). To identify $C[[t]]$ with the dual of $C[x]$, let $t_n = \frac{t^n}{n!}$, $n \in \mathbb{N}$. Then

$$t_m t_n = \binom{m+n}{m} t_{m+n}, \quad m, \; n \in \mathbb{N}. \tag{3.7}$$

The C-algebra $C[[t]]$, together with the basis $\{t_n\}$ is called the *umbral algebra.*

We can identify $C[[t]]$ with the dual C-module of $C[x]$ by taking $\{t_n\}$ to be the dual basis of $\{x^n\}$. In other words, t_k is defined by

$$t_k : C[x] \to C, \; x^n \mapsto \delta_{k,n}, \; k, \; n \in \mathbb{N}.$$

Rota and his collaborators removed the mystery of sequences of binomial type and Sheffer sequences by showing that such sequences have a simple characterization in terms of the umbral algebra.

Let f_n, $n \geq 0$, be a pseudo-basis of $C[[t]]$. That is, f_n, $n \geq 0$ are linearly independent and generate $C[[t]]$ as a topological C-module where the topology on $C[[t]]$ is defined by the filtration

$$F^n := \left\{ \sum_{k=n}^{\infty} c_k t_k \right\}.$$

A pseudo-basis f_n, $n \geq 0$, of $C[[t]]$ is called a *divided power* pseudo-basis if

$$f_m f_n = \binom{m+n}{m} f_{m+n}, \ m, \ n \geq 0.$$

Theorem 3.8 [26,25]

(a) *A polynomial sequence $\{p_n(x)\}$ is of binomial type if and only if it is the dual basis of a divided power pseudo-basis of $C[[t]]$.*

(b) *Any divided power pseudo-basis of $C[[t]]$ is of the form $f_n(t) = \frac{f^n(t)}{n!}$ for some $f \in C[[t]]$ with ord$f = 1$ (that is, $f(t) = \sum_{k=1}^{\infty} c_k t^k$, $c_1 \neq 0$).*

This theorem completely determines all polynomial sequences of binomial type. Algorithms to determine such sequences effectively have also been developed. See the book by Roman [24] for details.

λ-*umbral calculus.* Our first observation is that, with the operator

$$P : C[[t]] \to C[[t]], \ t_n \mapsto t_{n+1},$$

$C[[t]]$ becomes a Baxter algebra of weight zero, isomorphic to $\widehat{\text{III}}_C(C, 0)$. More generally, we have

Theorem 3.9 [12] *A sequence $\{f_n(t)\}$ in $C[[t]]$ is a divided power pseudo-basis if and only if the map $f_n(t) \mapsto t_n, n \geq 0$, defines an automorphism of the Baxter algebra $C[[t]]$.*

This theorem provides a link between umbral calculus and Baxter algebra. This characterization of the umbral calculus in terms of Baxter algebra also motivates us to study a generalization of binomial type sequences.

Definition 3.10 A sequence $\{p_n(x) \mid n \in \mathbb{N}\}$ of polynomials in $C[x]$ is a sequence of λ-binomial type if

$$p_n(x + y) = \sum_{k=0}^{n} \lambda^k \sum_{i=0}^{n} \binom{n}{i} \binom{i}{k} p_i(x) p_{n+k-i}(y), \ \forall y \in C, n \in \mathbb{N}.$$

When $\lambda = 0$, we recover the sequences of binomial type. Denote

$$e_\lambda(x) = \frac{e^{\lambda x} - 1}{\lambda}$$

for the series

$$\sum_{k=1}^{\infty} \frac{\lambda^{k-1} x^k}{k!} \, .$$

When $\lambda = 0$, we get $e_\lambda(x) = x$. We verify that

$$\mathfrak{q} := \{(e_\lambda(x))^n\}_n$$

is a sequence of λ-binomial type of $C[[x]]$. Let $C<\mathfrak{q}>$ be the C-submodule of $C[[x]]$ generated by elements in $\mathfrak{q}$.

Let $f(t)$ be a power series of order 1. Define

$$d_n(f)(t) = \frac{f(t)(f(t) - \lambda) \cdots (f(t) - (n-1)\lambda)}{n!}, \quad n \geq 0 \, .$$

Then $P : C[[t]] \to C[[t]]$, $d_n(f) \mapsto d_{n+1}(f)$ defines a weight λ Baxter operator on $C[[t]]$. Such a sequence is called a *Baxter pseudo-basis* of $C[[t]]$.

Definition 3.11 Fix a $\lambda \in C$. The algebra $C[[t]]$, together with the weight λ Baxter pseudo-basis $\{d_n(t)\}_n$, is called the λ-*umbral algebra*.

As in the classical case, we identify $C[[t]]$ with the dual algebra of $C<\mathfrak{q}>$ by taking $\{d_n(t)\}$ to be the dual basis of $\{(e_\lambda(x))^n\}$. We then extend the classical theory of the umbral calculus to the λ-umbral calculus. In particular, Theorem 3.8 is generalized to

Theorem 3.12 [12]

(a) *A pseudo-basis $\{s_n(x)\}$ of $C[[x]]$ is of λ-binomial type if and only if $\{s_n(x)\}$ is the dual basis of a Baxter pseudo-basis of $C[[t]]$.*

(b) *Any Baxter pseudo-basis of $C[[t]]$ is of the form $\{d_n(f)\}$ for some $f(t)$ in $C[[t]]$ of order 1.*

References

1. Anderson, E. S. *On the fluctuations of sums of random variables*, Math. Scand. **1** (1953), 263–285.
2. Andrew, G.E., Guo, L., Keigher, W., Ono, K. *Baxter algebras and Hopf algebras*, preprint.

3. Atkinson, F. V. *Some aspects of Baxter's functional equation,* J. Math. Anal. and Applications **7** (1963), 1–30.

4. Baxter, G. *An operator identity,* Pacific J. Math. **8** (1958), 649–663.

5. Baxter, G. *An analytic problem whose solution follows from a simple algebraic identity,* Pacific J. Math. **10** (1960), 731–742.

6. Cartier, P. *On the structure of free Baxter algebras,* Adv. Math. **9** (1972), 253-265.

7. Chen, K.T. *Integration of paths, geometric invariants and a generalized Baker-Hausdorff formula,* Ann. of Math. **65** (1957), 163–178.

8. de Bragança, S. L. *Finite dimensional Baxter algebras,* Studies in Applied Math. **LIV** (1975), 75–89.

9. Foata, D. *Étude algébrique de certains problèmes d'analyse combinatoire et du calcul des probabilités,* Publ. Inst. Statist. Univ. Paris **14** (1965), 81–241.

10. Guo, L. *Properties of free Baxter algebras,* Adv. Math. **151** (2000), 346–374.

11. Guo, L. *Ascending chain conditions in free Baxter algebras,* to appear in International Journal of Algebra and Computation.

12. Guo, L. *Baxter algebras and the umbral calculus,* to apprear in Adv. Applied Math.

13. Guo, L., Keigher, W. *On integration algebras,* preprint.

14. Guo, L., Keigher, W. *Baxter algebras and shuffle products,* Adv. Math. **150** (2000), 117–149.

15. Guo, L., Keigher, W. *On free Baxter algebras: completions and the internal construction,* Adv. Math. **151** (2000), 101–127.

16. Ireland, K., Rosen, M. *A Classical Introduction to Modern Number Theory,* Springer-Verlag, New York, 1982.

17. Keigher, W. *On the ring of Hurwitz series,* Comm. Algebra **25** (1997), 1845–1859.

18. Kingman, J. F. C. *Spitzer's identity and its use in probability thoery,* J. London Math. Soc. **37** (1962), 309–316.

19. Kolchin, E. R. *Differential Algebras and Algebraic Groups,* Academic Press, New York, 1973.

20. MacLane, S. *Categories for the Working Mathematician,* Springer-Verlag, New York, 1971.

21. Miller, J. B. *Some properties of Baxter operators,* Acta Math. Acad. Sci. Hungar. **17** (1966), 387–400.

22. Murru, Doctorate thesis, Italy, 1990.

23. Ree, R. *Lie elements and an algebra associated with shuffles,* Ann. Math. **68** (1958), 210–220.

24. Roman, S. *The Umbral Calculus*, Academic Press, Orlando, FL, 1984.
25. Roman, S., Rota, G.-C. *The umbral calculus*, Adv. Math. **27**(1978), 95–188.
26. Rota, G.-C. *The number of partitions of a set*, Amer. Math. Monthly **64** (1964), 498–504.
27. Rota, G.-C. *Baxter algebras and combinatorial identities I, II*, Bull. Amer. Math. Soc. **75** (1969), 325–329, 330–334.
28. Rota, G.-C. *Baxter operators, an introduction*, In: "Gian-Carlo Rota on Combinatorics, Introductory papers and commentaries", Joseph P.S. Kung, Editor, Birkhäuser, Boston, 1995.
29. Rota, G.-C. *Ten mathematics problems I will never solve*, Invited address at the joint meeting of the American Mathematical Society and the Mexican Mathematical Society, Oaxaca, Mexico, December 6, 1997. DMV Mittellungen Heft 2, 1998, 45–52.
30. Rota, G.-C. Kahaner, D., Odlyzko, A. *Finite operator calculus*, J. Math. Anal. Appl. **42** (1973), 685–760.
31. Rota, G.-C., Smith, D. A. *Fluctuation theory and Baxter algebras*, Istituto Nazionale di Alta Matematica, **IX** (1972), 179–201.
32. Spitzer, F. *A combinatorial lemma and its application to probability theory*, Trans. Amer. Math. Soc. **82** (1956), 323–339.
33. Thomas, G. P. *Frames, Young tableaux and Baxter sequences*, Adv. Math. **26** (1977), 275–289.
34. Wendel, J. G. *A brief proof of a theorem of Baxter*, Math. Scand. **11** (1962), 107–108.
35. Winkel, R. *Sequences of symmetric polynomials and combatorial properties of tableaux*, Adv. Math. **134** (1998), 46–89.